The biodiversity of many habitats is under threat and although seas cover the majority of our planet's surface, far less is known about the biodiversity of marine environments than that of terrestrial systems. It is also not clear whether many of the patterns known to occur on land also occur in the sea. Until we have a firmer idea of the diversity of a wide range of marine habitats and what controls it, we have little hope of conserving biodiversity, or determining the impact of human activities such as mariculture, fishing, dumping of waste and pollution. This book brings together key studies from the deep sea and open ocean to tropical shores and polar regions to consider how comparable the patterns and processes underlying diversity are in these different ecosystems. *Marine Biodiversity* will be a major resource for all those interested in biodiversity and its conservation.

MARINE BIODIVERSITY: PATTERNS AND PROCESSES

Acknowledgements

This book was the direct outcome of a conference on 'Marine Biodiversity: Causes and Consequences' sponsored by the Marine Biological Association of the UK and the Scottish Association for Marine Science. This took place at the University of York, UK, and included presentations by the authors contributing chapters to this book. Additional support to the conference was provided by the UK Institute of Estuarine and Coastal Sciences, the University of Hull, the Marine Conservation Society, The Natural History Museum, London, and the World Wide Fund for Nature.

Rupert F.G. Ormond
John D. Gage
Martin V. Angel

MARINE BIODIVERSITY

Patterns and Processes

Edited by
RUPERT F.G. ORMOND
Senior Lecturer in Marine Ecology, Biology Department,
University of York

JOHN D. GAGE
Senior Principal Scientific Officer and Senior Research Fellow,
Scottish Association for Marine Science,
Dunstaffnage Marine Laboratory, Oban

and

MARTIN V. ANGEL
Senior Principal Scientific Officer, George Deacon Division,
Southampton Oceanography Centre

CAMBRIDGE UNIVERSITY PRESS

PUBLISHED BY THE PRESS SYNDICATE OF THE UNIVERSITY OF CAMBRIDGE
The Pitt Building, Trumpington Street, Cambridge, United Kingdom

CAMBRIDGE UNIVERSITY PRESS
The Edinburgh Building, Cambridge CB2 2RU, UK http://www.cup.cam.ac.uk
40 West 20th Street, New York, NY 10011-4211, USA http://www.cup.org
10 Stamford Road, Oakleigh, Melbourne 3166, Australia

© Cambridge University Press 1997

This book is in copyright. Subject to statutory exception
and to the provisions of relevant collective licensing agreements,
no reproduction of any part may take place without
the written permission of Cambridge University Press.

First published 1997
Reprinted 1998, 1999

Printed in the United Kingdom at the University Press, Cambridge

Typeset in Times 10/13pt System Miles 33

A catalogue record for this book is available from the British Library

Library of Congress cataloguing in publication data

Marine biodiversity : patterns and processes / edited by Rupert F.G.
  Ormond, John D. Gage, and Martin V. Angel.
    p.   cm.
  Includes index.
  ISBN 0 521 55222 2
  1. Marine biology.   2. Biological diversity.   I. Ormond, Rupert.
  II. Gage, John D.   III. Angel, Martin Vivian.
  QH91.57.A1M3   1997
  574.5'2636—dc20   96-9334 CIP

ISBN 0 521 55222 2 hardback

Contents

List of contributors	*page* ix
Foreword: The value of diversity. Sir Crispin Tickell	xiii
1 Marine biodiversity in its global context. *M. Williamson*	1
2 Gradients in marine biodiversity. *J.S. Gray*	18
3 Pelagic biodiversity. *M.V. Angel*	35
4 Biological diversity in oceanic macrozooplankton: More than counting species. *A.C. Pierrot-Bults*	69
5 Large-scale patterns of species diversity in the deep-sea benthos. *M.A. Rex, R.J. Etter and C.T. Stuart*	94
6 Diversity, latitude and time: Patterns in the shallow sea. *A. Clarke and J.A. Crame*	122
7 High benthic species diversity in deep-sea sediments: The importance of hydrodynamics. *J.D. Gage*	148
8 Diversity and structure of tropical Indo-Pacific benthic communities: Relation to regimes of nutrient input. *J.D. Taylor*	178
9 Why are coral reef communities so diverse? *A.J. Kohn*	201
10 The biodiversity of coral reef fishes. *R.F.G. Ormond and C.M. Roberts*	216
11 The historical component of marine taxonomic diversity gradients. *J.A. Crame and A. Clarke*	258
12 Population genetics and demography of marine species. *J.E. Neigel*	274
13 Discovering unrecognised diversity among marine molluscs. *J. Grahame, S.L. Hull, P.J. Mill and R. Hemingway*	293
14 Ecosystem function at low biodiversity – the Baltic example. *R. Elmgren and C. Hill*	319

15 Land–seascape diversity of the USA East Coast coastal zone with particular reference to estuaries. *G.C. Ray, B.P. Hayden, M.G. McCormick-Ray and T.M. Smith* 337

16. The development of mariculture and its implications for biodiversity. *M.C.M. Beveridge, L.G. Ross and J.A. Stewart* 372

17. Protecting marine biodiversity and integrated coastal zone management. *J.S.H. Pullen* 394

18. Conserving biodiversity in North-East Atlantic marine ecosystems. *K. Hiscock* 415

Author index 429
Species index 437
Subject index 443

Contributors

Martin V. Angel
Institute of Oceanographic Sciences, Wormley, Godalming, Surrey, GU8 5UB, UK
Present address, Southampton Oceanography Centre, Empress Dock, Southampton, SO14 3ZH, UK
M.C.M. Beveridge
Institute of Aquaculture, University of Stirling, Stirling, FK9 4LA, UK
Andrew Clarke
British Antarctic Survey, High Cross, Madingley Road, Cambridge CB3 0ET, UK
J. Alistair Crame
British Antarctic Survey, High Cross, Madingley Road, Cambridge CB3 0ET, UK
Ragnar Elmgren
Department of Systems Ecology, Stockholm University, S-106 91 Stockholm, Sweden
Ron T. Etter
Department of Biology, University of Massachusetts, Boston, MA 02125, USA
John D. Gage
Scottish Association for Marine Science, Dunstaffnage Marine Laboratory, P.O. Box 3, Oban, Argyll, PA34 4AD, UK
J. Grahame
Department of Biology, The University of Leeds, Leeds, LS2 9JT, UK
John S. Gray
Department of Biology, University of Oslo, Pb 1064, 0316 Blindern, Norway

B.P. Hayden
Department of Environmental Sciences, University of Virginia, Charlottesville, VA 22903, USA
R. Hemingway
Department of Biology, The University of Leeds, Leeds, LS2 9JT, UK
Cathy Hill
Swedish Environmental Protection Agency, S-106 48 Stockholm, Sweden
Keith Hiscock
Joint Nature Conservation Committee, Monkstone House, Peterborough, PE1 1JY, UK
S.L. Hull
Science Section, University College Scarborough, Filey Road, Scarborough, YO11 3AZ, UK
Alan J. Kohn
Department of Zoology, University of Washington, Seattle, WA 98195, USA
M.G. McCormick-Ray
Department of Environmental Sciences, University of Virginia, Charlottesville, VA 22903, USA
P.J. Mill
Department of Biology, The University of Leeds, Leeds, LS2 9JT, UK
Joseph E. Neigel
Department of Biology, The University of Southwestern Louisiana, Lafayette, LA 70504, USA
Rupert F.G. Ormond
Tropical Marine Research Unit, Department of Biology, The University of York, York, YO1 5DD, UK
Annelies C. Pierrot-Bults
Institute for Systematics and Population Biology, University of Amsterdam, The Netherlands
J.S.H. Pullen
WWF UK, Panda House, Weyside Park, Catteshall Lane, Godalming, Surrey, GU7 1XR, UK
G.C. Ray
Department of Environmental Sciences, University of Virginia, Charlottesville, VA 22903, USA
Michael A. Rex
Department of Biology, University of Massachusetts, Boston, MA 02125, USA

C.M. Roberts
Department of Environmental Economics and Environmental Management, University of York, York YO1 5DD, UK
L.G. Ross
Institute of Aquaculture, University of Stirling, Stirling, FK9 4LA, UK
T.M. Smith
Department of Environmental Sciences, University of Virginia, Charlottesville, VA 22903, USA
J.A. Stewart
Institute of Aquaculture, University of Stirling, Stirling, FK9 4LA, UK
Carol T. Stuart
Department of Biology, University of Massachusetts, Boston, MA 02125, USA
John D. Taylor
Department of Zoology, The Natural History Museum, Cromwell Road, London SW7 5BD, UK
Crispin Tickell
Green College, Oxford, OX2 6HG, UK
Mark Williamson
Department of Biology, University of York, York, YO1 5DD, UK

Foreword

The value of diversity

The current debate about the diversity of life is relatively new. That on marine biodiversity has hardly started. It has been created not by some new appreciation of the marvels of nature – would that it were – but by the evident destruction of the natural world as we and our ancestors have known it. When discussion of an international treaty on biodiversity began, there were many who asked why we should worry, or who gave it low priority in comparison with such problems as human poverty or the hazards of climate change. The answers are both simple and complex.

Change and its impact

The sea is our mother and father. All life comes from it. It is still far richer in major groupings of animals than the land: of 34 animal phyla, 29 occur in the sea, and 14 of them only in the sea. The complexity of its species and ecosystems is immense. Yet, as on land, we are engaged in a process of extinguishing populations and ecosystems at something like 1000 times the natural rate. Globally, the present rate of species extinction is comparable to the extinctions at the end of the Cretaceous period 65 million years ago when the long dominance of the dinosaurs came to an end.

How is it that one animal species – our own – could have had such destructive effects on others? Many species have modified the environment to suit their needs. Beavers are a good example. But until the industrial revolution, the effects of human activities were local, or at worst regional, rather than global. All the great civilisations of the past cleared land for cultivation, introduced plants and animals from elsewhere, and caused lasting change. But the consequences of the industrial revolution have been more serious. On the one hand there has been a huge growth in human population; on the other there has been a huge growth in consumption of the world's resources, and saturation of its sinks. Higher standards of living inevitably involve higher consumption and more waste.

In broad terms there were around 10 million people at the end of the Ice Age, 1 billion in the lifetime of Thomas Malthus, 2 billion in 1930, and around 5.8 billion now. Short of catastrophe there will be around 8.5 billion in 2025. At the same time there has been an even steeper growth in urban populations, with all that implies for the resources surrounding centres, many along coasts. An observer from outer space, with a device for speeding up time, would see steadily increasing brown patches like freckles on the land surface of the earth. He would not yet see the increasing scum on the surface of the sea. High consumption of resources in rich countries and heavy pressure on resources in poor ones have already changed its face. We are better aware of the consequences on land than on the sea, and many miss the interaction between the two. Put in simple terms, we have been taking too many living resources out of the sea, discharging too many of our wastes into it, and failing to understand the marine environment with its myriad implications for particular habitats and life as a whole.

The story of the rise and impending fall of the world's fisheries – a complex of boom-and-bust cycles – is like one of those moral tales for Victorian children. At the turn of the century total catch was less than 5 million tons. As technology improved, the catch rose steeply. In 1989 it reached 86 million tons. By 1992 it had fallen to 80 million tons, and all the indications are that it is continuing to fall. At the same time fishermen are switching to new fish species as traditional ones diminish, sometimes to vanishing point: for example cod fishing off the Canadian Grand Banks, one of the richest locations in the world, has now ended. The new fishing technology, however successful in the short term, has proved destructive of all forms of life on the ocean floor, almost incredibly wasteful (shrimp trawlers discard between 50% and 90% of their catch), and profoundly damaging to other species and ecosystems in all their interdependencies (from puffins and sand eels in the Shetlands to other organisms up and down the food chain). Loss of total marine biomass is beyond calculation.

This destructive process has of course been accelerated by pollution of the sea from the land. Nearly every form of human waste – direct and indirect – finds its way into the sea sooner or later. The ocean is everyone's favourite sink. Like the land the sea has its deserts, its

Foreword

volcanoes, its mountains and its fertile plains, and it is of course the fertile plains, often along the continental shelf, which are worst affected. The steady flow of human wastes, from pathogens and nutrients derived from agriculture and sewage to industrial effluents such as oil, chemicals, plastics and a host of newly created chemicals quite alien to the natural environment, has already altered ecosystems worldwide. PCBs have even been found in Antarctic seals. The result again is beyond calculation.

Perhaps most alarming of all is our ignorance of the marine environment, both physical and biological. We do not yet fully understand, for example, what drives the oceanic circulation system, nor – more importantly – what causes it to change. A twitch in the behaviour of the Gulf Stream could put most of Britain back under ice. Oceanic ecosystems are equally mysterious. We add to them or subtract from them at our peril.

Removal of apparently unimportant creatures can often have unexpected effects, as is well known when fishermen in California tried to eradicate California sea otters, which they saw as their prime competitors for fish. The reduction in numbers of otters caused an explosion in the population of sea urchins on which the otters had fed. This in turn led to a dramatic loss of the kelp which not only formed the diet of the sea urchins, but was also a critical breeding habitat of the fish. So the fishermen found that eradication of sea otters led ineluctably to even fewer fish. Now, thankfully, the sea otters have been reintroduced. As E.O. Wilson has well said: 'The loss of a keystone species is like a drill accidentally striking a power line. It causes lights to go out all over'.

The ocean is closely linked to the atmosphere, and the exchanges between the two are crucial to life. They determine the distribution of heat from the equator to the poles, and the character of the world's climate. Part of the exchange is biological; normally phytoplankton play a key role in the regulation of atmospheric carbon. But we are changing the chemistry of the atmosphere. Acidification downwind of industry is widespread. Depletion of the ozone layer, which acts as a protective blanket against ultraviolet radiation, has the potential to harm all forms of life, both on land and near the surface of the ocean.

Above all, by adding carbon dioxide, methane and nitrous oxide to the atmosphere, we are also altering global climate with unforeseeable local consequences. According to reports from the Inter-governmental Panel on Climate Change, if we continue as we are, there could be an average rise in temperature of around 2 degrees Celsius by the end of the next century. These figures may seem small, but the effect on the climate may be dramatic. The results will be very different in different places. Temperature is likely to rise more steeply in polar and temperate regions than around the equator. Weather patterns could change drastically, with rain coming where it hardly came before, and droughts coming where there was previously ample rainfall. Further, to judge from recent evidence from ice-cores taken from Greenland, general warming could precipitate rapid instability as was the case some 120 000 years ago.

There could also be a rise in sea level of around 50 cm by 2100. Major uncertainties remain, but they are more about the magnitude and geographical distribution of change than about change itself. It could be less or more than predicted. Our natural assumption is that it will be less; but wise men – and insurance brokers – should certainly reckon with the possibility of more. Taken together, these factors indicate environmental change with effects on the diversity of life as rapid in geological times as any which have previously affected the earth. According to the Intergovernmental Panel, changes will be greater than have occurred naturally in the last 10 000 years, and the rise in sea level will be three to six times faster than in the last 100 years.

The value of biodiversity

How can we assess in our crude terms the value of biodiversity if our future is to be sustainable? The issue falls into four categories: ethical, aesthetic, direct economic and indirect economic.

On ethical grounds it is questionable whether we have the right to exterminate so many of our companions on the living planet whether they are of use to us or not. This is not a point that has caused Christianity much concern in the past. There are honourable exceptions; but most Christian thinkers have seen humans as separate from the rest of nature, which they believe was for their plunder or delectation. But respect for life as such has always been a central tenet of Buddhism and Taoism, among other systems of belief. There is an increasing awareness, especially in societies that have done most to destroy other forms of life, that humans have some kind of ethical responsibility for the welfare, or at least the continued existence, of our only known living companions in the universe.

The aesthetic aspects of nature usually go without saying, but they

Foreword

are very difficult to define. I have rarely felt more wonder than when diving on the Great Barrier Reef of Australia. There is, I believe, a profound human instinct that causes people to feel linked to the natural world of land and sea. Even the most hardened city dwellers need space and greenery in their work and play. The culture of every people is closely allied to its landscapes and their living inhabitants, and cannot be dissociated from them. Both ethical and aesthetic arguments are of enormous importance for the psychological health of any society, although it is very difficult to attach monetary values to them.

Our direct economic interest in biodiversity is more obvious. We need to maintain our own good health as well as that of the plants and animals, big and small, on which we depend for food. A large share (in some cases more than half) of the animal protein that people eat comes from the sea. Our economies simply could not work without the raw materials available from living organisms. These include not only foods but medicines. More than three-quarters of the population in poor countries depend on plant-based drugs, while in industrialised ones about a quarter of prescription drugs contain at least one compound that is or once came from higher plants. So far only 10% or less of flowering plants have been tested, but there have been some outstanding successes. For the future the sea may be the most promising source of new anti-viral and anti-tumour drugs.

As well as conserving diversity at the level of species, we also need to cherish the genetic diversity that occurs within them. Modern agricultural techniques, including fish farming, have led to an excessive dependence on a few miracle strains of even fewer plants and animals. Meanwhile the wild relatives of these strains are often lost when natural habitat is converted for other uses. Without a large natural genetic reservoir, we make our food supplies vulnerable to disease as the Irish potato growers in the last century and salmon farmers in this century learned to their cost.

It is not easy to weigh up the long-term potential value of unpolluted water full of fish against the short-term but rapidly diminishing value of a dump for nuclear, chemical or other less classifiable human rubbish. This same difficulty is greater when trying to assess indirect economic benefits provided by the diversity of life. At present we take as cost-free a broadly regular climatic system with ecosystems, terrestrial and marine, to match. We rely on forests and vegetation to produce soil, to hold it together and to regulate water supplies by preserving catchment basins, recharging groundwater and buffering extreme conditions. We rely on coral reefs and mangrove forests as spawning grounds for fish and wetlands, and on deltas as shock absorbers for floods.

Likewise we need nutrients to be recycled, and wastes to be disposed of. We rely upon the current balance of insects, bacteria and viruses; and we assume the health of plants and animals unless we find to the contrary. There is no conceivable substitute for these natural services. Economic values tend to be based on scarcity, and so far there has been no permanent shortage of such commodities as fertile land, clean water, clean air or the supply of genetic resources. Yet we cannot continue to assume that this natural bounty will continue.

The rise in public awareness

It is fair to say that until very recently most people were unaware of these issues most of the time. Even now there are huge variations between countries, within them, and between generations, and the problems of the sea have been a Cinderella. But public anxiety has greatly increased, often in a partial and muddled fashion. The UN Conference on the Environment in Stockholm in 1972 was followed by the creation of the UN Environment Programme. A protracted conference began to create a Law of the Sea. Meetings and papers multiplied. The subject took more and more space on the political and diplomatic agenda. Then came the UN Conference on Environment and Development in Rio de Janeiro in June 1992, which produced a Declaration, two Conventions and Agenda 21, or an agenda of environmental action for the next century. This was followed by the creation of a UN Commission on Sustainable Development, and the Law of the Sea, amended to take better account of environmental concerns, has at last come into effect.

The Convention on Biodiversity was the first of its kind. The idea of an international convention to protect it would have seemed bizarre only a few years ago. Yet the Convention was signed by 157 countries and has now been ratified by more than the minimum necessary, and come into effect. The Convention is not a beautiful text. It is written in the language of UN speak, and is full of ambiguities and hostages to fortune. The financing, through the Global Environment Facility, is incomplete and inadequate. Yet the Convention contains important matter. Whether and how this will apply in the marine area has yet to be seen.

The various conventions and conclusions of Rio laid new obligations on individual governments. In Britain the Government has produced papers on Sustainable Development, Climate Change, Forestry and Biodiversity, and created machinery for increasing public awareness of local as well as global environmental problems. There is even a national Biodiversity Action Plan. It is no good preaching to others if we practise otherwise at home. The Government also launched a project of its own: the Darwin Initiative for the Survival of Species. This supports co-operation between Britain and countries rich in biodiversity but poor in resources, and already has some excellent projects to its credit. Furthermore in the first and second annual reports of the Government Panel on Sustainable Development, a plan for a new Inter-Governmental Panel on the Oceans was proposed. This and other issues have now been taken up in the UN Commission on Sustainable Development, and the Global Environment Facility may well give it priority.

Changing minds

The Biodiversity Convention is an illustration of the way in which minds have changed. But we are still at the very beginning. A good place to start is the 1987 Report of the World Commission on Environment and Development. In this, sustainable development was defined as 'meeting the needs of the present generation without compromising the ability of future generations to meet their needs'. Moving towards inter-generational equity has many implications: among them stabilising population levels; protecting natural and international systems; seeing resources as a kind of capital stock; and recognising global interdependence.

Nowhere are our shortcomings clearer than in the conceptual as well as practical approach that we maintain to economics and economic management. Most economists – not so much dismal as dim – still work within a framework in which long-term environmental considerations play a minor if not peripheral part. Furthermore, economic growth, as enunciated by most politicians, is often cited as the only way out of our problems. But it takes no account of the impact of growth on the environment nor of its inevitable effect on natural resources, whether renewable or not, including those in the sea.

Economists usually argue that environmental considerations must be given a monetary value for them to be visible in the decision-making process. It is true that the benefits of any development must be sufficient to outweigh the costs. But cost involves pricing, and prices should always tell the truth. In addition to the traditional costs of research, process, production and so on, prices should reflect the costs involved in replacing a resource or substituting for it; and the costs of the associated environmental problems.

Considerations of biodiversity create even bigger problems for economists. Firstly there is the problem of irreversibility. Discount rates cannot accommodate this point. Nor can they accommodate the second problem: that we are forced to make decisions without more than the sketchiest idea of their ecological consequences. It is an understatement to refer to current levels of ignorance as uncertainty.

Thirdly, not all objects and processes are marketable. As we have seen, species are important in ways that have no direct or immediate effect on humans but are essential to the long-term health of the ecosystem of which they are part. How should we value an ecosystem? A common method used by economists to assign value is to ask individuals how much they would be willing to pay for a particular item or service. Imagine asking someone how much he would pay to keep atmospheric oxygen at its present level. Different peoples and different societies have different priorities. More generally how can we value the totality of diversity, including the range of habitats within an ecological community? Briefly we have to value the things that count rather than the things that can be counted. Nature should be recognised as having an existence beyond that of a warehouse of raw materials to be protected because they are useful to our particular animal species.

How can we find means to reflect this multidimensional value of ecosystems in a way that conveys meaning? Perhaps a hint of the answer lies in the recent work of the UN Development Programme in devising a Human Development Index to replace the old misleading pecking order of wealth based on Gross Domestic Product. By adding indices of child mortality, life expectancy, access to health services, literacy, political plurality and so on, a new and very different order has emerged (often to the indignation of countries who had come high on the previous list). The same principle could be applied to establish an Index of Planetary Wellbeing or Health that would bring in the values of biodiversity – ethical, aesthetic, directly and indirectly economic – and which would attribute the positive or negative contributions made by individual governments. That might help sharpen many minds.

Managing the oceans

The good health of the sea presents peculiar difficulties. Even incompetent governments have some control over the land of their nation states, and take legal responsibility for it. Most maps show a patchwork of colours of such states surrounded by the deep undifferentiated blue of the sea. Outside the exclusive economic zones of 200 miles, which it is in the interest of states to protect, comes the vast area of ocean that is the property of everybody and nobody. The Law of the Sea will do something to regulate marine activity in some areas and some respects and may at least create a framework for a future international regime. But it is not designed to cope with the issues of biodiversity that arise in a moving or migratory environment where fish carry no passports, waste travels with the currents, and the behaviour of water, like wind, affects the planet as a whole.

More than any problem on land, the problems of the sea require an integrated approach. Here the Biodiversity Convention is of only limited value. Such an approach must begin with the thousand ways in which human activity on land affects the sea: human population in all its forms, management of fresh water, wetlands, estuaries and coastlines, and changes in atmospheric chemistry. So far most debate has been over the protection of particular species, the disposal of particular wastes, or the study of particular phenomena that can be elucidated through research. The results have been pathetic. Fishing fleets are still far too big, and many governments covertly subsidise them (the FAO calculates that US $124 billion is spent every year to catch US $70 billion worth of fish). Technology is still senselessly destructive.

Perhaps we need some catastrophe, not too big but not too small, to persuade governments and peoples to look at ocean management in a more radical fashion. Steep decline in some fish stocks has not so far been sufficient. Nor has the poisoning of some inland seas and continental shelves. International agreements covering this or that aspect are no more than a beginning. They rarely relate to each other. With or without the stimulus of a shock, governments now need to put together something more ambitious, which extends old notions of sovereignty, takes account of the needs of future generations, creates a new authority with power to regulate and license, and sets long-term sustainability as its goal.

The problem is as always how to get from where we are to where we want to be. With ratification of the Law of the Sea, one idea would be to start a new negotiation within its framework to establish a regime for the open seas. Another would be for the United Nations agencies concerned to set up a body comparable to the Inter-governmental Panel on Climate Change to put together the science, look at the impacts of human activity on the sea, and explore possible responses to it. This in turn might lead to a separate convention like the Climate Change and Biodiversity Conventions signed at Rio in 1992. Yet another possibility would be a gradualist approach through the further development of regional agreements among riparian states covering specific biogeographical areas.

Whatever route is chosen, something of global scope to protect the global commons is essential. It may not happen soon but the logic for it is compelling. In the meantime governments, stimulated and helped by non-governmental organisations, should begin to work out the mechanisms – national, regional and international – for management of oceanic resources, and invest in the research still urgently needed for better, above all integrated, understanding of the element that accounts for over two-thirds of the surface of the planet.

Crispin Tickell
Green College
Oxford

Chapter 1
Marine biodiversity in its global context

MARK WILLIAMSON
Department of Biology, University of York, York, YO1 5DD, UK

Abstract

How important is marine biodiversity for understanding global biodiversity? This introductory chapter compares and contrasts marine with terrestrial (including freshwater) diversity by looking at the present state of knowledge and our ability to explain diversity patterns and richness.

Diversity is not only species number; phyletic, morphological, ecological and genetic diversity are important concepts, if difficult to measure satisfactorily. In phyletic diversity, marine systems are superior in the metazoa, inferior in green plants and fungi. Ecological diversity is probably the measure most needed, hardest to achieve and, for the moment, impossible to compare between sea and land. In species numbers, weak evidence suggests there are ten times as many multicellular terrestrial as marine species, but our ignorance, even in well-known areas, is great. In patterns, particularly the tropical–polar gradient, and in richness, terrestrial studies suggest that the nature of energy input, heterogeneity and specificity are all strongly involved. The detailed balance of different factors is undoubtedly different in different groups, and many factors are involved. The importance of heterogeneity and specificity in the sea is rightly a subject of active research.

In knowledge and in ignorance, marine studies are on a par with terrestrial. In academic activity and repute, marine studies are not given sufficient weight. The relatively low level of productivity of the sea and the differences in the scaling of heterogeneity may be important in the contrast with terrestrial systems. Further comparisons of marine and terrestrial diversity should improve our understanding of both.

1.1 Introduction

Most of the world's surface is sea, but most general works on biodiversity concentrate on the land. Is that because land organisms are more diverse or just better known? How important is marine biodiversity for understanding global biodiversity? What do studies of global biodiversity tell us about what should or could be studied with marine organisms?

In this review, I discuss some of the points that may be important and concentrate on four questions. Firstly, what do we know about biodiversity in general, and the status of marine organisms in particular? Secondly, can major patterns in biodiversity be explained? The pattern that is relevant here is the contrast between marine and terrestrial diversity, taking terrestrial to include (almost always) freshwater diversity. I will approach this contrast by looking at the intensively studied latitudinal gradient. Thirdly, can richness be explained? Finally, what do the answers to these questions say about the present status, in a global context, of marine biodiversity studies?

1.2 What do we know about diversity?

Although most studies on biodiversity discuss the number of species, other aspects are at least as important. To mention four, there is cladistic or phylogenetic diversity, morphological diversity, ecological diversity and genetic diversity. All four are harder to study, and to quantify, than species diversity, and all are independent of it.

With phylogenetic diversity, we are familiar with diagrammatic trees indicating the relationship of cellular organisms and showing three main branches for eubacteria, archaebacteria and eucaryotes (e.g. Embry et al., 1994 or Schlegel, 1994). Viruses do not appear. In such diagrams, the higher plants, animals and fungi occupy three small, closely related branches. At a molecular level, multicellular organisms may be very uniform, but such diagrams are a reminder that any one measure of biodiversity is insufficient. Animals (metazoa), are of course primarily marine; all phyla appear to have started in the sea (even the Onychophora, now purely terrestrial, but marine in the Cambrian). However, multicellular plants and fungi are primarily terrestrial; it is the marine forms that are derived. How many of the unicellular groups are primarily marine is unknown, as are most other aspects of their diversity (Margulis et al., 1989; Hawksworth & Colwell, 1992).

Morphological and ecological diversity are perhaps more relevant to the preservation of biodiversity, but it is far from clear how either should be measured. Claims that the morphological disparity of Cambrian arthropods is about the same (Briggs *et al.*, 1992a,b) or greater (Foote & Gould, 1992) than that of modern arthropods seem highly dependent on the characters chosen. If it were to be measured just by variation in wing structure then clearly modern disparity would be greater, but uninformative about anything except insects. It is not clear that the use of characters in these Cambrian studies is appreciably less arbitrary than using just wings, or just 16 S RNA. Wilson (1992), from a more holistic standpoint, also doubts the claims for great Cambrian morphological (and ecological) diversity.

Measuring ecological diversity is even harder, but probably nearer what is needed for conservation decisions. Harper & Hawksworth (1994) make the same point. As different ecological characters are incommensurable, any single index is more or less arbitrary. Ecological comparisons among similar organisms are conceivable; comparing the ecological diversity of marine and non-marine scarcely so. Genetic diversity has the merit of being measurable in nucleotide differences, but would only be informative if the relative importance of changes in different parts of the genome were weighted.

So it is not surprising that quantitative studies of biodiversity have, apart from a nod to phyletic diversity, been based almost entirely on species counts. It is well-known that most species have not been described. Counts of those that have been and estimates of those to come (or to become extinct before they are described) can be found, for example, in Groombridge (1992) and Wilson (1992). It is less well-known that the numbers can be changed appreciably by changes in the definition of a species. In 1899, R.B. Sharpe, using a definition comparable to that of most modern botanists, counted 18 939 species of birds (see Bock & Farrand, 1980). Modern estimates are about 9000 (9021 in Bock & Farrand, 1980, and over 9200 in Howard & Moore, 1991) and the variations are primarily due to differences of opinions between splitters and lumpers. New bird species are still being described, but only at a rate of about three or four a year (May, 1994). The number can be expected to increase as it is usually easier, for various reasons, to push for the conservation of a species than of a subspecies.

Birds are exceptionally well-known. In most groups the true number of species is not known to an order of magnitude. There has been much discussion of the number of insect species, as this is widely believed to be the group with the most species. It is certainly the group with the most described species, Wilson's (1992) figure being 751 000. Estimates for all insects are now in the range of from 3 to 10–30 million, the lower part of this range being more likely (Gaston, 1991; Groombridge, 1992). Estimates for deep-sea invertebrates range from 200 000 to 1–10 million (Grassle & Maciolek, 1992; May, 1992), so they could be either more or less speciose than insects. Estimates for the number of fungal species, mostly terrestrial, are comparable to those of deep-sea invertebrates, 0.2 to 1–1.6 million (May, 1991, 1994; Groombridge, 1992). No other set of multicellular taxa is thought to have as many as a million species, so it is likely, despite the wide extent of the oceans, that most multicellular species are terrestrial. The numbers of species of viruses, procaryotes and protists can not yet sensibly be estimated (Giovannoni et al., 1990; Groombridge, 1992; Fauquet, 1994).

Estimates of the world number of species usually depend on extrapolating from more or less accurate estimates for a small patch. It is possible to compare the number of recorded species of British insects with those of British marine crustacea (Fig. 1.1), both very well-known by global standards. There are about ten times as many insects as Crustacea, 22 056 compared with 2240. A difference of this size is robust against the obvious problems of taxonomy, area and recording. But Fig. 1.1 shows this comes from the richness of the four major insect orders: Hymenoptera, Diptera, Coleoptera and Lepidoptera. The distribution in the number of recorded species in the other insect orders is very much like the distribution of numbers in the Crustacea.

In a fashion common to papers on biodiversity, I shall extrapolate wildly to say that there are globally about ten times as many terrestrial and freshwater species as there marine species of eucaryotes. In support, this figure matches the published guesses (Wilson, 1992) for photosynthetic plants, about 250 000 *versus* 25 000, where again the difference is due to one particular taxon, the angiosperms. Briggs (1994) thinks the ratio of terrestrial to marine is nearer 60:1, but his total for all marine invertebrates is less than May's (1992, 1994) minimum estimate for the deep sea alone.

The ratio of species in different groups is probably a more reliable statistic than the total number, but it too is subject to change. Among British insects, the Coleoptera and Lepidoptera have always been much more popular than Diptera or Hymenoptera. Gaston (1991) showed that the rate of discovery in the first two orders flattened off in the mid-nineteenth century, while in the other two the number of species recorded continues to increase at a steady rate. (Gaston's figures are based on

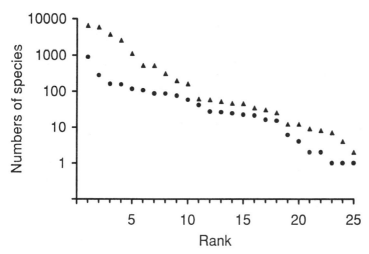

Fig. 1.1. A scree graph comparison of species diversity in British insects (▲) and British marine Crustacea (●). The ordinate is the logarithm of the number of species, the abscissa the rank order of those numbers. Data from Kloet & Hincks (1964–78), Emmet (1991) and Crothers (1997, with one datum from Dr M. Angel). The insect orders are, in rank order: Hymenoptera, Diptera, Coleoptera (including Strepsiptera), Lepidoptera, Homoptera, Heteroptera, Mallophaga, Collembola, Trichoptera, Thysanoptera, Neuroptera, Siphonaptera, Psocoptera, Ephemeroptera, Odonata, Plecoptera, Orthoptera, Anoplura, Diplura, Protura, Thysanura, Dictyoptera, Dermaptera, Mecoptera and Phasmida. The crustacean orders are, in rank order: Harpacticoidea, Amphipoda, Siphonostomatoidea, Decapoda, Podocopida, Poecilostomatoidea, Myodocopida, Isopoda, Mysidacea, Cyclopoida, Cumacea, Tanaidacea, Monstrilloidea, Calanoidea, Thoracica, Rhizocephala, Euphausiacea, Platycopida, Cladocera, Leptostraca, Mormonilloidea, Stomatopoda, Acrothoracica, Platycopioida and Misophrioida.

year of description rather than year of discovery in Britain). The numbers of both Hymenoptera and Diptera known in Britain have been increasing at a rate of about 30 species a year since the time of Linnaeus. Consequently, although equal numbers of Diptera and Coleoptera were known about a century ago (about 3500), there are now 60% more Diptera listed (about 6000 *versus* about 3800). However, even the Coleoptera and Lepidoptera are still adding species each decade that match the numbers of species in middlingly common orders. The recorded British Lepidoptera went up from 2357 in 1972 (Kloet & Hincks, 1964–78) to 2595 (Emmet, 1991) in 1991, averaging 125 per decade. It would be rash to draw conclusions from any differences less than times two when comparing distantly related groups. For instance, it is probably safe to say that there are about as many British Homoptera (1111 species in

Fig. 1.1, the fifth commonest order of insects) as there are British Harpacticoidea (900 species, the commonest Crustacea).

Although the numbers and ratios are so uncertain, they seem no more uncertain, on a logarithmic scale, in the sea than on land. They are also changing, due to improved collecting, improved taxonomy and natural causes. Both evolution and extinction are happening at rates as fast as ever before in the history of the planet. But the time scales are quite different. With extinction, the concern is with effects on scales of tens to thousands of years. With evolution, the scales are hundreds of thousands to tens of millions of years, four orders of magnitude slower. Wilson (1992) and Benton (1995) give graphs for the recorded history of both marine animals and terrestrial plants, which show the present (last 10 million years) high rate of increase, and the considerable variation in rate of change in the past. Both graphs show periods of around 200 million years when diversity was more or less static. The time series as a whole are certainly not stationary. Diversity has a strong historical component, and has apparently not been in equilibrium between origination and extinction in the Neogene and Quaternary, even without the effects of the current extinction crisis.

1.3 Can major patterns be explained?

The historical component no doubt figures largely in the differences between marine and terrestrial biodiversity; most lineages have been largely confined to one biome or the other for hundreds of millions of years. The differences arise from the action of ecological factors on species origination and extinction rates. Species can to some extent, and during the Pleistocene they were compelled to, migrate to find to an appropriate ecological environment. So it makes sense to consider whether and to what extent other major patterns in biodiversity (Brown, 1988) can be understood in terms of contemporary environmental factors.

The major pattern I shall consider is the well-known polar–tropical gradient in biodiversity, partly because it is shown in both marine and terrestrial systems. Other chapters in this book deal with this pattern in various marine systems. This pattern is remarkable for its pervasiveness, its lack of a generally agreed explanation, and for the plethora of explanations put forward. There are even those who dispute there is such a gradient. So I start with three empirical points and shall then turn to explanations. Firstly, there is such a gradient in diversity as a whole.

Secondly, many groups do not show this gradient, but they are more than compensated for by those that do. Thirdly, in detail the gradient is different in different groups and frequently not a uniform monotonic gradient. Understanding those points makes it easier to understand what is typical and what is unusual about marine biodiversity in a global context.

Fig. 1.2, which is a plot of trees/hectare in nearly equal sized plots, shows the major differences between temperate and tropical systems. Tropical communities are both richer and more even in their species diversity; the total number of species is greater, the scree plot of log number of individuals against log rank is flatter. Note that the total number of individuals is not significantly different in the two systems, a consequence of using a single life-form. The differences in diversity are reflected in all the species, not just in, say, the common or the rare ones.

There are many maps in the literature based on species in equal areas that show the polar–tropical gradient and its standard variations. On land, Africa usually has fewer species per unit area than South America or South-East Asia. So, when it is the richest, as in termites (Eggleton *et al.*, 1994), historical explanations are sought, but these are used to modify the primary explanation based on productivity or other contemporary factors. Similarly, in the oceans, it is normal for equal areas of the Indo-Pacific to be generally richer than the Atlantic, as for instance in Euphausiids (McGowan & Walker, 1993), often explained in terms of history (and total area). The same gradient is shown, albeit weakly and with much variance, in the deep-sea benthos (Rex *et al.*, 1993; and see Chapter 5). The importance of comparing like with like in diversity studies is evident in Brey *et al.* (1994).

Nevertheless, there are many groups that do not show a uniform gradient, i.e. that are not most diverse in the tropics. Penguins, being both marine and terrestrial, are a convenient and well-known example (Williamson, 1973). Birds as a whole are undoubtedly more diverse in the tropics than in temperate and arctic regions, though seabirds may not be. Various groups of insects are most diverse outside the tropics. A pattern with an extra-tropical maximum is shown in some parts of both the Hymenoptera and the Homoptera. In the Hymenoptera these are some (but not all) sections of the enormous parasitoid family Ichneumonidae, the herbivorous Symphyta (sawflies) and the Apoidea (bees) (Noyes, 1989). In the Homoptera, it is the Aphids and Psyllids (Dixon, 1985) that show this pattern.

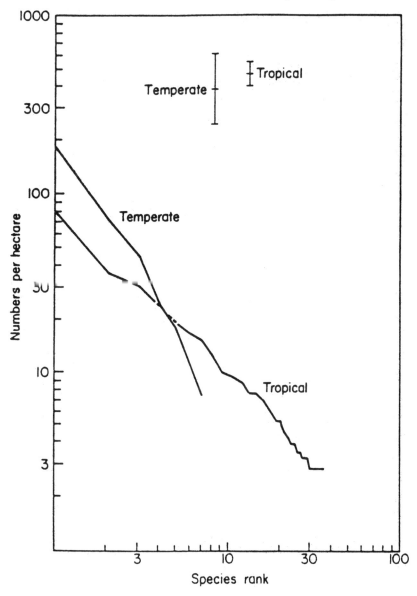

Fig. 1.2. A scree graph of the number of stems 10 cm (4 inches) in diameter in each species of tree in forests in Wisconsin and Brazil, on plots varying from 0.4–3 hectares. Numbers are plotted as log number per hectare, rank is also on a logarithmic scale. From Williamson (1973).

These patterns may just be marked examples of a general phenomenon. Rosenzweig & Abramsky (1993) argue that almost all tropical–polar clines, and for that matter clines in altitude, up mountains and into the deep sea, show a hump. The maximum diversity is not at the end of the cline, but at an intermediate point often much closer to the rich end. They give marine and terrestrial examples, and the phenomenon is well-known in the deep sea (Angel, 1993 and see Chapter 3; Gage & Tyler, 1991). As far as I am aware, none of the numerous general explanations of the polar–tropical gradient predicts this pattern. If humpiness is as common as Rosenzweig & Abramsky argue, then some new models are needed. Entomologists have made a start with possible explanations for the diversity of patterns seen in the Ichneumonidae (Gauld et al., 1992) based on the balance of advantage of different life-histories in relation to the diversity of other groups.

Another point not addressed by most explanations, is that the north–south pattern of the gradient is notably different in different groups. This is also evident, if not systematically documented, for marine groups. On land, equal area diversity plots are available for more groups in North America than anywhere else. North of Mexico, trees are most diverse in the east, in the southern Appalachians (Currie & Paquin, 1987); tiger beetles (Cicindelidae) in the centre of the continent, in Kansas (Pearson & Cassola, 1992; Pearson & Juliano, 1993); and birds in the western mountains (Cook, 1969; Williamson, 1981; Cotgreave & Harvey, 1994). The tiger beetles may show a hump in the USA, as there are fewer species in Texas, which is south of Kansas. In the Indian subcontinent they show two peaks, one, as expected, at the south of India and in Sri Lanka, the other in the Himalayas, in Nepal and Bhutan, and so there is a trough in central India (Pearson & Cassola, 1992).

The North American birds show the importance of both origination patterns and the detailed ecology of the group in explaining details of the geographical distribution of diversity. Although birds as a whole are richest in the west, individual families show different patterns (Cook, 1969). The Parulidae (wood warblers), are found throughout the continent up to the tree line but show a maximal diversity in the mid-eastern states, with a lesser centre of diversity in Mexico (Cook, 1969; Williamson, 1981). For birds as a whole, there is an interesting tongue of high species density stretching east–west across the contact zone of deciduous and coniferous forests (Williamson, 1981). Presumably this particular type and scale of habitat heterogeneity favours bird diversity in particular.

There are many explanations in the literature. Pianka (1994, and earlier editions) gives a well-known list of 10, which has been expanded to 14 in papers in the *American Naturalist* (Huston, 1979; Terborgh, 1985; Stevens, 1989; Pagel *et al.*, 1991), with a further explanation in Rohde (1992), who also gives a comprehensive list. Brown (1988) has a somewhat different, but largely overlapping, list of six factors. The comparison emphasises that many factors are correlated, and that it is difficult to decide what should be considered a distinct factor. Wilson (1992) favours energy, stability and area. Colwell & Hurtt (1994) put forward a null model. History, including origination rates, energy, constancy and heterogeneity are, I think, the factors most generally and enthusiastically championed, and there is a steady stream of papers debunking particular explanations in particular areas. For instance, Hoffman *et al.* (1994) show that in a zone covering 6 deg. of latitude, energy is negatively related, either by itself or in a multiple regression, to higher plant species richness. In the multiple regression, area and precipitation have significant positive coefficients. This negative relationship with energy, which has been seen in several studies (Huston, 1979) may fit with Rosenzweig & Abramsky's (1993) claim that there is in general a humped relationship between productivity and diversity.

It is obvious that no single explanation can be universally applicable, and it is likely that all the explanations have some validity in some circumstances. Rohde (1992) dismisses most explanations because they are inadequate on their own, but that is throwing out the baby with the bath water. Marine studies, particularly of deep-sea benthos, are likely to be important in teasing out the variation of the strength of different factors in different circumstances. We are some way from a sophisticated global model for any major group, let alone life as a whole. Nevertheless, there are some pointers to the factors that allow some groups to show remarkable richness in the tropics, though how these factors translate through population dynamics to co-existence is far from clear.

1.4 Can richness be explained?

If the variation in richness could be explained, it is possible that an understanding of the polar–tropical gradient, humpiness, the elevational gradient, and other major patterns would follow. In particular, the study of terrestrial richness might indicate why most marine systems are much less rich, if they are, or equally rich, if that turns out to be true.

There have now been sufficient quantitative studies on biodiversity that a start can be made on meta-analysis, seeing how concordant different studies are, and the extent to which combining weak conclusions leads to a stronger one. Rosenzweig & Abramsky (1993) tried to determine the relationship of biodiversity to productivity. In contrast, Wright *et al.* (1993) studied what factors have been implicated in variations in diversity. The weakness of their study is a weakness inherent in much modern ecology: most studies concentrate on a single hypothesis. Yet we all should know, from the statistical discipline of the design of experiments, that it is much more efficient and informative to study several factors simultaneously. So the review by Wright *et al.* is more a meta-study of what ecologists have thought important, than of what factors have been found to be important. The studies investigated are also largely terrestrial. In 53 studies I can only identify 5 that are freshwater (excluding marsh plants) and 3 that are marine, and of those 1 is on corals, 1 on mangroves and 1 on subtidal algae; there were no deep-sea or planktonic studies!

From 97 relationships, they found that the 41 relationships with energy-related factors were all statistically significant, as were all of the 4 for seasonality. In contrast, other factors gave mixed results. For instance habitat complexity etc gave nine significant to three not. So although there is a bias to studying energy and stability, there is good evidence that they are usually important. Counter-examples were noted above. The commonsense conclusion that the polar–tropical gradient in some way results from geophysical or astronomical effects on the environment, in seasonality and energy, is confirmed.

The question remains of how such physical effects are translated into variations in biodiversity. On land, the main contributors to richness are the insects and angiosperms, particularly trees, which set the structure of many terrestrial ecosystems. A simple model may encapsulate what happens, a model based on tree layering and two sorts of specificity in insects.

Terborgh (1985) pointed out that the simple geometrical facts of the angle of the sun and its variation through the year could explain the degree of layering seen in vegetation. When the sun is often vertical, as in the tropics, many layers, up to five, can coexist. When the sun is never vertical, and often at a low angle, trees grow as tall and narrow cones, in just one layer, as in the boreal forests. This variation in layering leads to an exponential increase in heterogeneity, as shown for instance by the abundance and variety of epiphytes. Roots are also a significant source of heterogeneity.

Terborgh's scheme is not enough by itself to explain the great richness of tropical rain forests. There are also non-forest terrestrial ecosystems, such as the South African fynbos (Cowling, 1992), which are remarkable for their plant diversity. (Fynbos is a small shrub heathland, with over 8500 plant species in less than 90 000 km^2, and surprisingly invasible by trees and shrubs.) The specificity of insects may add the necessary other factor. There are two common views on why insects, and in particular the Coleoptera, Diptera, Hymenoptera and Lepidoptera, are so diverse. These are that their metamorphic life-histories allow them to be either very specific plant herbivores, limited to just one tissue of a species, or to be very specific parasitoids (Wiegmann et al., 1993). The Lepidoptera exemplify the first, the Hymenoptera Parasitica the second. The Coleoptera, Diptera and the remaining Hymenoptera are more heterogeneous, but are still remarkable for their specialisations, particularly in the more species-rich sections.

A recent study in Costa Rica (Memmott et al., 1994) illustrates the point. This was a study of leaf-miners and their parasites, of insects that spend their larval lives in the thickness of a single leaf. Of 88 plant species, larval mines were found in 56. These plants supported 96 species of herbivore, which in turn supported 93 species of parasitoid. The parasitoids with one exception were Hymenoptera; the herbivores were a mixture of Lepidoptera, Coleoptera and Diptera. Any given species of miner occurred in only one host plant, although host plants could have between 1 and 12 species of miners. Similarly, 57 parasitoids were recorded as specific to a particular host, the others were recorded in from 2 to more than 20 hosts. This appearance of specificity is perhaps misleading, in that in large samples of over 100 mines, no parasitoid attacked fewer than four species, and there is a significant positive relationship of log sample size to log number of parasitoid species. Even so, specificity is clearly much higher than is normally found in larger, external, herbivores and predators.

All this leads to a model in which constancy and type of energy input lead to heterogeneity and a system in which many ecologically specific organisms can coexist. For this model to be worked out, the evolution of specificity should be included as well as the way specificity leads to co-existence. History, energy, constancy and heterogeneity are all essential components. How do these ideas impinge on marine studies? How can marine studies be used to clarify and modify these ideas?

These rich terrestrial groups suggest that the importance of heterogeneity and specificity have been undervalued, although the former, at least, appears in most lists of explanations. There is a contrast with the early views of the deep-sea benthos, which appeared to be unspecialised and living in a rather homogeneous habitat (Gage & May, 1993). However, the nature and scales of heterogeneity in the deep sea are now a subject of great interest (Gage & Tyler, 1991). In shallower waters the importance of specialisation, if not specificity, has been averred (Knowlton & Jackson, 1994).

1.5 Discussion: Where is marine diversity in a global context?

There are three aspects I want to end with; the general state of knowledge, the general academic regard for marine studies, and our understanding of the differences in marine and terrestrial systems.

In knowledge, or ignorance, I am impressed both by how much we know and how little we know. Well over a million species have been described, but that is perhaps 10% or less than those that await description. Even in the best-known phyla new species are still found at a significant rate, although there are a few orders that are apparently completely known. Genetic studies are turning up sibling species in many groups, and the problem of the definition of species remains. In all these respects, marine studies seem not appreciably different from terrestrial ones. The species to be discovered will always be, on average, smaller (Pine, 1994), harder to find, and with poorer diagnostic characters than those already known. There is still a great deal to be learnt of the distribution and abundance of species in all groups. Studies of phylogenetic diversity are making rapid progress thanks to new techniques. The more basic studies on ecological diversity await a satisfactory methodology.

In academic studies, and in the general perception of biodiversity, marine studies are doing poorly. This book may help to redress the balance. The crisis of extinction is more immediate and more obvious in terrestrial systems, so the political impetus for studying biodiversity is concentrated there. Perhaps as a result, academic papers on biodiversity ignore much that is relevant and useful in the sea. The review by Wright *et al.* (1993) mentioned above, which includes only 3 marine out of 53 studies is an example. The splendid volume of Ricklefs & Schluter (1993), of which Wright *et al.* is a part, has only 4 chapters out of 30 on marine topics. It could be argued that if, as I have suggested above, only about 10% of multicellular species are marine, these proportions are about right. That would be to ignore the importance of being able to compare marine and terrestrial systems in understanding the causes and consequences of biodiversity, and the very different view of biodiversity that is got from studying corals, plankton and deep-sea benthos than from studying tropical rain forests, migratory birds and arctic–alpine systems.

The major puzzle in biodiversity is, for me, the problem of the origin and maintenance of richness, and of the ecological structures that that implies. In solving this, comparisons of rich marine and terrestrial studies will be important. A major question is how diverse is the deep-sea benthos, a question addressed in various ways in Chapters 2, 5 and 7. Many of the statistics are derived by rarefaction and I would be happier if rarefaction were done by area (or volume) rather than by number of individuals (Brewer & Williamson, 1994). For this discussion I will presume, as before, that there are ten times as many terrestrial multicellular species as marine ones.

If that is so, some statistics by Cohen (1994) are relevant. He estimates that the productivity of the oceans is about 69 tonnes C a^{-1} km^{-2}, and for the continents 330 tonnes C a^{-1} km^{-2}, a difference of a little less than five times, which reduces to twice if the total area is taken. That is an appreciably lower factor than that I have assumed for species diversity. On the other hand, Cohen's estimate of biomass goes the other way, with 2 GT (giga-tonnes) C in the oceans compared with 560 GT on the continents, a factor of 280. Maybe both productivity and biomass limit the total diversity that can be supported, but that seems unlikely given the geological history, discussed above, showing that diversity has never been stationary. But some contemporary relationship between productivity, biomass, size distributions (May, 1994) and diversity seems likely (see also Chapter 8). This is an important and rather neglected area of biodiversity, and one in which a mixture of marine and terrestrial studies will be essential.

The relationship of heterogeneity to diversity is another area needing clarification (May, 1994). The logo of one of the conference sponsors, the Scottish Association for Marine Science, leads me to put forward a suggestion that has crossed my mind before, that the strength of heterogeneity at different scales is different on land and in the sea. On land travelling a kilometre or so (horizontally) often covers appreciable habitat diversity. In the sea, and I write as a one-time plankton man, that distance is negligible because of mixing. But at larger scales, three orders of magnitude or more, it could be argued that the sea is more heterogeneous, and that in usable vertical heterogeneity the sea is again the greater. That is all very speculative, but nevertheless underlines the importance of marine studies in getting a balanced view of global biodiversity.

Acknowledgements

I am grateful to Dr Martin Angel, Dr Kevin Gaston, Dr Peter Hogarth and Professor Robert May for discussion, information and advice.

References

Angel, M.V., 1993. Biodiversity in the pelagic ocean. *Conservation Biology*, **7**, 760–72.
Benton, M.J., 1995. Diversity and extinction in the history of life. *Science*, **268**, 52–8.
Bock, W.J. & Farrand, J., Jr, 1980. The number of species and genera of recent birds: A contribution to comparative systematics. *American Museum Novitates*, **2703**, 1–29.
Brewer, A. & Williamson, M., 1994. A new relationship for rarefaction. *Biodiversity and Conservation*, **3**, 373–9.
Brey, T., Klages, M., Dahm, C., Gorny, M., Gutt, J., Hain, S., Stiller, M. & Arntz, W.E., 1994. Antarctic benthic diversity. *Nature*, **368**, 297.
Briggs, D.E.G., Fortey, R.A. & Wills, M.A., 1992a. Morphological disparity in the Cambrian. *Science*, **256**, 1670–3.
Briggs, D.E.G., Fortey, R.A. & Wills, M.A., 1992b. Morphological disparity in the Cambrian. *Science*, **258**, 1817–18. (Reply to Foote & Gould).
Briggs, J.C., 1994. Species diversity: Land and sea compared. *Systematic Biology*, **43**, 130–5.
Brown, J.H., 1988. Species diversity. In *Analytical Biogeography*, ed. A.A. Myers & P.S. Giller, pp. 57–89. London: Chapman and Hall.
Cohen, J.E., 1994. Marine and continental food-webs: Three paradoxes? *Philosophical Transactions of the Royal Society, Series B*, **343**, 57–69.
Colwell, R.K. & Hurtt, G.C., 1994. Nonbiological gradients in species richness and a spurious Rapoport effect. *American Naturalist*, **144**, 570–95.
Cook, R.E., 1969. Variation in species density of North American birds. *Systematic Zoology*, **18**, 63–84.
Cotgreave, P. & Harvey, P.H., 1994. Associations among biogeography, phylogeny and bird species diversity. *Biodiversity Letters*, **2**, 46–55.
Cowling, R. (ed.), 1992. *The Ecology of Fynbos*. Cape Town: Oxford University Press.
Crothers, J., 1997. *A Key to the Major Groups of British Marine Invertebrates*. Aidgap. Preston Montford: Field Studies Council. (In press).
Currie, D.J. & Paquin, V., 1987. Large scale biogeographic patterns of species richness of trees. *Nature*, **329**, 326–7.
Dixon, A.F.G., 1985. *Aphid Ecology*. Glasgow: Blackie.
Eggleton, P., Williams, P.H. & Gaston, K.J., 1994. Explaining global termite diversity: Productivity or history? *Biodiversity and Conservation*, **3**, 318–30.

Embry, T.M., Hirt, R.P. & Williams, D.M., 1994. Biodiversity at the molecular level: The domains, kingdoms and phyla of life. *Philosophical Transactions of the Royal Society, Series B*, **345**, 21–33.

Emmet, A.M., 1991. Chart showing the life history and habits of the British Lepidoptera. In *The Moths and Butterflies of Great Britain and Ireland*, ed. M.A. Emmet and J. Heath, vol. **7(2)**, pp. 61–301. Colchester: Harley Books.

Fauquet, C.M., 1994. Taxonomy and classification—general. In *Encyclopedia of Virology*, ed. R.G. Webster & A. Granoff, pp. 1396–1410. London: Academic Press.

Foote, M. & Gould, S.J., 1992. Cambrian and Recent morphological disparity. *Science*, **258**, 1816–17.

Gage, J.D. & May, R.M., 1993. A dip into the deep seas. *Nature*, **365**, 609–10.

Gage, J.D. & Tyler, P.A., 1991. *Deep-sea Biology: A Natural History of Organisms at the Deep-sea Floor*. Cambridge: Cambridge University Press.

Gaston, K.J., 1991. The magnitude of global insect species richness. *Conservation Biology*, **5**, 293–96.

Gauld, I.D., Gaston, K.J. & Janzen, D.H., 1992. Plant allelochemicals, tritrophic interactions and the anomalous diversity of tropical parasitoids: The 'nasty' host hypothesis. *Oikos*, **65**, 353–7.

Giovannoni, S.J., Britschgi, T.B., Moyer, C.L. & Field, K.G., 1990. Genetic diversity in Sargasso Sea bacterioplankton. *Nature*, **345**, 60–3.

Grassle, J.F. & Maciolek, N.J., 1992. Deep-sea richness: Regional and local diversity estimates from quantitative bottom samples. *American Naturalist*, **139**, 313–41.

Groombridge, B. (ed.), 1992. *Global Biodiversity, Status of the Earth's Living Resources*. A report compiled by the World Conservation Monitoring Centre. London: Chapman and Hall.

Harper, J.L. & Hawksworth, D.L. (eds), 1994. Biodiversity: Measurement and Estimation. (Symposium). *Philosophical Transactions of the Royal Society, Series B*, **345**, 1–136. (Authors of the Preface, pp. 1–12).

Hawksworth, D.L. & Colwell, R.R. (eds), 1992. Microbial Diversity 21: Biodiversity Amongst Microorganisms and Its Relevance. (Symposium). *Biodiversity and Conservation*, **1**, 221–345. (Authors of the Introduction, pp. 221–6).

Hoffman, M.T., Midgley, G.F. & Cowling, R.M., 1994. Plant richness is negatively related to energy availability in semi-arid southern Africa. *Biodiversity Letters*, **2**, 35–8.

Howard, R. & Moore, A., 1991. *A Complete Checklist of the Birds of the World*. London: Academic Press.

Huston, M., 1979. A general hypothesis of species diversity. *American Naturalist*, **113**, 81–101.

Kloet, G.S. & Hincks, W.D., 1964–78. *A Check List of British Insects*, 2nd edn. Royal Entomological Handbooks for the Identification of British Insects, vol. **11**, parts 1–5. London: Royal Entomological Society of London.

Knowlton, N. & Jackson, J.B.C., 1994. New taxonomy and niche partitioning on coral reefs: Jack of all trades or master of some? *Trends in Ecology and Evolution*, **9**, 7–9.

Margulis, L., Corliss, J.O., Melkonian, M. & Chapman, D.J., 1989. *Handbook of Protoctista*. Boston: Jones & Bartlett.

May, R.M., 1991. A fondness for fungi. *Nature*, **352**, 475–6.

May, R.M., 1992. Bottoms up for the oceans. *Nature*, **357**, 278–9.

May, R.M., 1994. Biological diversity: Differences between land and sea. *Philosophical Transactions of the Royal Society, Series B*, **343**, 105–11.
McGowan, J.A. & Walker, P.W., 1993. Pelagic diversity patterns. In *Species Diversity in Ecological Communities*, ed. R.E. Ricklefs & D. Schluter, pp. 203–14. Chicago: Chicago University Press.
Memmott, J., Godfray, H.C.J. & Gauld, I.D., 1994. The structure of a tropical host–parasitoid community. *Journal of Animal Ecology*, **63**, 521–40.
Noyes, J.S., 1989. The diversity of Hymenoptera in the tropics with special reference to Parasitica in Sulawesi. *Ecological Entomology*, **14**, 197–207.
Pagel, M.D., May, R.M. & Collie, A.R., 1991. Ecological aspects of the geographical distribution and diversity of mammalian species. *American Naturalist*, **137**, 791–815.
Pearson, D.L. & Cassola, F., 1992. Worldwide species richness patterns of tiger beetles (Coleoptera: Cicindelidae): Indicator taxon for biodiversity and conservation. *Conservation Biology*, **6**, 376–91.
Pearson, D.L. & Juliano, S.A., 1993. Evidence for the influence of historical processes in co-occurrence and diversity of tiger beetle species. In *Species Diversity in Ecological Communities*, ed. R.E. Ricklefs & D. Schluter, pp. 194–202. Chicago: Chicago University Press.
Pianka, E.R., 1994. *Evolutionary Ecology*, 5th edn. New York: HarperCollins.
Pine, R.H., 1994. New mammals not so seldom. *Nature*, **368**, 593.
Rex, M.A., Stuart, C.T., Hessler, R.R., Allen, J.A., Sanders, H.L. & Wilson, G.D.F., 1993. Global-scale latitudinal patterns of species diversity in the deep-sea benthos. *Nature*, **365**, 636–9.
Ricklefs, R.E. & Schluter, D. (eds), 1993. *Species Diversity in Ecological Communities*. Chicago: Chicago University Press.
Rohde, K., 1992. Latitudinal gradients in species diversity: The search for the primary cause. *Oikos*, **65**, 514–27.
Rosenzweig, M.L. & Abramsky, Z., 1993. How are diversity and productivity related? In *Species Diversity in Ecological Communities*, ed. R.E. Ricklefs & D. Schluter, pp. 39–65. Chicago: Chicago University Press.
Schlegel, M., 1994. Molecular phylogeny of eukaryotes. *Trends in Ecology and Evolution*, **9**, 330–5.
Stevens, G.C., 1989. The latitudinal gradient in geographical range: How so many species coexist in the tropics. *American Naturalist*, **133**, 240–56.
Terborgh, J., 1985. The vertical component of plant species diversity in temperate and tropical forests. *American Naturalist*, **126**, 760–76.
Wiegmann, B.M., Mitter, C. & Farrell, B., 1993. Diversification of carnivorous parasitic insects: Extraordinary radiation or specialized dead end? *American Naturalist*, **142**, 737–54.
Williamson, M., 1973. Species diversity in ecological communities. In *The Mathematical Theory of the Dynamics of Biological Populations*, ed. M.S. Bartlett & R.W. Hiorns, pp. 325–35. London: Academic Press.
Williamson, M., 1981. *Island Populations*. Oxford: Oxford University Press.
Wilson, E.O., 1992. *The Diversity of Life*. New York: W.W. Norton & Company.
Wright, D.H., Currie, D.J. & Maurer, B.A., 1993. Energy supply and patterns of species richness on local and regional scales. In *Species Diversity in Ecological Communities*, ed. R.E. Ricklefs & D. Schluter, pp. 66–74. Chicago: Chicago University Press.

Chapter 2
Gradients in marine biodiversity

JOHN S. GRAY
Department of Biology, University of Oslo, Pb 1064, 0316 Blindern, Norway

I found to my great surprise at this enormous depth − not, as might be presumed according to Forbes' hypothesis − a poor and oppressed Fauna, but on the contrary a richly developed and varied animal life; so that my father (Michael Sars) was able in 1869 to increase the catalogue of the forms of animal life observed at the depths of 200−300 fathoms, by the addition of not less than 335 species (in all 427) of which nearly all were taken at one locality, namely the fishing place Skraaven in Lofoten.

We may therefore, ... presume that the Cretaceous formation is continued undisturbed at this present day in the depths of the Ocean ... while the Fauna at smaller depths, and especially the shore-Fauna, would be relatively short time, by reason of telluric and physical revolutions, be forced entirely to change its character.

G.O. SARS 1872.

Abstract

It has been claimed that just as on land the marine domain shows a cline of increasing diversity from poles to tropics. Recent data however, suggest that this may only hold for the northern hemisphere. On land there is a gradient of decreasing diversity with altitude. A marine counterpart has been suggested, but here with an increase in diversity with increasing depth. The original data on which this gradient was postulated are examined critically in the light of new data. It is shown that the species numbers found in the original data are not representative of shallow coastal areas. Recent data suggest that coastal diversity can be as high as that of the deep sea. However, there are few data sets available and many more are needed before gradients can be established.

Marine biologists usually use the rarefaction method to compare the diversity of samples of different sizes. It is known that the degree of dominance influences the estimates of diversity from rarefaction curves, yet this fact does not seem to be widely appreciated. Examples are given of the magnitude of error in diversity estimates that can be made.

Finally, the pressing need for better knowledge of the diversity of coastal systems, in light of the threats posed to loss of diversity, is emphasised.

2.1 Introduction

There are strong relationships between sampling scale and the processes that influence diversity (Huston, 1994). At small scales all species are presumed to interact with each other and to be competing for similar limiting resources. This has been called within-habitat or alpha diversity (Fisher *et al.*, 1943; Whittaker, 1960, 1967). At slightly larger scales habitat and/or community boundaries are crossed and sampling covers more than one habitat or community. This scale has been called between-habitat or beta diversity (Whittaker, 1960, 1975, 1976). At an even larger scale (regional scale) where evolutionary rather than ecological processes operate the pattern has been called landscape or gamma diversity (Whittaker, 1960; Cody, 1986). Huston (1994) has reviewed these definitions. It is clearly important, therefore, to specify what scale (and hence type of diversity) is being studied especially when dealing with patterns of diversity.

The best-known diversity pattern in the marine domain is probably that on a regional scale of generic diversity in coral genera and species. Highest diversity is found in the Malaysian archipelago and decreases across the Pacific Ocean to the east and to the Indian Ocean in the west, with lowest diversity in the Caribbean (Stehli & Wells, 1971). Clearly this pattern is of gamma scale diversity. Similar patterns have been shown for mangroves, and gastropod snails (see Huston, 1994). The reason for this high level of diversity in the Indo-Pacific region is thought to be the result not only of a long period of evolutionary stability, but also to the high diversity of islands and archipelagos that have given rise to periods of isolation and allopatric speciation followed by reunification and sympatric speciation. Throughout geological time there have been periods of massive extinction followed by rapid evolution (Kauffman & Fagerstrom cited in Huston, 1994).

At the largest diversity scale terrestrial systems show characteristic gradients with high diversity in the tropics and at low altitudes, and low diversity at the poles and at high altitudes. Similar gradients and diversity patterns have been postulated for the marine environment with high diversity in tropical seas and low diversity in polar seas. Not surprisingly attempts were made to see if such patterns applied to the marine domain

20   *Gradients in marine biodiversity*

and from the Arctic to the tropics there appeared to be an increase in species diversity of hard substratum epifauna (Thorson, 1957). Stehli & Wells (1971) showed that bivalve molluscs at species, genus and family levels show increased diversity towards the tropics in the Indo-Pacific. Recent data from the deep sea (Rex *et al.*, 1993) purports to confirm these findings. However, in the latter data there is much scatter and if one removes the data from the Norwegian Sea, which have low diversity, the trends are far less clear if evident at all. The Norwegian Sea is a recently glaciated area and it is not surprising that diversity is low. In the southern hemisphere the gradient is less clear.

Clarke (1992) suggested that although there may be an increase in diversity from Arctic to tropics this is not clearly the case in the southern hemisphere. He showed that the Antarctic has high diversity for many taxa. Likewise data from Australia show that in coastal soft sediment areas 800 species have been recorded from just 10 m^2 of sediment in Bass Strait and that 700 species occur in Port Phillip Bay sediments (Poore & Wilson, 1993). These values are as high as the highest values for soft sediments found anywhere. Furthermore, it is known that seaweed (macroalgal) diversity is higher in temperate latitudes than the tropics and lowest at the poles (Silva, 1992). It seems probable that there is a cline of increasing diversity from the Arctic to the tropics but the cline from the Antarctic to the tropics is far less well established if it occurs at all. However, there is clearly a need to better document diversity patterns in other areas of the southern hemisphere such as the African and American continents.

These large-scale patterns are discussed in Chapter 11, and will not be discussed here. In this brief review I will concentrate on within-habitat (alpha) and between-habitat (beta) diversity.

One of the best known within-habitat diversity patterns is the gradient in the fauna of soft sediments from low intertidal species diversity to high deep-sea species diversity (Hessler & Sanders, 1967). First I will give a critical analysis of the data and present data that show that the patterns may not be as has been assumed and then I will argue that much theorising has been done on the basis of a false interpretation.

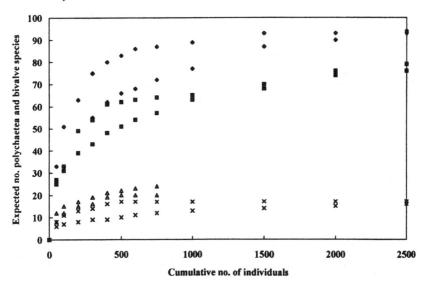

Fig. 2.1. Sanders' original data giving the basis for the shallow–deep sea increase in diversity. The expected number of species are calculated using the method described by Sanders (1968). ♦, tropical shallow; ■, deep sea; ▲, tropical estuary; ×, boreal continental climate.

2.2 Shallow water–deep sea sediment benthos diversity gradient

Sanders (1968) collected samples with a semi-quantitative anchor-dredge from a variety of different habitats from the boreal and the tropics and from estuaries to the deep-sea slope. Fig. 2.1 shows the data plotted as rarefaction curves (see below).

Sanders (1968) also gave a figure that showed the percentage of the total fauna represented by the polychaete–bivalve fraction. From this it is possible to estimate the total number of species. This gives the data shown in Table 2.1.

The boreal shallow area has low diversity with the deep-sea slope and has low diversity compared with the tropical shallow. Thus two gradients had to be explained for which Sanders (1968) erected the stability–time hypothesis. The basic idea is that in shallow (and boreal) areas the environment is physically unstable, changing frequently and irregularly so that species, over evolutionary time, have had to adapt primarily to the changing physical environment and not to other species. In shallow (and boreal) areas competition can be severe but does not lead to niche specialisations since the time between disturbances is

Table 2.1. *Species diversity taken with a benthic anchor-dredge*

Area	Total number species of polychaetes and bivalves	Total no. species
Boreal shallow (20 m)	10–19	12–32
Deep sea (487–2086 m)	62–76	86–99
Tropical estuary	20–23	24–28
Tropical shallow water	93–95	110–176

Source: From Sanders (1968).

shorter than that required to produce specialisations. Sanders' hypothesis postulates that this results in shallow areas having few species with large overlapping niches. In deep-sea (and tropical) areas by contrast the environment is very stable over evolutionary time so that species do not need to adapt to the environment, but instead have adapted to each other by becoming highly specialised. Here the result is that there are many species that co-occur and have small non-overlapping niches (Grassle & Sanders, 1973).

Alternative hypotheses have been erected to explain the shallow–deep sea gradient: the disturbance hypothesis (Dayton & Hessler, 1972), the area hypothesis (Abele & Walters, 1978), the spatial heterogeneity hypothesis (Jumars & Gallagher, 1982) and the dynamic equilibrium model (Huston, 1994). For a review of these hypotheses, see Huston (1994).

Sanders (1969) distinguished between the shallow, coastal fauna of the east and west coasts of the USA showing that at Friday Harbor, San Juan Island, Washington State the rarefaction curves were much higher than those of the boreal shallow (Buzzard's Bay, MA). He explained this by the fact that the temperature range on the west coasts of continents is much lower than that on the east coasts (6.5–12°C below 10 m at Friday Harbor and −1.5 to 23°C in Buzzard's Bay). Sanders called the west coast fauna the Boreal Maritime Climate and distinguished this from the Boreal Continental Climate on the east coast. Fig. 2.2 shows the data.

Although Sanders (1969) claimed that 'the diversity found in the maritime climate shallow boreal samples is as high as those of the deep sea stations and approaches, and perhaps reaches, the levels encountered in the more diverse tropical shallow-water habitats' his data do not show

J. S. Gray   23

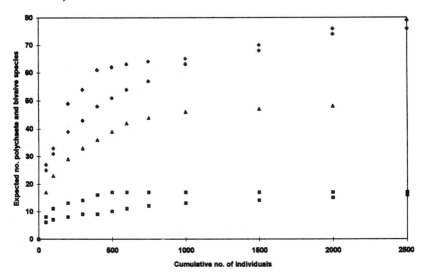

Fig. 2.2. Comparison of diversity (rarefaction curves) of the benthic sediment-living fauna of the deep sea with that of the boreal climate (San Juan Island, WA, USA) with the boreal maritime climate (Buzzard's Bay, MA, USA) from Sanders (1969). ♦, deep sea; ■, boreal continental climate; ▲, boreal maritime climate.

this. The diversity is considerably less than that of the deep-sea stations (maximum species number 80 whereas there were only 49 in the boreal maritime climate).

Much new data have been obtained since Sanders' (1968) study but only the deep-sea component of the putative gradient has been considered (e.g. Rex, 1981). Rex's data are used to illustrate the fact that the coast–deep sea soft-sediment gradient does not increase linearly but tends to show maxima between 2000 and 3000 m on the continental slope and decreases on the abyssal plain. Fig. 2.3 shows the patterns obtained.

What is most surprising about Fig. 2.3 is the extremely low expected number of species found in Rex's study and the large scatter in the data. Yet hypotheses have been erected to explain this dome-shaped curve (e.g. Huston's (1994) dynamic equilibrium model) without questioning whether the pattern described really represents the shallow–deep sea gradient.

All of the hypotheses advanced to explain the shallow–deep gradient have assumed that the data presented by Sanders are representative of a general pattern of low species diversity in shallow coastal areas. Surprisingly no-one has questioned whether or not this is the case. This

Gradients in marine biodiversity

(a)

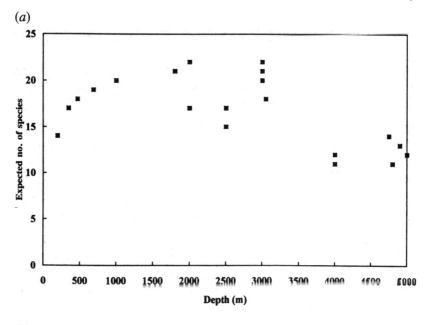

(b)

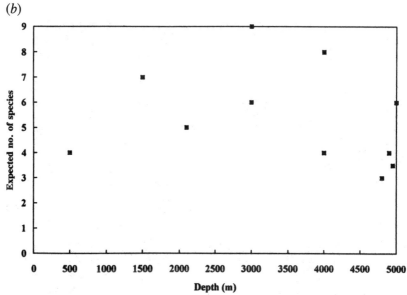

Fig. 2.3. Expected species numbers against depth (from Rex, 1981). (a) Polychaeta. (b) Protobranchia. (c) Gastropoda.

(c)

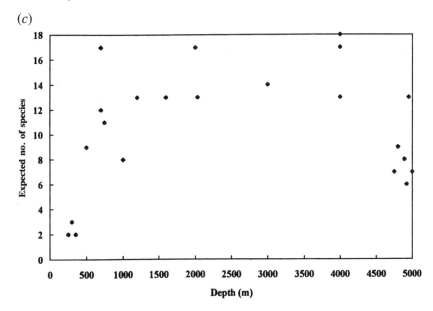

is all the more remarkable since there are large numbers of studies that have been done in coastal areas that show that the low number of species shown in Table 2.1 and Figs 2.1–2.3 (less than 100) are not at all typical for the diversity of coastal sediments.

As an illustration of typical modern data from a coastal area, Fig. 2.4 shows a rarefaction curve for the polychaete–bivalve fraction of data from a coastal environment, the relatively polluted inner Oslofjord, depth range 20–120 m (Aschan & Skullerud, 1990). In this data set there is a total of 148 species of which the polychaete–bivalve fraction comprises 70.1% of the species and 97.5% of the total number of individuals. Compared with Sanders' data (Figs 2.1 and 2.2) it shows a much higher number of species for a comparable number of individuals. Yet these data are from a disturbed environment where species diversity is known to be lower than in unperturbed areas such as the outer Oslofjord.

There are many recent studies of soft-sediment fauna in boreal areas that show high numbers of species from soft-sediment habitats, e.g. Pearson (1975), Gray et al. (1991) and Olsgard & Hasle (1993) to name but a few. Thus Sanders' data are clearly not representative of shallow water species diversity.

One could argue that quantitative techniques are much improved since the 1960s and thus it is to be expected that species numbers are higher in recent studies. Yet in fact the earliest quantitative studies of the

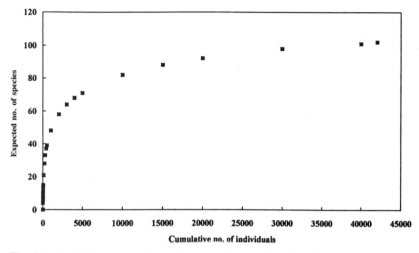

Fig. 2.4. Rarefaction curve for benthic data taken with a Day Grab from the inner part of Oslofjord, Norway (based on Aschan & Skullerud, 1990).

benthos of soft-sediments in coastal areas of Denmark and Norway (which were used by C.G.I. Peterson in the early 1900s to produce the first description of assemblages of species), described 264 taxa (Stephenson et al., 1982). These data were from 193 sites. Not all species were determined and methods were much less rigorous than today so that species diversity is certainly much higher. It is however, possible to compare in a fairly rigorous and quantitative way recent data from the deep-sea and shallow environments.

Grassle & Maciolek (1993) have presented data that 'indicate a much greater diversity of species in the deep sea than previously thought'. They thus argue that deep-sea diversity is much higher than even Sanders thought, thereby strengthening even further the idea that species diversity increases from shallow to deep areas. From depths of between 1500 m and 2100 m along a transect of 176 km off the east coast of the USA a total of 798 species and 90 677 individuals was found from a total area sampled of 21 m^2. Rarefaction curves were calculated for a reduced data set covering nine stations at the 2100 m contour, which gave around 625 species and 65 000 individuals and is shown in Fig. 2.5.

Using data obtained from monitoring of oil activities in the Norwegian continental shelf for a north to south transect from the Heidrun field to the Tommeliten field in the North Sea (a distance of 1200 km), a total of 620 species and 39 582 individuals was found (Fig. 2.5). The depth range covered was from 70–305 m.

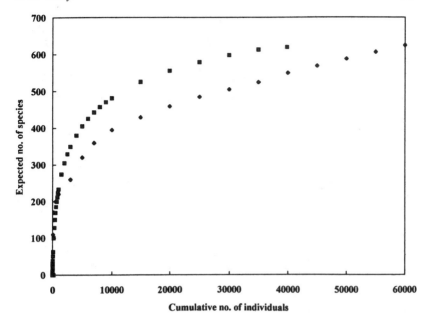

Fig. 2.5. Hurlbert's rarefaction curves of the fauna of the Norwegian shelf (Gray, 1994) and the deep sea off the east coast of the USA (redrawn from Grassle & Maciolek, 1993). ♦, deep sea; ■, Norwegian shelf.

Fig. 2.5 shows that the Norwegian shelf data have a similar number of species from a lower number of individuals. From the rarefaction curve for the deep-sea data for 40 000 individuals approximately 550 species were found, which is less than the 620 species from the Norwegian continental shelf. Thus, on the basis of Fig. 2.5, the continental shelf of Norway has as high, if not higher, species diversity than the new data from the deep sea. This raises the question of whether the paradigm of high deep-sea diversity needs to be revised?

It might be argued that along a transect of 1200 km compared with only 176 km in the deep sea one is sampling between-habitat (beta) diversity rather than within-habitat (alpha) diversity since the sediment and depth are not uniform. There is no clear definition of what constitutes a within-habitat comparison. One clear example of a study of alpha (within-habitat) diversity is that of Warwick et al. (1986) who showed the influence of the feeding behaviour of a single large polychaete on the abundance and community structure of meiofauna. The scale of this study, however, was within a radius of 15 cm.

Can one argue that soft sediments are one habitat and all comparisons are thereby within-habitat, whereas if one compares the diversity of sediments with that of hard surfaces and, say, a sea-grass bed, then that constitutes a between-habitat comparison? It is known that a variety of sediment properties affect the number and kinds of species inhabiting a given area. Marine benthic community structure and hence species diversity is known to vary with median grain size, with sorting (Gray, 1974; Rhoads, 1974) and with sediment heterogeneity (Etter & Grassle, 1992). Likewise depth is known to be an important variable influencing species diversity and distribution patterns (see Gage & Tyler, 1991 for a full discussion of depth-related patterns). However, if both depth and sediment properties co-vary then it is difficult to distinguish between correlation and causal factors.

For example Etter & Grassle (1992) found relationships between species diversity and depth and between species diversity and sediment heterogeneity. For the fauna of the deep-sea slope off the coast of the east USA sediment particle size diversity had an important role in determining the number of species within a community. Bathymetric patterns were largely attributable to changes in sediment characteristics with depth. Thus, if comparative studies are to be made, influences of depth and habitat heterogeneity on species diversity must be determined.

In Grassle & Maciolek's (1993) study the deep-sea slope sediment was relatively uniform sandy mud to clayey muds. Although both depth and type of sediment varied across the Norwegian shelf, Gray (1994) found no relationship between either number of species and depth or number of species and median particle diameter or sorting (a measure of sediment heterogeneity). Sorting may not, however, be a good indicator of sediment heterogeneity. R.J. Etter (personal communication) has replotted the data reported in Etter & Grassle (1992) and finds no relationship between species diversity and sorting. No re-analysis of the Norwegian sediment data has been done so whether there is a relationship between species diversity and sediment heterogeneity on the Norwegian continental shelf is not known.

Gray (1994) discussed in more detail the validity of making a comparison between the Norwegian continental shelf data and that of the deep-sea slope. What is clear is that the data shown by Sanders (Figs 2.1 and 2.2) are not representative of the species diversity of the coastal area as a whole and the paradigm of low coastal and high deep-sea diversity needs to be re-examined.

Confirmatory evidence of the high diversity of coastal areas has come from Poore & Wilson's (1993) data showing that in Australia 800 species have been recorded from just 10 m² of sediment in Bass Strait and that 700 species occur in Port Philip Bay sediments. These data support the contention that the shelf has equal, if not higher, diversity than the deep sea and strengthen the need for a focusing of efforts on coastal diversity.

2.3 Problems with rarefaction curves

Finally I will consider a methodological problem that does not seem to be widely appreciated when considering marine diversity. Following Sanders, diversity in the marine environment is usually expressed as a curve of species number against the number of individuals, so-called rarefaction curves. Sanders constructed the rarefaction method in order to be able to compare samples of different size. The method is based on estimating the expected number of species that will be found at a reduced sample size assuming that the species dominance patterns are identical to that of the full sample size. The expected number of species $(E(s)_n)$ for a smaller number of individuals (n) can thus be estimated so that sample sizes can be compared. Sanders' method has been shown to overestimate the expected number and has been revised by Hurlbert (1971). This method is widely used to compare diversity of samples in the marine domain, but has not been used with terrestrial or freshwater data.

A major problem with this method, which has been recognised by Gage & Tyler (1991) and Gage & May (1993), is the assumption of a common dominance pattern at reduced sample sizes. Gage & Tyler (1991) show that for data from six box-core samples from the Rockall trough, 80 species were found from 1600 individuals. A rarefaction curve was plotted. If one wishes to compare a sample size of 400 individuals the curve gives an $E(s)_n$ value of 50 species. Yet if the dominance pattern was in fact perfectly even the estimate should have been 80 species, while for a perfectly uneven sample the estimate is 20 species. Thus the estimate is 50 species with a range from 20 to 80, a huge difference! Whilst this is a theoretical argument it is extremely important since should the dominance pattern vary from that of the total sample then the rarefaction estimates of diversity will be incorrect.

In order to obtain data comparable to that shown by Grassle & Maciolek (1993) the diversity study of the Norwegian continental shelf was done by pooling samples in a sequence starting with sample 1 of the most northerly field. The cumulative number of species was plotted against the cumulative number of individuals for the sequentially pooled samples. In order to assess whether the assumption of the rarefaction

Table 2.2. *Dominance changes with increasing sample size. Values are % of total abundance*

A. Pooled samples within sites and within fields

Dominant species	Tommeliten	Ekofisk	Gullfaks	Snorre	Heidrun
1st	13.7	15.9	23.5	12.4	6.5
2nd	7.0	11.0	11.4	10.5	6.0
3rd	6.2	8.0	7.1	8.0	5.8
10th	2.7	2.1	1.6	2.0	3.1

B. Pooled samples within fields

Dominant species		Dominance %
1st:	Mean	14.4
	Range	6.5–23.5
2nd:	Mean	9.3
	Range	6.0–11.5
3rd:	Mean	7.0
	Range	5.8–8.0
10th:	Mean	2.3
	Range	1.6–3.1

C. Pooled samples over all fields

Dominant species	Dominance %
1st	7.1
2nd	5.1
3rd	3.5
10th	1.7

method that the dominance pattern for the total cumulative number of species and individuals was the same for smaller samples, dominance patterns were determined over different sample sizes. Dominance patterns were calculated for pooled samples within sites within fields, for pooled sites within fields and finally for pooled fields. Table 2.2 shows the results.

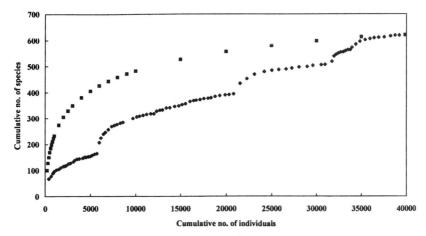

Fig. 2.6. Cumulative species number (♦) obtained by summing samples across the gradient and Hurlbert's rarefaction curves (■) for the benthic fauna of the Norwegian continental shelf.

It is the final (Table 2.2C) data that are used as the basis for estimating $E(s)_n$. Table 2.2 shows clearly that dominance decreases with increasing sample size. This is a general property of all samples that we have studied. The result is that the use of Hurlbert's (1971) rarefaction method to estimate species number at reduced sample size will over-estimate the species richness. This is shown clearly in Fig. 2.6, where for a sample size of 10 000 individuals the rarefaction curve estimates approximately 500 species yet the cumulative curve shows that in the actual sample only just over 300 species were in fact obtained. The extent of the over-estimate will be proportional to the degree of dominance, the higher the dominance the greater the over-estimate. It is clear that rarefaction methods should be used with extreme caution if the aim is to compare diversity of small sample sizes, since these are likely to have very different dominance patterns to large sample sizes, as pointed out by Gage & May (1993).

2.4 Conclusion

Much controversy has been generated by Grassle & Maciolek's (1993) claim, based on extrapolation from their data, that there may be 10 million undescribed species in the deep ocean. This suggestion was based on an extrapolation from the data available on the number of new species found from just 21 m^2 off the east coast of the USA to the whole area of the deep sea. May (1993) has strongly criticised the suggestion and his estimate is that there are probably only 500 000 species undescribed. This debate has focused the research agenda for marine diversity heavily towards the need to describe the unknown species from the deep sea. The data presented above and that given by Poore & Wilson (1993) suggest that Sanders' original paradigm may be incorrect.

There is clearly a need to obtain more quantitative data in a comparative way so that marine diversity gradients can be properly compared. There are many data sets from coastal sediments that could be used in this context. The influential IUBS/SCOPE/UNESCO study *From Genes to Ecosystems: A Research Agenda for Biodiversity* (Solbrig, 1991) suggests that marine high diversity hot spots should be studied and emphasises that this requires a much greater effort than is the case for land areas.

Within coastal areas there are a wide variety of different habitats. These range from hard rocky surfaces with a rich encrusting flora and fauna and interstitial fauna within for example clumps of mussels, to kelp forests in boreal areas with a very high diversity of fauna within holdfasts (e.g. Moore, 1972 found over 350 species on the species-poor east coast of the UK), to sea-grass beds with very rich associated fauna (McRoy & Helfferich, 1977), to mangal (MacNae, 1968; Walsh, 1974) and coral reefs (Loya, 1972; Sheppard, 1980; Huston, 1985) with their associated flora and faunas. Thus in terms of between-habitat diversity the coast is likely to be much richer than the deep sea. It is the coastal area that is subjected to man's influence from habitat destruction, either directly via construction works or indirectly by changing hydrology and sediment transport in rivers, from pollution from land-based sources and countless other disturbances including climate change (Gaudie, 1990). Thus it can be argued that the research agenda for marine diversity should concentrate primarily on the coastal regions particularly in tropical and subtropical areas where man's influences are greatest.

References

Abele, L.G. & Waters, K., 1979. Marine benthic diversity: A critique and an alternative explanation. *Journal of Biogeography*, 6, 115–26.

Aschan, M. & Skullerud, A.M., 1990. Effects of changes in sewage pollution on soft-bottom macrofauna communities in the inner Oslofjord, Norway. *Sarsia*, 75, 169–80.

Clarke, A., 1992. Is there a latitudinal diversity cline in the sea? *Trends in Ecology and Evolution*, 7, 286–7.

Cody, M.L., 1986. Diversity, rarity, and conservation in Mediterranean-climate

regions. In *Conservation Biology: The Science of Scarcity and Diversity*, ed. M.E. Soule, pp. 122–52. Sunderland, MA: Sinauer.
Dayton, P.K. & Hessler, R.R., 1972. The role of biological disturbance in maintaining diversity in the deep sea. *Deep Sea Research*, **19**, 199–208.
Etter, R.J. & Grassle, J.F., 1992. Patterns of species diversity in the deep-sea as a function of sediment particle size diversity. *Nature*, **360**, 576–8.
Fisher, R.A., Corbet, A.S. & Williams, C.B., 1943. The relationship between the number of species and the number of individuals in a random sample of an animal population. *Journal of Animal Ecology*, **12**, 42–58.
Gage, J.D. & May, R.M., 1993. A dip into the deep sea. *Nature*, **365**, 609–10.
Gage, J.D., & Tyler, P.A., 1991. *Deep Sea Biology*. Cambridge: Cambridge University Press.
Gaudie, A., 1990. *The Human Impact on the Natural Environment*. Oxford: Blackwell Scientific.
Grassle, J.F. & Maciolek, N.J., 1993. Deep-sea species richness: Regional and local diversity estimates from quantitative bottom samples. *American Naturalist*, **139**, 313–41.
Grassle, J.F. & Sanders, H.L., 1973. Life histories and the role of disturbance. *Deep Sea Research*, **22**, 643–59.
Gray, J.S., 1974. Animal-sediment relationships. *Oceanography and Marine Biology; An Annual Review*, **12**, 223–61.
Gray, J.S., 1994. Is the deep sea really so diverse? Species diversity from the Norwegian continental shelf. *Marine Ecology Progress Series*, **112**, 205–9.
Gray, J.S., Clarke, K.R., Warwick, R.M. & Hobbs, G., 1991. Detection of initial effects of pollution on marine benthos: An example from the Ekofisk and Eldfisk oilfields, North Sea. *Marine Ecology Progress Series*, **66**, 285–99.
Hessler, R.R. & Sanders, H.L., 1967. Faunal diversity in the deep sea. *Deep Sea Research*, **14**, 65–78.
Hurlbert, S.H., 1971. The nonconcept of species diversity: A critique and alternative parameters. *Ecology*, **52**, 577–86.
Huston, M.A., 1985. Patterns of species diversity in relation to depth at Discovery Bay, Jamaica. *Coral Reefs*, **4**, 19–25.
Huston, M.A., 1994. *Biological Diversity: The Coexistence of Species on Changing Landscapes*. Cambridge: Cambridge University Press.
Jumars, P. & Gallagher, E.D., 1982. Deep sea community structure: Three plays on the benthic procenium. In *Rubey Colloquium Environment of the Deep Sea*, ed. W.E. Ernst & J.G. Morin, pp. 217–55. Los Angeles, CA: Rubey Colloquium.
Loya, Y., 1972. Community structure and species diversity of hermatypic corals at Eilat, Red Sea. *Marine Biology*, **13**, 100–23.
MacNae, W., 1968. A general account of the fauna and flora of mangrove swamps and forests in the Indo-west Pacific Region. *Advances in Marine Biology*, **6**, 73–270.
May, R.M., 1993. Bottoms up for the oceans. *Nature*, **357**, 278–9.
McRoy, C.P. & Helfferich, C. (eds), 1977. *Sea-grass Ecosystems. A Scientific Perspective*. New York: Dekker.
Moore, P.G.M., 1972. The kelp fauna of Northeast Britain III. Qualitative and quantitative ordinations, and the utility of a multivariate approach. *Journal of Experimental Marine Biology and Ecology*, **16**, 257–300.
Olsgard, F. & Hasle, J., 1993. Impact of waste from titanium mining on benthic fauna. *Journal of Experimental Marine Biology and Ecology*, **172**, 185–213.
Pearson, T.H., 1975. The benthic ecology of Loch Linnhe and Loch Eil, a sea-loch

system on the west coast of Scotland. I. Changes in the benthic fauna attributable to organic enrichment. *Journal of Experimental Marine Biology and Ecology*, **20**, 1–41.

Poore, G.C.B. & Wilson, G.D.F., 1993. Marine species richness. *Nature*, **361**, 597–8.

Rex, M.A., 1981. Community structure in the deep sea benthos. *Annual Review of Ecology and Systematics*, **12**, 331–53.

Rex, M.A., Stuart, C.T., Hessler, R.R., Allen, J.R., Sanders, H.L. & Wilson, G.D.F., 1993. Global-scale latitudinal patterns of species diversity in the deep-sea benthos. *Nature*, **365**, 636–9.

Rhoads, D., 1974. Organism-sediment relations on the muddy sea floor. *Oceanography and Marine Biology Annual Review*, **12**, 263–300.

Sanders, H.L., 1968. Marine benthic diversity: a comparative study. *American Naturalist*, **102**, 243–82.

Sanders, H.L., 1969. Benthic marine diversity and the stability-time hypothesis. In *Diversity and Stability in Ecological Systems*, ed. G.M. Woodwell & H.H. Smith, pp. 71–81. Brookhaven Symposium No. 22. Brookhaven, NY: Biology Department, Brookhaven National Laboratory.

Sars, G.O., 1872. On some remarkable forms of animal life from the great depths off the Norwegian coast. I. University of Christiana (Oslo).

Sheppard, C.R.C., 1980. Coral cover, zonation and diversity on reef slopes of Chagos Atolls and population structure of major species. *Marine Ecology Progress Series*, **2**, 193–205.

Silva, P.C., 1992. Geographic patterns of diversity in benthic marine algae. *Pacific Science*, **46**, 429–37.

Solbrig, O.T. (ed.), 1991. *From Genes to Ecosystems: A Research Agenda for Biodiversity*. Sponsored by Cambridge, MA: IUBS.

Stehli, F.G. & Wells, J.W., 1971. Diversity and age patterns in hermatypic corals. *Systematic Zoology*, **20**, 115–26.

Stephenson, W., Willimas, W.T. & Cook, S.D., 1972. Computer analyses of Petersen's original data on bottom communities. *Ecological Monographs*, **42**, 387–415.

Thorson, G., 1957. Bottom communities (sublittoral and shallow shelf). In *Treatise on Marine Ecology and Palaeoecology*, vol. 1. *Ecology*, ed. J.W. Hedgpeth, pp. 461–534. Memoirs of the Geological Society of America. New York: Geological Society of America.

Walsh, G.E., 1974. Mangroves: A review. In *Ecology of Halophytes*, ed. R.J. Reimold & W.H. Queen, pp. 51–174. New York: Academic Press.

Warwick, R.M., Gee, J.M., Berge, J.A. & Ambros, W.G., 1986. Effects of the feeding activity of polychaete *Streblosoma bairdii* (Malmgren) on meiofaunal abundance and community structure. *Sarsia*, **71**, 11–16.

Whittaker, R.H., 1960. Vegetation of the Siskiyou Mountains, Oregon and California. *Ecological Monographs*, **30**, 279–338.

Whittaker, R.H., 1967. Gradient analysis of vegetation. *Biological Reviews*, **42**, 207–64.

Whittaker, R.H., 1975. *Communities and Ecosystems*. 2nd edn. New York: Macmillan.

Whittaker, R.H., 1977. Evolution of species diversity in land communities. *Evolutionary Biology*, **10**, 1–67.

Chapter 3
Pelagic biodiversity

MARTIN V. ANGEL
Institute of Oceanographic Sciences, Wormley, Godalming, Surrey, GU8 5UB (now at Southampton Oceanography Centre, Empress Dock, Southampton, SO14 3ZH, UK)

Abstract

This chapter looks at the factors controlling large-scale biogeographical patterns of pelagic taxa, the present biogeography of pelagic communities, and the origins of pelagic biodiversity. Oceanic versus neritic biodiversity is considered before the existing patterns of distribution of bathyetric communities are discussed.

3.1 Introduction

Biodiversity, according to the Biodiversity Convention, is defined as 'the variability among living organisms from all sources including *inter alia*, terrestrial, marine, and other aquatic ecosystems and the ecological complexes of which they are a part; this includes diversity within species, between species and of ecosystems'. Biodiversity therefore includes all variability in the natural world, at all scales in time and space, and at all levels of organisation from the genetic variability within the organisms to the structure of ecosystems. In trying to understand the causal factors determining the variability, the general approach has been to look for correlations between the patterns in the biological characteristics and other abiotic and biotic factors, the inference being that positive and negative correlations indicate causal relationships.

The end of the nineteenth and beginning of the twentieth centuries saw the great transoceanic expeditions that laid the foundations for modern biogeographical studies of pelagic faunas and floras (e.g. Ekman, 1953). But as more detailed studies were conducted, incompatibilities emerged between the interpretations based on studies of large-scale distributions and those of variability over finer space and time scales. The factors that appeared to be important in determining the observed patterns varied between different samplers and between different sampling scales, both in time and space. Over 30 years ago terrestrial botanists became aware of the importance of scale (e.g. Greig-Smith, 1964), but it was not until Haury et al. (1978) published their classic paper that the significance of scale was finally brought home to biological oceanographers, although an attempt had been made earlier to apply the botanical surveying techniques to benthic ecology (Angel & Angel, 1967). Haury et al. (1978) succeeded in expressing the variability of zooplankton biomass at different space/time scales by using a Stommel diagram (Fig. 3.1). Stommel (1963) had shown there is an approximately linear relationship between the size and longevity of eddy structures in the oceans. So any pattern in biological distributions with similar space/time relationships to these hydrodynamic features seem likely to have been determined by the hydrodynamics. Patterns with very different space/time characteristics were interpreted as being generated either by biological processes such as behaviour (e.g. diel vertical migration) and reproduction (e.g. responses to upwelling) or non-eddy related features such as frontal structures.

The concept of biodiversity is particularly challenging in open ocean biology because it requires the integration of very disparate sciences that focus on different scales within the space/time continuum:

1. Marine palaeontology addresses the geological and evolutionary time-scales, which cannot be associated with spatial scale in a Stommel diagram, because of the limited size of the Earth. The long-term processes include the tectonic changes in the shape and morphology of the oceanic basins, one-off vicariance events caused by the opening or closure of sea-ways, and planetary cycles such as the Milankowitch–Croll (100 ky), the obliquity cycle (44 ky) and the precessionary cycle (23 ky) have driven evolutionary processes through changes in climate and sea-level fluctuations.
2. Oceanic biogeography, in describing the large-scale distribution patterns of species and communities within the World's oceans, endeavours to identify the causes of the patterns. Ideally it should link the palaeontological record with present-day distributions, and indeed interpretations of past climates based on microfossil distributions in sediments depend very heavily on recent distribution patterns. However, there is usually a major gap in accessible time-scales, since the shortest time-scale distinguishable

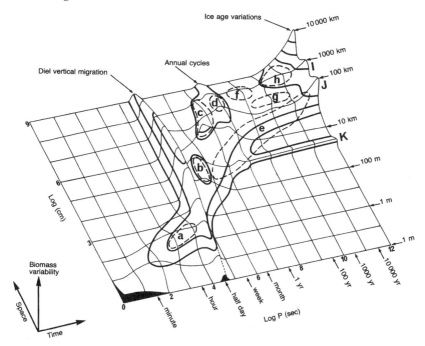

Fig. 3.1. A Stommel diagram of the variability of zooplankton biomass in time and space provides a conceptual framework for understanding the causal mechanisms of this variability. There is a strong correlation between biomass variability and the eddy structure of the water, but certain physical features (e.g. fronts) and biological responses (e.g. diel vertical migration) generate variability outside the predominantly linear envelope of variance caused by the eddying (redrawn from Haury et al., 1978).

from most oceanic sediments is of the order of a millennium, whereas the longest time series of direct observations are limited to a very few decades (e.g. the Continuous Plankton Recorder Survey (Warner & Hays, 1994) and CalCOFI (California Cooperative Oceanic Fisheries Investigation) programme). Hence the excitement generated by interpretations of data from varied deposits in the hypoxic basins off California where analyses of 1500 years of records have revealed persistent 65–70 year cycles.

3. Pelagic ecology studies processes within local regions and within mesoscale structures (i.e. eddies approximately 50–200 km or less). While these scales are highly relevant to determining how the local structure of the pelagic assemblages is maintained, they generally fail to match the scales studied by biogeographers. Discussions on diversity have often failed to distinguish processes

determining the evolutionary origins of species richness with those regulating species richness and dominance in contemporary populations.
4. Population biology has made little impact on oceanic ecology except in fisheries oceanography (Sinclair, 1988), probably because its development has been inhibited until recently by a lack of appropriate methodology. Now technological developments in molecular biology can be expected to open up research into the importance of metapopulations in the open ocean.
5. Taxonomy and systematics while addressing the fundamental problems of natural classification, that reflect the evolutionary relationships, are not specifically concerned with ecological processes that regulate populations, communities and assemblages.

3.2 Factors controlling large-scale patterns

There are three main categories of factor that either have been, or are currently, important in determining the large-scale biogeographical patterns of pelagic taxa:

1. Regional or climatic factors that impose latitudinal gradients on the distributions, through, for example, variations in solar radiation which play a large role in determining the surface wind field and the thermohaline circulation of the oceans. These currently control the distributional patterns.
2. The evolutionary and geological origins of the pelagic communities, influenced by the long-term changes in the morphology of oceans' basins resulting from sea-floor spreading and plate tectonics, and also the various planetary factors associated with the Earth's orbit. These have determined the species pool from which the present patterns are being generated.
3. The local biotic and abiotic conditions and ecological processes that currently maintain the biodiversity at all scales and levels of organisation. In terrestrial ecology these conditions are often determined at local to very fine scales by persistent landscape and microhabitat features such as soil type, aspect and even biological structure. While there are fine-scale environmental features in the pelagic ocean, these are for the most part too ephemeral to have any evolutionary significance.

In trying to integrate these different approaches it is probably simplest to begin by describing the patterns seen in present biogeographical patterns, because these cover scales over which all disciplines have some commonality.

3.3 Present biogeography of pelagic communities

A simplistic approach to understanding how communities are structured is that evolutionary events have led to a pool of species within a region from which a subset is 'selected' by factors that may operate 'bottom-up' or 'top-down'. Bottom-up control operates through the primary producers determining the structure of the food web, whereas top-down operates *via* selective predation curbing the dominance by a limited number of species; the classic example is Paine's (1980) study of the importance of starfish predation on an exposed rocky shore in maintaining high species richness. In contrast to terrestrial habitats in the pelagic ocean, there is an absence of semi-permanent fine- to meso-scale structure limiting distributional ranges (i.e. equivalent to soil and rock type on land), nor is there any three-dimensional biological structuring that can generate microclimates (i.e. equivalent to forests or coral reefs). The main control on distributional ranges is the basin-scale current circulation patterns on which are superimposed the effects of latitudinal gradients.

The basin-scale circulation patterns are determined by the rotation of the Earth, in combination with the differential heating by solar radiation between the poles and the equator. These processes generate variations in atmospheric pressure and hence the wind patterns, which drive the ocean current gyres. At mid-latitudes there are major current gyres that are clockwise in the northern hemisphere and anti-clockwise in the southern hemisphere. These currents redistribute heat polewards from the equatorial regions especially *via* the eastern boundary currents, whereas along the western continental margins cold currents flow towards the equator. Along the equator in the Pacific and the Atlantic there are westward flowing surface currents that are underlain by eastward flowing counter-currents. In the Indian Ocean there is a more complex situation resulting from the seasonal reversal of the monsoon winds.

The two polar oceans are markedly different, both in their geographical attributes and their geological histories. The Arctic Ocean is truly polar and is almost totally enclosed by the Eurasian and American land-masses, so that exchange of water with the other oceans is quite limited. Its ice cover originated only about 3 million years ago. It has a layer of low-salinity water at the surface that stabilises the water column and tends to keep nutrient concentrations in the euphotic zone rather low. So the productivity of the Arctic Ocean is generally relatively low. The ocean is almost entirely covered by pack-ice, much of which is composed of multi-year old ice. The area of pack-ice in the Arctic fluctuates annually by only about 10%.

The Southern Ocean surrounds the polar continent of Antarctica, and is fully open to all the other major oceans. There is a strong circumpolar eastward-flowing current to the south of the Antarctic Convergence. Nutrient concentrations remain high in the vicinity of the Polar Front, creating a circumpolar zone of high productivity during the austral summer. Most of the pack-ice melts in the summer so that its area fluctuates annually by about 70%. The retreating ice edge is a zone of high productivity, so that annual productivity in the Antarctic is much higher than in the Arctic.

Sea ice forms at −1.9°C; it contains relatively little salt, so the surface water becomes more saline, making it much denser than the surrounding water. Consequently it sinks into the interior of the ocean. In most regions it mixes within the water column and so fails to reach the bottom. But, in the Weddell Sea and also to the west of Greenland, in the northern hemisphere (for reasons that are still not fully understood) it penetrates to the bottom. There it is identifiable by its temperature and salinity characteristics as Bottom Water, which can be traced spreading throughout the World's oceans. Since before it sinks it is in gaseous equilibrium with the atmosphere, it therefore contains high concentrations of both dissolved oxygen and carbon dioxide. Bottom Water formation not only ensures that there is enough dissolved oxygen throughout most of the deepest waters of the World's oceans to support aerobic respiration, but is also an important sink for atmospheric carbon dioxide. As the Bottom Water sinks at high latitudes, water is displaced upwards elsewhere. This thermohaline circulation slowly stirs the oceans and carbon-14 distributions suggest that the major oceans turn-over every 250–500 years (Stuiver et al., 1983). Thus, if there were no factors limiting the distributions of pelagic species, they would become cosmopolitan within less than a millennium, well within the evolutionary time-scale of speciation.

At mid-latitudes in the centres of the major gyres, other water masses form. Evaporation from the sea surface exceeds precipitation (rain) and the surface water becomes saltier and heavier than the surface waters on both its poleward and equatorward sides. As a result along the boundaries there are convergences where these heavier salty waters sink beneath the neighbouring lighter fresher waters. The salt and heat properties of these waters are conservative, and so can be used to track the spread and gradual mixing of the water. Other tracers, such as their nutrient contents and dissolved anthropogenic inputs, such as tritium and chlorofluorocarbons (CFCs: e.g. Smethie, 1993), are now being used to confirm the pathways and measure the rates of spread. Pelagic species that are unable to regulate their horizontal distributions and lack the ability to navigate will be dispersed by these large-scale hydrodynamic processes. So the distribution of the water masses can be used as a proxy for the distribution of biological assemblages ('proxy' is the term used by palaeontologists to describe a parameter that is readily measurable and that can be used as an indicator of another parameter or rate process).

It was in the 1940s that the early pelagic biogeographers first reported correlations between the distributions of pelagic species and the water masses. McGowan (1971, 1974) produced the seminal papers that provided an integrated concept for communities of plankton in the Pacific. This was elaborated for a number of individual taxonomic groups (e.g. euphausiids, globally by Brinton, 1975; and pelagic fishes in the Atlantic, Backus *et al.*, 1977). Syntheses were published by van der Spoel & Pierrot-Bults (1979) and van der Spoel & Heyman (1983). More recently White (1994) has produced a synthesis of biogeographical and palaeoceanographic information for the Pacific that links the co-evolution of the water masses and the pelagic assemblages that inhabit them.

However, ranges of individual species seldom match the water mass distribution, although community structure does show a reasonable relationship. Attempts at understanding what is happening at the boundaries between the biogeographical regions, where the changes in the assemblages must be associated with changes in the food-web structure and the flow of materials through the pelagic ecosystem, have been thwarted by the ill-defined nature of the boundaries. This is partly because the boundary is often a complex of fronts and eddies, and partly because many of the horizontal gradients in physical and chemical properties are weak, relative to some of the vertical gradients, and are too weak to limit the distributions of most species. Many species can survive as 'expatriates' after being carried far beyond the range within which they can successfully reproduce and maintain their populations.

Locally these expatriates may continue to play important roles in ecological interactions (food-web structure and material fluxes), although their populations are only sustained by immigration. Thus the assemblage of species sampled at a place may not represent a community in the classical Eltonian sense. In the pelagic ocean, boundaries tend to be broad transition zones (ecotones) with few, if any species being completely bounded by them (van der Spoel, 1994). Even so, there is a broad range of surface water properties that replicate the patterns seen in the pelagic communities (Reid et al., 1978), so it is neither easy to identify which are the dominant factors that determine the biological distributions, nor is it likely that every species is being limited by the same factor (or factors).

Nutrient concentrations (nitrate, phosphate and silicate) in subthermocline waters differ between the different water masses, partly because there are substantial regional differences in the fluxes of sedimenting particles that resupply the nutrients through remineralisation, and partly because of differences in the age of the deep water (i.e. the time since it was at the surface) and how long the process of resupply by remineralisation has gone on. Fig. 3.2 shows a chart of mean annual silicate concentrations at a depth of 150 m (Levitus et al., 1993) and illustrates the striking regional variations in nutrient concentrations (note the

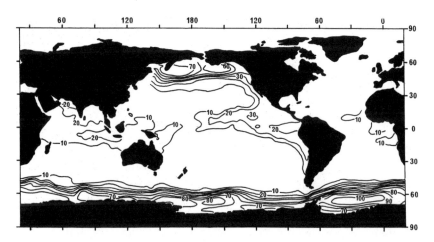

Fig. 3.2. Distribution of mean annual silicate concentrations (in μmol/l) at 150 m depth in the World's oceans, showing that there are major differences between different ocean basins. These patterns relate to the main circulation features and a comparison with Fig. 3.3 shows that they are also repeated in the biogeographical distribution of pelagic species and assemblages (redrawn from Levitus et al., 1993).

impoverishment of the temperate North Atlantic relative to the North Pacific, the Eastern Tropical Pacific and the Southern Ocean). Silicate is essential for diatom growth and so its availability strongly influences both the quantity and the quality of primary (and export) production. Hence the structure of food webs both within the surface waters and deep within the water column will be influenced by these variations in superficial nutrient availability.

At latitudes lower than 40° the surface waters are generally permanently thermally stratified. The warm surface waters of the euphotic zone contain low concentrations of nutrients that are only resupplied by localised stirring by eddies and the exceedingly slow process of diffusion. Most production is supported by nutrients recycled within the euphotic zone, and the phytoplankton populations and production tend to be dominated by picoplankton (very tiny cells <2 µm). Since these cells are too small to be filtered from the water by the mode of suspension feeding seen in copepods, most are consumed by tiny protozoan grazers and mucus-web feeders. As a result most (up to 90%) of organic production is usually recycled within the euphotic zone and relatively little gets exported into deep water. The production cycle is only weakly seasonal, usually showing two- to threefold increases in wintertime.

Along the equator in the Pacific and the Atlantic the easterly winds generate a divergence that brings nutrient-rich subthermocline water up into the euphotic zone, so that there is a narrow band of high productivity along the equator. Similarly along the eastern boundaries of the major oceans (off California, Peru, Northwest and Southwest Africa) alongshore trade winds seasonally drive surface waters off-shore, which draws nutrient-rich waters up to the surface, creating highly productive coastal upwelling zones. The type of blooms stimulated depends on the nutrient status of the upwelled water, and this influences the structure of the local food web. Fig. 3.2 illustrates how the upwelling along the western boundary of Africa and Europe introduces water with lower concentrations of dissolved silicate (and other nutrients) up into the euphotic zone, than the comparable upwellings along the western boundary of the Americas.

At latitudes higher than 40°, wintertime cooling of the surface waters results in convective overturn and mixing, resupplying nutrients back into the surface waters. However, phytoplankton growth is held in check until the surface waters become thermally stratified when turbulent mixing ceases continually to mix the phytoplankton down below the photosynthetic compensation depth (i.e. the maximum depth at which production by photosynthesis exceeds the losses from respiration). In the Atlantic, remote-sensing of seasonal surface chlorophyll concentrations has revealed a clear latitudinal banding in seasonality (Campbell & Aarup, 1992). At temperate latitudes 40°–60°N the production cycle is seasonally bimodal producing 10 to 100-fold increases in phytoplankton biomass during a large spring bloom and a smaller autumn bloom. In the North Pacific and Southern Ocean although production increases in the Spring, there is very little increase in the standing crop of phytoplankton. At latitudes higher than 60° these two seasonal production maxima converge to form a single summertime bloom. These mostly latitudinal variations in the production cycle are important in influencing the lifecycles, seasonal succession and distributions (all aspects of biodiversity) of the pelagic assemblages.

In the North Atlantic the dramatic increase in phytoplankton production is not checked until virtually all the available nutrients in the euphotic zone are exhausted (Savidge *et al.*, 1992). Consequently the bloom crashes. The surface nutrient concentrations remain depleted until the equinoctial storms begin to erode the seasonal thermocline, resupplying nutrients to the euphotic zone and, in many years, stimulating a short-lived autumnal bloom. The autumn bloom is curtailed by light limitation, as winter sets in and vertical mixing extends deeper and deeper. In some regions of the North Atlantic this deep convective mixing can extend to depths of 500–800 m. Throughout the winter, nutrients are in plentiful supply in the euphotic zone, but phytoplankton growth is once again kept in check by the constant turbulent mixing.

In the North Pacific and the Southern Ocean ecosystems respond very differently. Although there are springtime increases in productivity, the concentrations of nutrients in the surface waters never fall low enough to limit photosynthetic activity. The reason for this absence of a spring bloom is currently the subject of active debate. There are two conflicting hypotheses as to what inhibits bloom development. One suggestion is that the lack of some chemical factor (notably iron) is limiting the growth of the phytoplankton. The other is that herbivorous grazing is able to match the rate at which the phytoplankton populations are growing so that there is no dramatic increase in phytoplankton biomass and nutrients are recycled as fast as they are utilised (de Baar, 1994). There is also a tantalising correlation between those areas where nutrient concentrations are maintained and never become limiting, and where there are

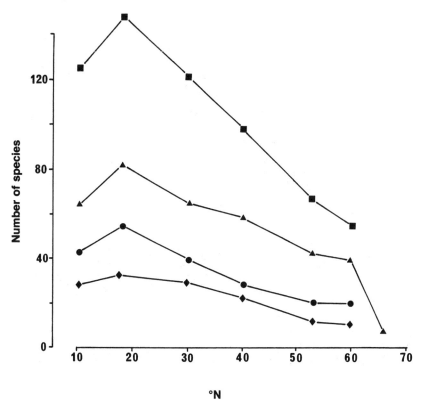

Fig. 3.3. Total numbers of species of four pelagic taxa caught at six stations along 20°W in the North-East Atlantic; at each station 14 day and 14 night samples were collected systematically from the top 2000 m of water column (Angel, 1991). ■, Fish; ▲, Ostracoda; ●, Decapoda; ♦, Euphausiids.

high silicate concentrations in the deep water (Fig. 3.2), which implies that silicate availability may somehow be implicated.

Associated with these patterns are two trends in the structure of the pelagic communities. Firstly there is a clear latitudinal decrease in the size of the pelagic species pool for several (perhaps all?) taxa from low latitude to high latitude (Fig. 3.3). Secondly there is a poleward increase in the mean adult body size of the planktonic assemblages, which has important implications for ecosystem function. Larger body size is associated with the longer life-cycles (Denman, 1994) that are needed if reproductive effort is to be timed to coincide with the seasonal production peak. The need to survive the months when food availability is sharply reduced has been achieved in a number of (non-exclusive) ways:

1. By having a body size that is large enough to be able to store sufficient reserves not only to survive until the following spring but also to fuel the following season's reproductive effort. Larger animals produce larger faecal pellets. Since the larger a particle is the faster it sinks, it is less likely to be intercepted and recycled before it sinks out of the euphotic zone. So the size of the pelagic populations may have a significant influence on the proportions of organic matter that get exported into deep water.
2. By having a flexible feeding strategy and switching from feeding directly on phytoplankton to detritus or even carnivory. This directly affects species richness, because there are relatively more generalist taxa at high latitudes there is a concomitant reduction in resource partitioning (and hence in the numbers of species). Moreover, the palaeontological record shows that in shallow-living taxa speciation and extinction rates are lower in generalist genera than in specialist genera. If this is equally true for pelagic species then high latitude species can be expected to be more tolerant of environmental fluctuations and to have persisted longer through geological time. Moreover, extinction events may have had relatively less impact on high latitude taxa than low latitude taxa.
3. By having a resting phase either as an egg, or as a juvenile or adult in a state of diapause (which in pelagic species is often linked with a seasonal migration into deep water where predation pressure is lower). At very high latitudes the productive season may be too short for many of the species to complete their life-cycles within a single year. So life-cycles tend to be extended and involve several resting phases (e.g. Miller & Clemons, 1992). Thus samples from one time of year and/or over a limited depth range are unlikely to reveal the full suite of species inhabiting that water column.

In contrast mid-latitude epipelagic species inhabiting the central waters of the oligotrophic gyres are generally smaller in size (see below), probably in response to the lower, but more consistent levels of production and the smaller sizes of the particles of organic matter suspended in the water. They have several generations per year. Since there are more species in the assemblages and much greater evenness in population sizes, not only is the partitioning of resources likely to be greater but also there is likely to be greater specialisation in feeding and life-history characteristics. McGowan (1990) showed that the composition and relative abundances of the species in these communities remains extremely stable from year to year; even in a year when there was a doubling of mean biomass, the rank order of abundance of the species remain unchanged. Diversity indices expressing sample evenness and dominance, as well as species richness, are also stable (Angel, 1991).

Large-scale patterns that have emerged from objective analyses of data from widely-spaced stations in the north-eastern Atlantic (e.g. Angel & Fasham, 1975; Fasham & Foxton, 1979), have highlighted the significance of the 40°N boundary and the correlation between the distributions of the different assemblages (but not ranges of the individual species!) and the water mass structure along 20°W in the North Atlantic. One water mass that failed to generate a signal in the community structure was the Mediterranean Outflow Water. Hypothetical reasons to explain this absence of a specific assemblage might be based on either history, oceanographic process or methodology. Firstly, for example, the characteristics of the water mass may have changed so much since the last glaciation, that there has been insufficient time for a community response to have occurred. Secondly the presence of the water mass may be either too intermittent in time and space for a characteristic assemblage to develop, or there is no return flow whereby species can maintain themselves within the outflow system. A third explanation is simply that the sampling protocol was inadequate to detect the assemblage. A fourth possible explanation is that water mass distributions are not very good proxies for the distribution of the biological populations.

Diel vertical migration plays a significant role in both local and regional variability (Angel, 1989b). These migrations are probably both a response and a contribution to the latitudinal variability and, because of bathymetric differentials in current speed and direction, the migrations provide a mechanism whereby quite small species can migrate horizontally. In spring and summer at temperate latitudes, a greater proportion of the standing crop (and the overall biomass of migrants) is involved in the migrations, than in the centres of the oligotrophic gyres (Angel & Hargreaves, 1992). However, diel migrations occur throughout the year in the oligotrophic regions and over much larger bathymetric ranges, whereas at higher latitudes diel migrations are mostly curtailed during the winter, when many species undertake seasonal migrations into deep water and enter diapause. Many species also stop diel migrations during the summer when primary production is low. Vertical migration plays a small but significant role in the export of organic matter from the euphotic zone, rapidly transmitting the impact of processes in the euphotic zone deeper into the water column and influencing processes and diversity deep down (see below).

3.4 The origins of pelagic biodiversity

Metazoans first appeared about 600 million years ago in the ocean, it was about a further 150 million years before they started to invade land. So it is not very surprising that more phyla occur in the oceans than in either terrestrial or freshwater environments. Pearse & Buchsbaum (1987) produced a table of the distribution of metazoan phyla by major habitat. Of the 33 metazoan phyla they listed, 28 phyla are found in the oceans, 13 of which occur nowhere else. Compare this with the 13 phyla found in terrestrial environments of which only one, the Onychophora, is thought to be endemic (even that is now questionable). The remaining four phyla are all symbionts and two of these are restricted to marine organisms. However, this high disparity of oceanic faunas is not a characteristic of holopelagic species, which include representatives of only 14 phyla, none of which is restricted to the pelagic realm. Since this is only one more phylum than occurs in terrestrial habitats, it seems to have been as hard to become a holopelagic species as it was to invade land.

The other striking feature of Pearse & Buchsbaum's table is the species-poorness of pelagic taxa, despite the oceanic environment offering >180 times the living space offered by terrestrial or freshwater habitats (Cohen, 1994). For example, only about 2200 pelagic copepods have been described from marine and brackish waters, compared with over a million known insect species. However, since there are estimated to be only 3500–4500 species of marine phytoplankton compared with an estimated 250 000 species of terrestrial green plant on land, this is not a surprising result, if control of diversity is bottom-up. It still leaves unresolved the reason for there being so few primary producers, and why the stock of living biomass per unit volume in the oceans is only 1500 kg carbon/km^3 compared with 75 000 kg carbon/km^3 on land. Cohen (1994) lists many other parameters that illustrate marked differences between structure and function of food webs in the oceans and on land, and may contribute to the low species diversity of oceanic faunas and floras. But before discussing further why there are so few pelagic species in the ocean, some consideration needs to be given to speciation and extinction in the oceans.

Pelagic biogeographers were slow to recognise the significance of

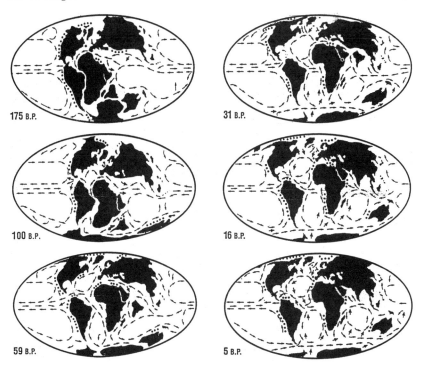

Fig. 3.4. Reconstruction of the distribution of the continents and the main surface circulation of the oceans through the last 175 million years following the fragmentation of the supercontinent Pangea. The distribution of old genera and families still retain the imprint of the ancient circulation patterns. Vicariance events such as the interruption of circum-equatorial circulation and sea-level changes associated with the glaciations and the desiccation and reflooding of the Mediterranean have had major impacts on patterns of speciation and redistribution (modified from Parish & Curtis, 1982). B.P., million years before present.

plate tectonics. Parish & Curtis (1982) produced a series of outline maps of the disposition of the continental land masses at various phases of the fragmentation of the supercontinent of Pangea (Fig. 3.4). They were able to hindcast the likely patterns of circulation at each phase, based on the predicted wind fields and the constraints of the basin morphology. By analogy with present day biogeographical patterns, each of these ancient current gyres was occupied by a different assemblage of species. The early stages of the fragmentation coincided with two of the five major extinction events recorded in the fossil record since metazoans first appeared some 600 million years ago. Since then the number of marine families has grown steadily (Sepkoski, 1992). Even the last extinction event at the end of the Cretaceous, when about 65–70% of shallow-living marine species (i.e. those for which there is a good fossil record) went extinct, caused only a temporary pause in the rate of increase in number of families. However, for some taxa recovery was never complete, Knoll (1989) records that across the Cretaceous–Tertiary (K-T) boundary the number of genera of calcareous nanoplankton decreased from 65 to 25; after an initial partial recovery to 40 during the early Eocene, numbers declined again to 30 by the beginning of the Oligocene. The K-T extinction event (65 million years ago) appears to have preceded the start (50 million years ago) of the cooling of ocean bottom waters from around 10°C to the present-day 2°C (Shackleton, 1982). Zachos et al. (1989) present compelling evidence that the extinction coincided with a sharp breakdown of the normal depth-related gradient in the $^{13}C : ^{12}C$ ratio, together with significant decreases in the fluxes of barium and carbonate that persisted for about half a million years. These proxy indicators imply that global marine productivity halved during this era. Distributions of microfossils show that the pelagic species that survived became cosmopolitan possibly as a result of an intensification of the thermohaline circulation – resulting in the so-called 'Strangelove Ocean' (Broecker & Peng, 1982). These cataclysmic events totally disrupted the biogeography of the World Ocean, and it took some time for the patterns to be gradually re-established in response to the environmental characteristics of the large-scale current gyres. White (1994) showed how the phylogenies of some monogeneric taxa matched the cladogram of water-mass history derived from Ocean Drilling Program (ODP) data. Thus as the water-masses gradually became distinct as a result of the reorganisation of the continental land-masses and the cooling of the deep ocean, so the distributions of pelagic taxa became reorganised into their present patterns. If he is correct then the foundations for present biogeographical distributions were established after the K-T event. Some biographers consider that memories of the intervening circulation patterns are detectable in the zoogeographical distribution patterns of the older taxa of euphausiids and hydromedusae (van der Spoel et al., 1990).

A key question to resolve is how rapidly has speciation occurred in pelagic environments. Can speciation occur sympatrically in pelagic habitats, or is isolation a pre-requisite? Palumbi (1992) suggested that sympatric speciation can occur given a large enough habitat in which gene flow is very slow. However, if speciation can only occur allopatrically and isolation is a prerequisite, then vicariance events have probably played a key role in evolution in marine species. Vicariance events are usually one-off phenomena associated with either tectonic phenomena or climatic changes (resulting in sea-level fluctuations), which have either opened up or closed off interoceanic connections. There have been at least four major tectonic events during the Cenozoic that are known to have had a major impact on ocean circulation patterns and hence on biogeographical distributions –

1. The deepening of the Tasman Seaway and
2. the opening of the Drake Passage, both of which contributed to the establishment of the circumpolar current in the Southern Ocean. These seem to have led to the progressive cooling of deep waters (see above), and also to a substantial deepening of the carbonate lysocline (the depth at which calcium carbonate dissolves in seawater).
3. The closure of the Indonesian Seaway, and
4. the elevation of the Panamanian Isthmus, both of which created barriers to the dispersion of pantropical taxa, interrupting circumequatorial gene flow.

Another series of tectonic events that must have had a major impact on global biogeography were the episodes of desiccation and catastrophic reflooding of the Mediterranean during the Miocene. The reflooding, in particular, must have resulted in rapid falls in global sea level of several tens of metres, which will have caused extensive disruption of shallow inshore ecosystems, but may not have had so much impact on open ocean communities.

In addition to these major tectonic events, the climatic oscillations induced by the eccentricity (Milankovitch–Croll cycles), obliquity and precessionary characteristics of the Earth's orbit during the last few million years, have resulted in the glaciation cycles. Global sea levels oscillated by around 100 m, falling during the expansion of the ice-caps and rising again as the ice melted during the warm interglacials. Both the waxing and waning of the ice-caps and the filling and emptying of the ocean basins have also resulted in major isostatic changes in relative sea levels. In some regions particularly in the region of the East Indies, these sea-level fluctuations were extensive enough to close sea-ways and isolate deep basins during glaciations and then re-open them during interglacials. The frequency of these oscillations, with the precessionary periodicity of approximately 23 ky, appears to have been long enough for speciation to occur on land but what about in marine habitats?

Fleminger (1986) ascribed the high species richness of coastal copepods of the genera *Pontella*, *Labidocera* and *Undulina*, and the marked changes in the species composition of inshore pelagic assemblages that occur across Wallace's Line (a line described on the basis of terrestrial biogeography) through the Indonesian islands, to such episodes of fluctuating sea levels.

Climatic changes during the glacial cycles have also isolated populations and possibly contributed to speciation. For example, at the last glacial maximum in the North Atlantic, subpolar waters extended as far south as the Straits of Gibraltar – glacial debris transported by icebergs is found in surficial sediments off southern Portugal. Several species presently more typical of cool subpolar and temperate waters found their way into the Mediterranean and have persisted as glacial relicts. These relict populations are totally isolated from their parental stocks, never experience waters cooler than 12.7°C (a temperature that the parent stocks only encounter during summertime when their vertical migration takes them into the epipelagic zone at night) and experience a different productivity regime. Some of the relict species, such as the euphausiid *Meganyctiphanes norvegica* and the myctophid fish *Benthosema glaciale*, also have relict populations isolated in the cool waters associated with the coastal upwelling region off northwest Africa. In the Mediterranean there are also putative relicts from when the Tethys Sea linked the tropical Indo-Pacific and the Atlantic; these include the copepods *Oithona hebes*, *Centropages aucklandicus* and *Oncaea dentipes* (Furnestin, 1979). Although how these species have persisted more or less unchanged through the salinity crises of the Miocene is an unresolved question. The specific status of these relicts needs to be investigated using molecular biological techniques to examine whether or not these populations have indeed remained unchanged or whether their apparent morphological similarity hides substantial genetic differences (i.e. they have become cryptic species). If the relicts from the last glaciation have already diverged, it would indicate that oceanic taxa have the potential to speciate quite rapidly, and that in the absence of isolation, ocean circulation is dynamic enough to maintain sufficient gene-flow to inhibit speciation. However, if neither they nor the Tethyan relicts have failed to speciate, then the time-scales for speciation in the ocean are orders of magnitude longer than those recorded in terrestrial and freshwater habitats.

In the pelagic environment the dominant scale of variability is mesoscale. Mesoscale eddies with spatial scales of 50–200 km can persist for time-scales of months to 2 years. They can disperse pelagic species and larvae (particularly those that do not vertically migrate) within and far beyond their reproductive range (e.g. Fairbanks *et al.*, 1980). The eddies impose a strong short-term selective pressure on those populations that get caught up within them and so contribute to the maintenance of wide genetic variability and gene-flow within populations of individual species. Stochastic factors determine the composition and structure of the populations caught up within each eddy so although the sequential changes that occur follow a fairly predictable succession, each eddy tends to be a unique 'experiment' with subtle variations occurring in the relative dominance and differential rates of gain and loss between populations of migrant species and non-migrant species. These eddies are an important source of sampling variability, but their time characteristics are far too short for them to function as an isolating mechanism for allopatric speciation to occur. In terrestrial ecology, the populations of many species are subdivided into series of metapopulations between which genetic exchange is limited. In pelagic habitats, by facilitating gene-flow over long distance, turbulent mixing and advection in mesoscale eddies serves to prevent the formation and maintenance of similar metapopulations in the open ocean. It is only where there are persistent hydrographic features associated with bottom topography in shelf seas, along continental slopes, and maybe associated with sea-mounts that there may be selection of life-history characteristics that serve to retain (and possibly to isolate) populations within semi-closed circulation features (Sinclair, 1988).

Indeed in those open ocean areas where eddies containing pelagic assemblages of mixed origins are common and are advecting species from one water-mass to another (as occurs in Gulf Stream rings) a large fraction of a sample may consist of expatriates, which while actively interacting ecologically (competing for food and other resources and acting as predators and/or prey) are unable to maintain themselves in that body of water by reproduction. The sample may either include other species that are not actively involved in the local ecological processes because they are in a state of diapause, or it may exclude others, which at that particular season are only present as eggs, or have vertically migrated beyond the range of the sampling, and yet play a substantial role in the processes at other times of the year. This calls into question whether the concept of community, as used in terrestrial ecology especially for plants, can be applied to pelagic ecosystems. It also calls into question the significance of sample diversity without a full context of time and spatial information.

3.5 Oceanic *versus* neritic diversity

In the pelagic ocean the biogeographical zonation (see Fig. 3.5) shows that only regionally-scaled environmental factors are linked with geographical location. Most local factors are linked with properties of the water body that are not only independent of locality but are also ephemeral at evolutionary time-scales. Thus locality, which is so important in terrestrial ecology, is relatively unimportant in the pelagic environment. However, in shelf seas locality is important (Ray, 1991). The physical environment is strongly influenced by the geomorphology of the coast and shelf through its interactions with both abiotic and biotic factors (e.g. exposure to surf, the interactions between any tides and coastal morphology, the presence of hard or soft bottoms and riverine inputs). There are major quantitative and qualitative differences between the pelagic ecosystems of open ocean and continental shelf seas. For example, over deep water the pelagic ecosystem processes are independent of what happens on the sea-floor, whereas over the shelf there are always influences from the sea-bed extending right to the surface. As a result neritic communities are geographically more constrained than open ocean communities, limited off-shore by the continental margin and along-shore by features associated with the coastal morphology. Shelf sea ecosystems are often linear, with the long-shore dimension being very much greater than the across-shelf dimension. The faunas and floras of the eastern continental margins differ from those of the western margins, because transoceanic exchange is often (but not always) inhibited by the distances involved. Many neritic species have planktonic dispersive larval stages whose persistence in the plankton ranges from a few days to several months. The durations of these stages and the behaviour of the larvae are often adaptive in maintaining their populations within relatively restricted geographical ranges.

Several species are known to have life-history characteristics that are finely tuned to the physical circulation in a region. In coastal upwelling regions, the copepod *Calanoides carinatus* (Smith, 1984) and the euphausiids *Euphausia lucens* and *Nyctiphanes capensis* (Barange & Pillar, 1992) undergo ontogenetic migrations that place the larvae in counter-currents. This enables them to 're-seed' the upwelled water so that the species stays within the upwelling system. It has been suggested that the life history of *Calanus finmarchicus* is adapted to exploit the slope current along the shelf-break to the west of Britain so that it maintains its position in the North-East Atlantic (e.g. Diel & Tande,

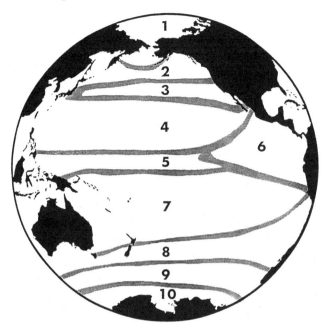

Fig. 3.5. Pelagic biogeographic zones of the Pacific Ocean (White, 1994). Comparable zones occur in the Atlantic and the Indian Oceans that also reflect the main oceanographic features and the large-scale circulation gyres. These zones have developed since the extinction event at the end of the Cretaceous era.

1992). This must also be true for many coastal and inshore species on oceanic islands.

In pelagic assemblages these interactions result in *local* shelf communities being appreciably poorer in species than the oceanic communities immediately off-shore (Fig. 3.6: Hopkins *et al.*, 1981). However, because neritic habitats are spatially more restricted than open ocean habitats, species richness of neritic faunas is by far the richer on the global scale. A comparison of the numbers of known species of mysids and euphausiids and their zoogeographical ranges exemplifies this pattern. There are 86 known species of euphausiids (Baker *et al.*, 1990) – assuming that there are no cryptic species amongst them – the majority of which are either cosmopolitan at latitudes <40°, or have distributions that are latitudinally banded at high latitude. Apart from four neritic species all euphausiids occur in the open ocean. In contrast the vast majority of the 971 known species of mysid (Müller, 1993) are either neritic or associated with continental slopes. Many are meroplanktonic rather than holoplanktonic, so maybe the comparison is artificial. How-

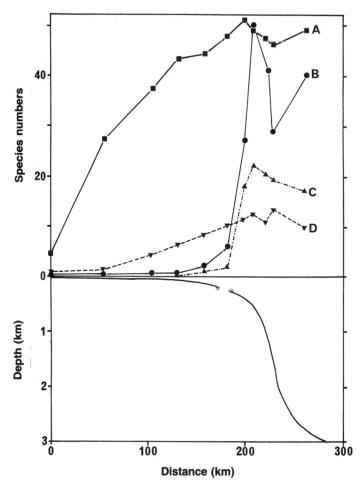

Fig. 3.6. Changes in the numbers of pelagic species taken across the shelf, shelf-break and continental slope off Florida in the Gulf of Mexico (A, Copepoda; B, mid-water fish; C, Decapod crustaceans; D, Euphausiids), showing how for these four pelagic taxa the numbers of species increase off-shore and often peak at or close to the shelf-break (modified from Hopkins et al., 1981).

ever, mysids generally have much more restricted biogeographical ranges than euphausiids, although the few open ocean pelagic species are just as widely distributed as the euphausiids. Fig. 3.7 shows the oceans divided into broad biogeographical regions with the ratio of mysid : euphausiid species known from each. There is a much smaller number of mysid species known from the southern hemisphere than from the subtropical and tropical regions, and also from the northern

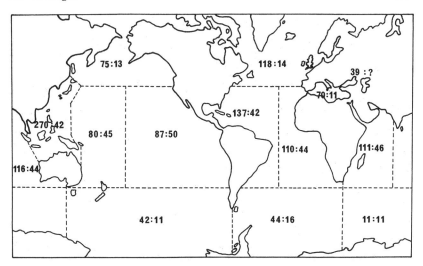

Fig. 3.7. Numbers of mysid : euphausiid species presently known from each of 12 broad zoogeographical regions.

hemisphere. This north–south difference may result either from an undersampling of the Southern Ocean, or from there being fewer and less varied coastal areas around the Southern Ocean than at similar latitudes in the northern hemisphere. The very high numbers of mysid species around Japan and the South China Sea may also either be a true reflection of a richer mysid fauna (maybe as a result of the sea-level fluctuations during the Pleistocene glacial cycles), or it may be an artefact resulting from the greater taxonomic effort by Japanese taxonomists, for whom the group has held a particular fascination. The data for the euphausiids also show a clear latitudinal trend, with half the known species occurring in each of the subtropical/tropical regions.

3.6 Existing patterns of community distributions – bathymetric

Most secondary production in the open ocean is supported by primary production in the euphotic zone. The chemosynthesis associated with hydrothermal vents, hydrocarbon and cold saline seeps provides oases of high standing crop in very small localised areas of the sea-floor, but there is little evidence of substantial export of organic carbon up into the overlying water column. Export production from the euphotic zone is influenced by a range of physical (water movements, especially those resulting in vertical mixing and upwelling, the light regime, and *in situ* temperatures and water stratification), chemical (nutrient availability, notably nitrogen, phosphate, silicate and maybe iron, and their concentration ratios) and biological factors (the species composition of the phytoplankton, the food-web structure and the grazing pressure). There is also evidence that active export by diel vertical migrants can often be significant (Angel, 1989a). Export in the form of dissolved organic matter (DOM) may also be important but remains poorly quantified; a recent estimate from off Bermuda suggested that the export of DOM was roughly equivalent to the export of particulate organic carbon (POC: Carlson et al., 1994). In-shore riverine inputs of DOM can be very important. However, it is the POC flux that is by far the most important for the deeper-living species.

Particulate export through the base of the euphotic zone is dependent not only on the amount of primary production occurring in the overlying water, but also on the types and size spectrum of the particles produced, mechanisms of aggregation and disaggregation, the particulate grazing, microbial degradation and chemical remineralisation processes and the physical factors influencing sedimentation rates. The sedimentary flux is often episodic. Diatom blooms in particular tend to crash very suddenly once the nutrients are depleted, so that all the cells sink out of the euphotic zone at rates in excess of 100 m/day (e.g. Pitcher et al., 1991). At the base of the euphotic zone the export flux is extensively modified but, thereafter, the rates of change of the flux appear to be predictable. Thus linear regressions of the log-transformed standing crop of micronekton and micronekton with depth have very similar slopes over a wide latitudinal range in the northeastern Atlantic (Angel, 1994). There is a clear trend for the average size of the pelagic populations to increase with depth, at least to depths of 1000 m; for example in the vicinity of the Azores Front the biomasses per unit volume of water of macroplankton and micronekton converged with depth from ratios of 10–21 : 1 in the euphotic zone to 1 : 1 at around a depth of 1000 m (Fig. 3.8: Angel, 1989b). At a highly oligotrophic site (31°30'N, 25°30'W), *in situ* sieving of macroplankton showed the relative proportion of the catch retained by 1 mm mesh became larger than the fraction passing through but being retained by 0.32 mm mesh at depths of around 500–600 m (Fig. 3.9: Angel, 1989a). Similar size changes occur in some individual taxonomic groups; e.g. Angel (1979) showed that the size spectra of planktonic ostracods sampled extended to larger sizes in a step-wise fashion across the boundaries of the main bathymetric zones.

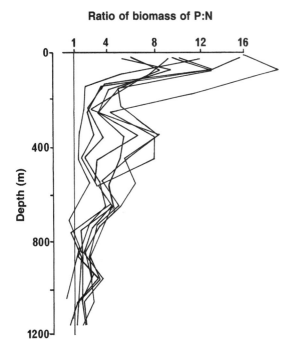

Fig. 3.8. Daytime profiles of the macroplankton (P) : micronekton (N) ratios of biomass per 1000 m^3 sampled at eight stations (three in November 1980, five in June 1981) in the North Atlantic in the vicinity of the Azores Front. At all the stations the ratios approach unity at a depth of about 1000 m, indicating a consistent trend for mean body size of the total community to increase with depth (modified from Angel, 1989b).

These classic zones, epipelagic, shallow and deep mesopelagic, bathypelagic and abyssopelagic were first identified on the basis of the characteristic morphologies of the animals inhabiting them. The coloration of the animals and their use of bioluminescence, which correlate with these bathymetric zones, are dependent on the colour, intensity and vertical distribution of the *in situ* light, and in most cases function as camouflage against visual detection. Such adaptations are only effective over limited depth ranges. For example, in clear oligotrophic waters, fishes with mirror-sides and well developed ventral photophores, such as *Argyropelecus* spp. are almost entirely restricted by day to the shallow mesopelagic zone at 300–700 m (Badcock & Merrett, 1976). The deepest depth to which macroplankton migrates is around 700 m, whereas migrations by micronekton species generally extend down to a maximum depth of 1000–1200 m (i.e. the boundary between the deep mesopelagic

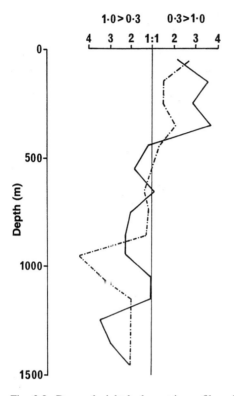

Fig. 3.9. Day and night bathymetric profiles of the ratios of biomass (displacement volumes) of macroplankton size-fractions sampled at 31°30'N, 25°30'W from 100 m strata to a maximum depth of 1500 m, and sieved *in situ* using concentric cod-ends of decreasing mesh size. The larger size-fraction passed through a mesh of 4.5 mm but was retained by the 1 mm mesh, the smaller size-fractions passed through the 1 mm mesh but was retained by the 0.32 mm mesh. The profiles show how during the day the biomass of the small size-fraction organisms was the greater down to depths of 400 m. At night it was still larger but its dominance was reduced as a result of diel vertical migration. Below 400 m the large size-fraction was either similar to or larger than the small size-fraction (Angel, 1991). —, day; — · — · —, night.

and the bathypelagic). However, adults of the myctophid *Ceratoscopelus* commute daily between the epipelagic zone at night to daytime depths of 1600–1700 m off the Azores.

In one sense the epipelagic zone must be the richest in species because the early larval stages of so many pelagic taxa are spent in the euphotic zone. During these early stages there is a need to maximise food intake that outweighs the greater risk of predation at shallow depths. As the larvae grow the risk of their being detected and consumed by visually hunting predators increases, and so they start an ontogenetic migration into deeper water. Since visual acuity is related to light intensity and the size of the target, the larger an organism grows the deeper it needs to move during the day, if it is to remain safe from visual predation. There is a trade-off between the benefits of staying shallow where there is greater availability of food but an increased risk of predation, and the greater safety at depths where food is more limited.

Euphausia species have a different strategy and undertake an ontogenetic migration in the opposite mode. Their eggs are heavy and sink into deep water. Once they have hatched the larvae start to swim back up towards the surface. The newly-hatched larvae are non-feeding relying on their stores of lipid for development. By the time they reach their first feeding stage, they have arrived back in the near-surface layers where food is available. This developmental migration enables the euphausiids to optimise the investment of resources in each individual embryo.

Analyses of catches from plankton and micronekton nets with mesh sizes of 320 μm and 4.5 mm respectively, show that the total numbers of species caught per sample increases with depth reaching a maximum at around 1000 m, even though the standing crop declines by an order of magnitude (Angel & Baker, 1982). Fig. 3.10 illustrates how the species richness increases with depth for one group, the planktonic ostracods. Moreover, since the mean size of the animals also increases with depth (see Fig. 3.8), the numbers of individuals per unit volume decreases by more than an order of magnitude.

Many of the environmental gradients associated with increasing depth are either almost invariant properties of water (hydrostatic pressure, spectral changes in light intensity, angular distribution of light) or are linked to large-scale processes (salinity, deep-water temperature, nutrient and dissolved gas concentrations). So most of these properties have bathymetric profiles that have remained consistent over evolutionary time-scales. Hence physiological, morphological and behavioural adaptations to restricted zones within these gradients have become very finely tuned by natural selection. Species adapted to daytime mesopelagic depths are not only tightly restricted to those bathymetric zones, but also they are probably highly effective at out-competing any new species that attempt to invade deep-water habitats. Even in those seas, such as the Mediterranean, where for geological reasons the deep-living fauna has been wiped out and appears not to have been replaced by migration, invasion of deep-water habitats has proved to be an almost insurmountable problem. In the case of the Mediterranean, the physiological chal-

62 Pelagic biodiversity

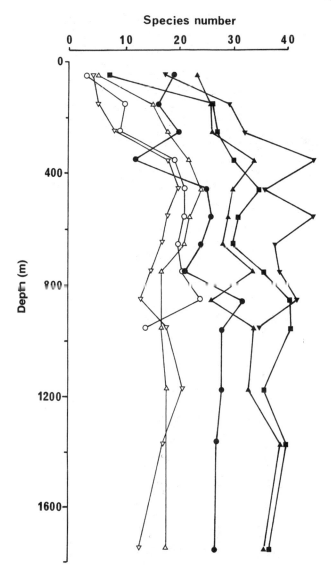

Fig. 3.10. Profiles of the numbers of planktonic ostracod species throughout the surface 2000 m along 20°W in the North-East Atlantic, showing how at latitudes <40°N there are more species at all depths than at higher latitudes. The change in average species richness appears to occur at the southern boundary of seasonal turn-over in the near-surface waters and the regions where there is a springtime peak in phytoplankton production (Angel, 1991). Latitude: ▲, 11°N; ▼, 18°N; ■, 30°N; ●, 40°N; ○, 44°N; △, 53°N; ▽, 60°N.

lenge is accentuated by the relatively low productivity of the surface waters and the warmth of the deep water (12.7°C), which keeps metabolic rates high thus stimulating the rapid microbial degradation of what little organic material is exported from the euphotic zone.

The zoogeographical ranges of taxa appear to change with depth. Tropical and subtropical species that inhabit the near surface layers throughout, or at some phase of, their life-cycle tend to be much more panoceanic than those inhabiting bathypelagic depths, possibly as a result of the existence of a shallow-water connection between the Pacific and the Atlantic across the Panamanian isthmus until about 2 million years ago. Bathypelagic species have been separated in their ocean basins for very much longer, but at abyssopelagic depths of >3000 m many of the species appear to be cosmopolitan. Is this because speciation rates are much slower or because gene-flow is more rapid as the result of the exchanges of deep water? Other distribution patterns that were extensively discussed by the early biogeographers are bipolarity and submergence. There are several pelagic taxa that have closely related sibling species in the Arctic and Southern Ocean, for example, the sibling pairs of planktonic ostracods *Conchoecia borealis* and *C. australis*, and *Conchoecia obtusata* and *C. antarctica*; both pairs were originally described as being conspecific. Their apparent similarity would suggest that relatively recently in the geological past there has been an exchange of faunas between the two polar oceans. Submergence is one way in which this exchange may have occurred. The classic example of submergence is the chaetognath *Eukrohnia hamata*. At latitudes >60° this species occurs in near-surface waters, whereas at lower latitudes it tends to be restricted to the cold-water sphere, living below 100 m at 40°, and at around 1000 m on the equator.

3.7 Summary

The physical and chemical environment of the pelagic ocean is structured by planetary forcing, ocean–atmosphere interactions and hydrodynamic processes. These generate major circulation cells that are clockwise in the northern hemisphere and anti-clockwise in the Southern. Where ice forms and in regions where evaporation from the sea-surface exceeds freshwater inputs from rainfall, large volumes of water are formed with characteristic and conservative properties of temperature and salinity (water-masses), which can be tracked as they follow the internal circulation of the ocean. These water-masses provide a proxy for biogeographical distributions of typical assemblages of pelagic species, but rarely for any individual taxa. The boundary regions between watermasses are nearly always diffuse and leaky, partly because of the effects of mesoscale eddies, but also because very many pelagic taxa have no difficulty in persisting for weeks to months as expatriates if they are advected into regions where they are unable to reproduce.

There are clear latitudinal and bathymetric gradients in species richness of pelagic communities. However, the patterns of species richness do not show a straightforward correlation with productivity. Circumstantial evidence points to the importance of the seasonality of organic input as a factor in determining local diversity. Species richness decreases and dominance increases as organic inputs become more seasonally pulsed.

The geological history of a region can impose significant restraints on the pool of species contributing to local and regional diversity. Changes in the morphology of the ocean basins since the fragmentation of the supercontinent Pangea have resulted in important changes in ocean circulation, exchanges between the different oceans and the characteristics of the deep ocean. ODP results imply that at the K-T event biogeographical zones were disrupted and those species that survived became cosmopolitan. As the water-mass composition of the oceans recovered, so present-day biogeographical patterns became established. There were still one-off vicariance events, such as the uplifting of the Panamanian Isthmus and the drying out and reflooding of the Mediterranean that had a major impact on pelagic distributions.

The mechanisms leading to speciation in the pelagic oceans, and the potential speciation rates, remain debatable; it has been argued that sympatric speciation occurs in very large environments in which there is slow gene-flow. But, if speciation is mostly allopatric, then sea-level fluctuations or the latitudinal oscillations in climatic zones during the glacial cycles resulting in relict faunas in semi-enclosed seas, may provide isolating mechanisms.

The great richness in phyla (disparity) displayed by marine faunas is restricted to the benthos. In pelagic communities disparity is similar to that of terrestrial faunas but species richness, while being high locally, is much lower at regional and global scales. This low species richness is the result of the ephemeral characteristics of the predominantly mesoscale structure of the physical environment. Mesoscale eddies, while enhancing gene-flow and reducing isolation at geological time-scales, at short time-scales provide a habitat diversity that can be expected to favour the maintenance of high genetic diversity in most pelagic species. Fluctuations in sea-level combined with redistribution of latitudinal zonation during Holocene glacial/interglacial cycles has led to isolation and speciation in those regions where the morphology of the continental margins is complex with shallow sea-ways and deep shallow-silled basins such as the East Indies.

Open ocean communities are often locally richer in species than neighbouring shelf communities, but the interaction between the morphology of ocean margins and physical processes leads to a finer-scaled structuring of neritic habitats. So regionally and globally the shelf and in-shore communities are richer in species than the open ocean communities.

References

Angel, H.H. & Angel, M.V., 1967. Distribution pattern analysis: In a marine benthic community. *Helgoländer wissenschaftliche Meeresuntersuchungen*, **15**, 445–54.

Angel, M.V., 1979. Studies on Atlantic halocyprid ostracods; their vertical distributions and community structure in the central gyre region along latitude 30°N from Africa to Bermuda. *Progress in Oceanography*, **8**, 1–122.

Angel, M.V., 1989a. Vertical profiles of pelagic communities in the vicinity of the Azores Front and their implications to deep ocean ecology. *Progress in Oceanography*, **22**, 1–46.

Angel, M.V., 1989b. Does mesopelagic biology affect the vertical flux? In *Productivity of the Oceans: Past and Present*, ed. W.H. Berger, V.S. Smetacek & G. Wefer, pp. 155–73. Chichester: John Wiley.

Angel, M.V., 1991. Variations in time and space: Is biogeography relevant to studies of long-time scale change? *Journal of the Marine Biological Association of the United Kingdom*, **71**, 191–206.

Angel, M.V., 1994. Spatial distribution of marine organisms: Patterns and processes. In *Large Scale Ecology and Conservation Biology*, ed. P.J. Edwards, R. May & N.R. Webb, pp. 59–109. Oxford: Blackwell Scientific Publications.

Angel, M.V. & Baker, A. de C., 1982. Vertical standing crop of plankton and micronekton at three stations in the north-east Atlantic. *Biological Oceanography*, **2**, 1–30.

Angel, M.V. & Fasham, M.J.R., 1975. Analysis of the vertical and geographic distribution of the abundant species of planktonic ostracods in the north-east Atlantic. *Journal of the Marine Biological Association of the United Kingdom*, **55**, 709–37.

Angel, M.V. & Hargreaves, P.M., 1992. Large scale patterns in the distributions of planktonic and micronektonic biomass in the N.E. Atlantic. *ICES Journal of Marine Science*, **49**, 403–11.

de Baar, H.J.W., 1994. Von Liebig's law of the minimum and plankton ecology. *Progress in Oceanography*, **33**, 347–86.

Backus, R.H., Craddock, J.E., Haedrich, R.L. & Robison, B.H., 1977. Atlantic mesopelagic zoogeography. In *Fishes of the Western North Atlantic*, vol. **7(1)**,

Sears Foundation for Marine Research, pp. 266–87, New Haven, CT: Yale University Press.
Badcock, J. & Merrett, N.R., 1976. Midwater fishes in the eastern North Atlantic – 1. Vertical distribution and associated biology in 30°N, 23°W, with developmental notes on certain myctophids. *Progress in Oceanography*, **7**, 3–58.
Baker, A. de C., Boden, B.P. & Brinton, E., 1990. *A Practical Guide to the Euphausiids of the World*. London: Natural History Museum Publications.
Barange, M. & Pillar, S.C., 1992. Cross-shelf circulation, zonation and maintenance mechanisms of *Nyctiphanes capensis* and *Euphausia hanseni* (Euphausiacea) in the northern Benguela Upwelling system. *Continental Shelf Research*, **12**, 1027–42.
Brinton, E., 1975. Euphausiids of southeastern Asian waters. *Naga Reports*, **4**, 1–287.
Broecker, W.S. & Peng, T.H., 1982. *Tracers in the Sea*. New York: Eldigio Press.
Campbell, J.W. & Aarup, T., 1992. New production in the North Atlantic derived from seasonal patterns of surface chlorophyll. *Deep-Sea Research*, **39**, 1669–94.
Carlson, C.A., Ducklow, H.W. & Michaels, A.F., 1994. Annual flux of dissolved organic carbon from the euphotic zone in the northwestern Sargasso Sea. *Nature*, **371**, 405–8.
Cohen, J.E., 1994. Marine and continental food webs: three paradoxes? *Philosophical Transactions of the Royal Society of London, Series B*, **343**, 57–69.
Denman, K.L., 1994. Scale-determining biological-physical interactions in oceanic food webs. In *Aquatic Ecology: Scale, Pattern and Process*, ed. P.S. Giller, A.G. Hildrew & D.G. Raffaelli, pp. 377–402. Oxford: Blackwell Scientific Publications.
Diel, S. and Tande, K., 1992. Does the spawning of *Calanus finmarchicus* in high latitudes follow a reproducible pattern? *Marine Biology*, **113**, 21–31.
Ekman, S., 1953. *Zoogeography of the Sea*. London: Sidgwick & Jackson Ltd.
Fairbanks, R.G., Wiebe, P.H. & Bé, A.W.H., 1980. Vertical distribution and isotopic composition of living planktonic Foraminifera in the western north Atlantic. *Science*, **207**, 61–3.
Fasham, M.J.R. & Foxton, P., 1979. Zonal distribution of pelagic Decapoda (Crustacea) in the eastern north Atlantic and its relation to the physical oceanography. *Journal of Experimental Marine Biology and Ecology*, **37**, 225–53.
Fleminger, A., 1986. The Pleistocene equatorial barrier between the Indian and Pacific Oceans and a likely cause for Wallace's line. *UNESCO Technical Papers in Marine Science*, **49**, 84–97.
Furnestin, M.-L., 1979. Aspects of the zoogeography of the Mediterranean. In *Zoogeography and Diversity in Plankton*, ed. S. van der Spoel & A.C. Pierrot-Bults, pp. 191–253. Utrecht: Bunge Scientific Publishers.
Greig-Smith, P., 1964. *Quantitative Plant Ecology*. London: Butterworth Press.
Haury, L.R., McGowan, J.A. & Wiebe, P.H., 1978. Patterns and processes in the time-space scales of plankton distributions. In *Spatial Pattern in Plankton Communities*, ed. J. Steele, pp. 277–328. New York: Plenum Press.
Hopkins, T.L., Milliken, D.M., Bell, L.M., McMichael, E.J., Hefferman, J.J. & Cano, R.V., 1981. The landward distribution of oceanic plankton and micronekton over the west Florida continental shelf as related to their vertical distribution. *Journal of Plankton Research*, **3**, 645–59.

Knoll, A.H., 1989. Evolution and extinction in the marine realm: Some constraints imposed by phytoplankton. *Philosophical Transactions of the Royal Society of London, Series B*, **325**, 279-90.
Levitus, S., Conkright, M.E., Reid, J.L., Najjar, R.G. & Mantyla, A., 1993. Distribution of nitrate, phosphate and silicate in the world oceans. *Progress in Oceanography*, **31**, 245-74.
McGowan, J.A., 1971. Ocean biogeography of the Pacific. In *The Micropalaeontology of Oceans*, ed. B.M. Funnel & W.R. Riedel, pp. 3-74. Cambridge: Cambridge University Press.
McGowan, J.A., 1974. The nature of oceanic ecosystems. In *The Biology of the Oceanic Pacific*, ed. C. Miller, pp. 9-28. Corvallis: Oregon State University Press.
McGowan, J.A., 1990. Species dominance-diversity patterns in oceanic communities. In *The Earth in Transition*, ed. G.M. Woodwell, pp. 395-421. Cambridge and New York: Cambridge University Press.
Miller, C.B. & Clemons, M.J., 1992. Revised life history analysis for large grazing copepods in the ocean subarctic Pacific. *Progress in Oceanography*, **20**, 275-92.
Müller, H.-G., 1993. *World Catalogue and Bibliography of the Recent Mysidacea*. Wetzlar: Laboratory for Tropical Ecosystems.
Paine, R.Y., 1980. Food webs: Linkage, interaction strength and community infrastructure. *Journal of Animal Ecology*, **49**, 667-85.
Palumbi, S.R., 1992. Marine speciation on a small planet. *Trends in Ecology and Evolution*, **7**, 114-18.
Parish, J. & Curtis, R.L., 1982. Atmospheric circulation, upwelling and organic-rich rocks in the Mesozoic and Cenozoic eras. *Palaeoceanography, Palaeoclimatology and Palaeoecology*, **40**, 31-66.
Pearse, J. & Buchsbaum, A., 1987. *Living Invertebrates*. Oxford: Blackwell Scientific Publications.
Pitcher, G.C., Walker, D.R., Mitchell-Innes, B.A. & Moloney, C.L., 1991. Short-term variability during an anchor station study in the southern Benguela upwelling system: Phytoplankton dynamics. *Progress in Oceanography*, **28**, 39-64.
Ray, G.C., 1991. Coastal-zone biodiversity patterns. *BioScience*, **41**, 490-8.
Reid, J., Brinton, E., Fleminger, A., Venrick, E.L. & McGowan, J.A., 1978. Ocean circulation and marine life. In *Advances in Oceanography*, ed. H. Charnock & Sir G. Deacon, pp. 65-130. New York and London: Plenum Press.
Savidge, G., Turner, D.R., Burkill, P.H., Watson, A.J., Angel, M.V., Pingree, R.D., Leech, H. & Richards, K.J., 1992. The BOFS 1992 Spring Bloom Experiment: Temporal evolution and spatial variability of the hydrographic field. *Progress in Oceanography*, **29**, 235-81.
Sepkoski, J.J., 1992. Phylogenetic and ecologic patterns in the Phanerozoic history of marine biodiversity. In *Systematics, Ecology and the Biodiversity Crisis*, ed. N. Eldredge, pp. 77-100, New York: Columbia University Press.
Shackleton, N.J., 1982. The deep-sea sediment record of climate variability. *Progress in Oceanography*, **11**, 199-218.
Sinclair, M., 1988. *Marine Populations: An Essay on Population Regulation and Speciation*. Seattle and London: Washington Sea Grant Program, University of Washington Press.
Smethie, W.M., 1993. Tracing the thermohaline circulation in the western north Atlantic using chlorofluorocarbons. *Progress in Oceanography*, **31**, 51-99.
Smith, S.L., 1984. Biological interactions of active upwelling in the northwestern

Indian Ocean in 1964 and 1979, and a comparison with Peru and northwest Africa. *Deep-Sea Research*, **31**, 951–67.

Stommel, H., 1963. Varieties of oceanographic experience. *Science*, **139**, 572–6.

Stuiver, M., Quay, P.D. & Ostlund, H.G., 1983. Abyssal water carbon-14 distribution and age of the World ocean. *Science*, **220**, 849–51.

van der Spoel, S. 1994. The basis for boundaries in pelagic biogeography. *Progress in Oceanography*, **34**, 121–33.

van der Spoel, S. & Heyman, R.P., 1983. *A Comparative Atlas of Zooplankton: Biological Patterns in the Oceans*. Utrecht: Bunge Scientific Publishers.

van der Spoel, S. & Pierrot-Bults, A.C. (eds), 1979. *Zoogeography and Diversity in Plankton*. Utrecht: Bunge Scientific Publishers.

van der Spoel, S., Pierrot-Bults, A.C. & Schalk, P.H., 1990. Probable mesozoic vicariance in the biogeography of Euphausiacea. *Bijdragen tot de Dierkunde*, **60**, 155–62.

Warner, A.J. & Hays, G.C., 1994. Sampling by the continuous plankton recorder. *Progress in Oceanography*, **34**, 237–56.

White, B.N., 1994. Vicariance biography of the open-ocean Pacific. *Progress in Oceanography*, **34**, 257–82.

Zachos, J.C., Arthur, M.A. & Dean, W.E., 1989. Geochemical evidence for suppression of pelagic marine productivity at the Cretaceous/Tertiary boundary. *Nature*, **337**, 61–4.

Chapter 4
Biological diversity in oceanic macrozooplankton: More than counting species

ANNELIES C. PIERROT-BULTS
Institute for Systematics and Population Biology, University of Amsterdam, The Netherlands

Abstract

Pelagic systems are the largest on earth. Compared with the terrestrial habitat these systems, which are composed of taxa with vast ranges of distribution, do not show high species richness. In general species richness in pelagic animal groups decreases from lower to higher latitudes. Speciation processes seem to be slow in the ocean. However, the genetic diversity within a pelagic species might be very high because of the very large distributional ranges where geographic differences occur in environmental conditions. There might be a number of cryptic species as yet undetected.

There is a suggestion of a strong link between the general oceanic circulation and the productivity regime, and pelagic distributions in the different regions. The polar production cycle is characterised by a single peak in summer, the temperate cycle has a spring and autumn bloom and the (sub)tropical regime has relatively low continuous production with a slight peak in winter. The main abundance and biomass of Chaetognatha is in the upper 200 m of the water column and north of 40° latitude. Vertically and horizontally there seems to be an inverse relationship between productivity and biological diversity. The relationship might not be a direct one but might rather reflect regions with high mixing and regions with fewer perturbations.

4.1 Introduction

Pelagic ecosystems are the largest on earth. The oceans cover about 70% of the surface of the globe to an average depth of 4 km. Pelagic systems are composed of taxa with very wide-ranging distributions, for example many taxa of oceanic macrozooplankton are distributed from

40°N–40°S in all three oceans. There is no biotope on land that is comparable in terms of living space.

On the higher taxonomic levels the marine realm is very rich in diversity. Of the 34 animal phyla, 29 occur in the marine habitat of which 14 are exclusively marine (Pearse & Buchsbaum, 1987; Funch & Kristensen, 1995). However, of the 29 phyla that are present in the ocean, only 12 are represented in the pelagic systems, moreover there is not a single phylum endemic to them. So it seems that in animal evolution it has been as difficult to move from the bottom up into the pelagic, as moving from the sea onto the land (Angel, 1993). Compared with the terrestrial habitat oceanic pelagic systems do not show high species richness. According to May (1994) only about 15% of all recorded animal and plant species live in the sea, and 56% of the terrestrial taxa are insects. The Plant Kingdom is represented by an estimated 3500–4500 phytoplankton species in the global ocean (Sournia et al., 1991), whereas it is estimated that there are >250 000 species of higher plants on land. Crustacea (phylum Arthropoda) are the most numerous group in marine macrozooplankton, but their estimate species richness is lower by about three orders of magnitude than the number of insect species on the land.

Speciation is thought to be a relatively slow process in the open ocean (Pierrot-Bults & van der Spoel, 1979). The large-scale oceanic circulation, and the gradual changes in the oceanic water-masses both in time and space, inhibit the development of firm barriers to dispersal of pelagic organisms. Also planktonic vertical migrations (diel, ontogenetic and seasonal) are mechanisms that serve to keep the populations well mixed. This lack of isolation results in there being relatively few species with very extensive geographical ranges in the pelagic environment of the global ocean, although local assemblages can be relatively rich in species.

However, recent molecular research indicates that there might well be a number of cryptic species in the sea, both in phytoplankton (Medlin et al., 1991) as well as in zooplankton (Bucklin et al., 1996). In many zooplankton groups there are fewer species at any given place in neritic regions than in the upper 200 m of open ocean environments. However, in the global ocean neritic species are the more numerous because they have far more restricted ranges. Coastal morphology, for example estuaries and inlet seas, and its interactions with hydrodynamic processes (tides, currents and sea-level fluctuations) can generate the isolation that induces speciation. Hence neritic species tend to show local endemism, while very few oceanic species are endemic to such localised regions. For example in chaetognaths (Bieri, 1991; Pierrot-Bults & Nair, 1991) there are on average about 10 species in the top 200 m of the water column in any given place in the subtropical open ocean, while in the neighbouring neritic environment there may be only one or two. In total, however, the number of neritic species is about 18 in the subtropical oceans.

Species richness shows latitudinal trends, with zooplankton species richness declining from the tropics to higher latitudes, both for oceanic and for neritic species. The biomass maximum of zooplankton (and micronekton at night) is in the upper 200 m of the water column, whereas maximum species richness occurs at about 800–900 m (Pierrot-Bults, 1982; Angel, 1993). The main productive areas of the world's oceans are at the edges of the central gyres at higher latitudes and at upwelling areas off the continental margins (Fig. 4.1). These are regions where there is mixing of water-masses, and deeper-living organisms are brought more to the surface by upwelling. Both factors result in a greater mixing of faunas that enhances species richness locally. This is very clearly demonstrated in the figures of Reid *et al.* (1978) showing species richness in Pteropoda, Euphausiidae and Chaetognatha. Boundary zones where there is either mixing between water-masses or there are seasonal shifts in their distributions show a high species diversity. However, the maintenance of this high diversity requires continuous influx from other areas, and the breeding range of many of the species lies elsewhere.

The final important influence on local species-richness is the role being played by anthropogenic introductions of exotic species, notably by the transport of large quantities of ballast-water and imports of stock for aquaculture (Omori *et al.*, 1994). Although not all of these exotic species are able to establish themselves, those that do often come without the diseases and parasites that may have been limiting their competitive success in their native waters. So although all those that become established initially enhance species-richness some can out-compete or otherwise reduce the viability of the native species, and may also adversely affect exploited populations.

4.2 Species richness in Chaetognatha compared with other macrozooplankton groups

The phylum Chaetognatha currently consists of about 80 pelagic and 20 benthic species. Chaetognatha are ubiquitous in the ocean – occurring from coast to coast and from the surface to the bottom. Typically the number of chaetognath species caught at a given time at a station in the North Atlantic between 0–40° latitude in the top 2000 m of the water

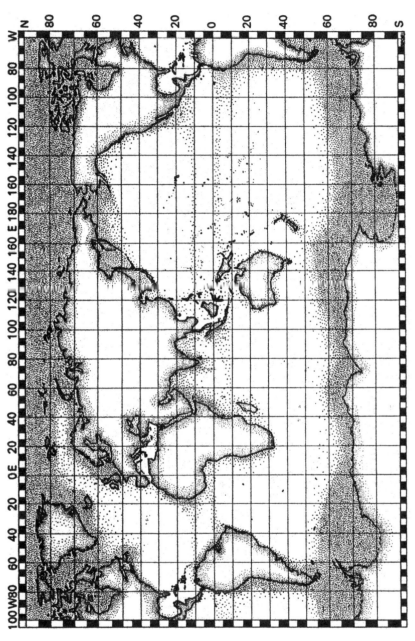

Fig. 4.1. Regional differences in productivity in the oceans. Dotted areas represent regions of high productivity; blank areas represent regions of low productivity.

Table 4.1. *Species richness in the Atlantic, Pacific and Indian Oceans for Chaetognatha, Pteropoda and Euphausiidae*

	Chaetognatha			Pteropoda			Euphausiidae		
	A	P	I	A	P	I	A	P	I
Arctic	1	1		3	4		5	5	
Subarctic	9	6		20	17		10	10	
40°N–40°S	25	34	29	120	90	95	25	30	30
Subantarctic	9	8	8	17	17	17	10	10	10
Antarctic	4	4	4	11	11	11	5	5	5

A, Atlantic; P, Pacific; I, Indian.

column is 18–20. So the species richness is not very high. The highest species richness is at about 900 m depth where there is mixing of the meso- and bathypelagic fauna.

As stated above, species richness in pelagic taxa decreases from low to high latitudes, with the greatest decline at around 40° latitude. Table 4.1 and Figs 4.2(*a*)–(*c*) summarise the latitudinal changes in species richness of three pelagic groups i.e. Chaetognatha, Pteropoda (S. van der Spoel, personal communication) and Euphausiidae (Reid *et al.*, 1978) in the three major oceans, illustrating the fact that these latitudinal trends are the same in representatives of the three phyla Chaetognatha, Mollusca and Arthropoda. Numbers of (sub)tropical epipelagic oceanic species are quite low. In Chaetognatha there are 14 epipelagic species in the Indo-Pacific and only 8 in the Atlantic (Fig. 4.2(*a*)). Most of the species richness in the (sub)tropical regions is contributed by meso- and bathypelagic species (Fig. 4.3). The same is true for pteropods; at (sub)tropical latitudes the epipelagic pteropod faunas consist of 38, 37 and 33 species in the Atlantic, Pacific and Indian Ocean, respectively (Fig. 4.2(*b*)), whereas the total (S. van der Spoel, personal communication) numbers of species for the (sub)tropical region are 120 (107), 90 (77) and 95 (76) for the western (eastern) waters of the Atlantic, Indian and Pacific Oceans, respectively. There are considerable differences between the western and eastern regions of the three oceans in Pteropoda, but in Chaetognatha these differences are less pronounced. Only in the Pacific is there a clear difference, with 30 species occurring in western waters and 25 species in eastern (Fig. 4.2(*a*)).

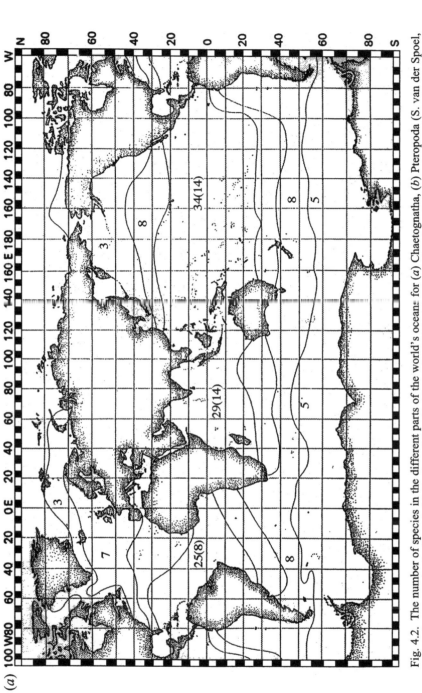

Fig. 4.2. The number of species in the different parts of the world's oceans for (a) Chaetognatha, (b) Pteropoda (S. van der Spoel, personal communication), and (c) Euphausiidae (Reid et al., 1978). Figures in brackets indicate the number of meso- and

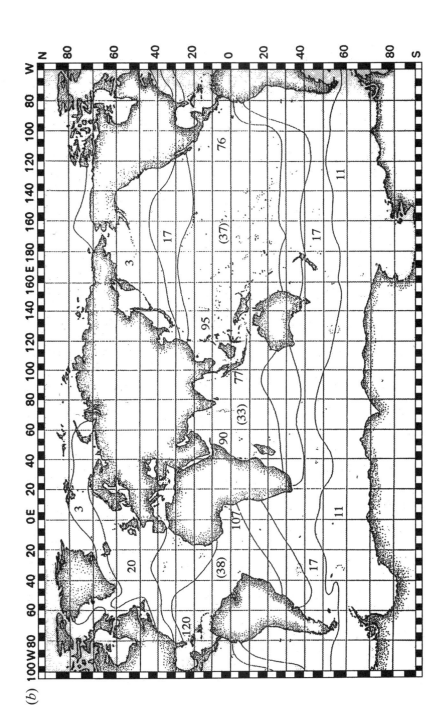

(b)

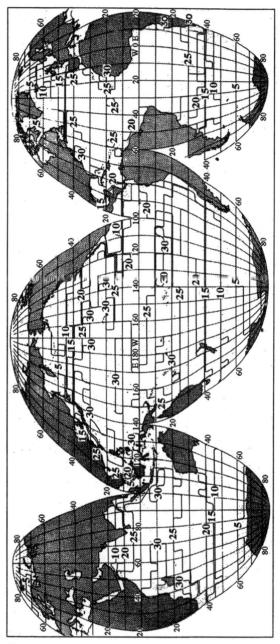

(c)

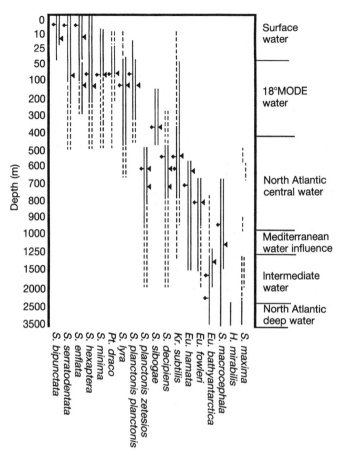

Fig. 4.3. The vertical distribution of Chaetognath species in the North Atlantic, horizontal bars indicate layers of maximum abundance (after Pierrot-Bults, 1982). ▼, day; ♦, night; = = = =, single specimens.

4.3 Relations between distribution patterns and biological diversity

4.3.1 Oceanic species

Oceanic pelagic species generally have very extensive distributional ranges. For example a typical distribution pattern for (sub)tropical epipelagic (from 0–200 m depth) and shallow mesopelagic species (from 200–500 m depth) is a tropical–subtropical range in all three oceans from 40°N to 40°S, as is found for the chaetognath *Pterosagitta draco*

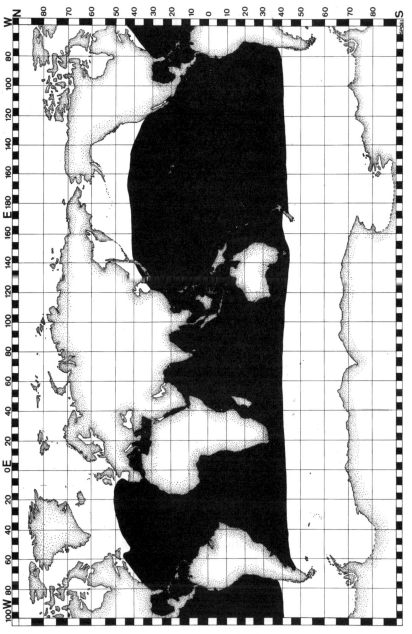

Fig. 4.4. The distribution of *Pterosagitta draco*, an example of a 40°N–40°S distribution in all three oceans.

Krohn, 1853 (Pierrot-Bults & Nair, 1991: Fig. 4.4) or the copepod *Clausocalanus paupulus* Farran, 1926 (Fleminger & Hulsemann, 1973), which shows a similar distribution. There are no barriers to gene-flow between the populations of these species in the different oceans. One consequence of these wide ranges is that there is relatively little difference in the species composition of shallow-living planktonic species between the different oceans. For example, the list of chaetognath species in the top 1200 m of the water column in the North-West Atlantic near Bermuda published by Pierrot-Bults (1982) is almost identical to the list for the South-West Pacific published by Fagetti (1972). The only specific difference in the epipelagic Chaetognatha within the 40°N–40°S range is the replacement of *Sagitta serratodentata* Krohn, 1853 in the Atlantic by *S. pacifica* Tokioka, 1940: these are closely related species that are classified together in a species-group with more restricted distributions (Figs 4.5(*a*) and (*b*)).

Deep-mesopelagic (500–1000 m depth) and bathypelagic (>1000 m depth) species, such as the chaetognath *S. macrocephala* Fowler, 1905, have even more extensive ranges, i.e. from the subarctic to the subantarctic.

Genetic heterogeneity within such wide-ranging species seems likely to be high because within such very large distributional ranges, factors influencing selection are likely to vary extensively. The thermohaline circulation and mixing must be rapid enough over evolutionary timescales to inhibit speciation. But do these widespread species such as the chaetognath *S. pacifica* (Fig. 4.5(*a*)) show more infraspecific genetic variation than species restricted to a single water-mass (such as *S. pseudoserratodentata* Tokioka, 1938 which is restricted to the north Pacific: Fig. 4.5(*b*))? If so, are these differences sufficient for the variants to be considered to represent cryptic species or do they still only represent infraspecific variation?

4.3.2 Neritic species

Neritic species generally have more restricted ranges than oceanic ones. Roughly there is a warm-water (40°N–40°S) and a cold-water pattern (latitudes >40°), with eastern and western coastal waters having distinctive faunas, showing a similar latitudinal succession to their oceanic counterparts. Speciation in neritic taxa has been, and can be, greatly

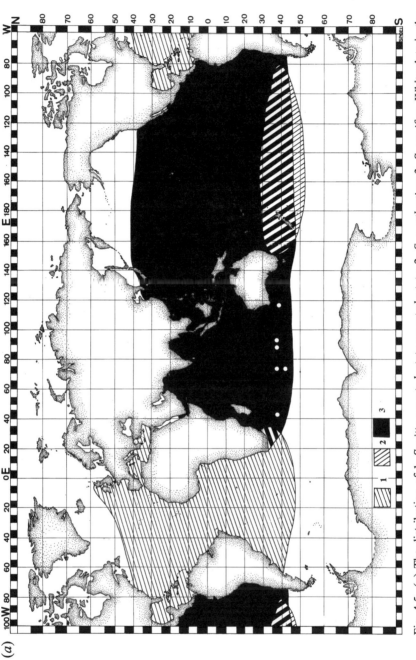

Fig. 4.5. (a) The distribution of 1, Sagitta serratodentata serratodentata; 2, S. s. atlantica; 3, S. pacifica. White dots in the South Indian Ocean represent transport of S. serratodentata to the South Pacific. (b) Distribution of 1, S. tasmanica; 2, S. bierii;

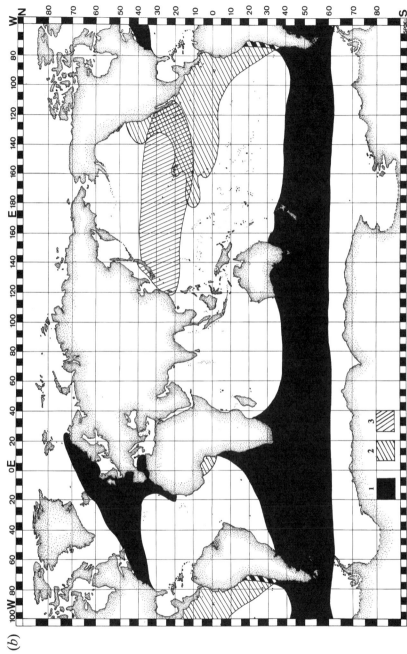

influenced by sea-level fluctuations and surface cooling by coastal upwelling. These phenomena, by creating abiotic barriers to distributions, can isolate populations long enough for speciation to occur, as demonstrated by Fleminger (1986) for copepods around the Indo-Malayan region, and by Tokioka (1979) for 13 Indo-West Pacific inlet-water species of Chaetognatha. The ranges of neritic species are often discontinuous and their morphological variation is greater than in oceanic species. For example, the neritic North Atlantic chaetognath species *S. setosa* J. Müller, 1847 ranges from the Baltic to the North Sea, and occurs both in the Mediterranean and in the Black Sea (Fig. 4.6). Within this species concept are two taxa originally described as *S. batava*, Biersteker & van der Spoel, 1966 (from the Scheldt estuary, The Netherlands) and *S. euxina*, Moltschanoff, 1909 (from the Black Sea), which are now considered to be local genetic variants that are not sufficiently differentiated to separate as distinct species (Pierrot-Bults, 1976). *Sagitta friderici* Ritter-Zahony, 1911 is another species with a disjunct distribution. It occurs along the coasts of West Africa and South America. The populations seem to be geographically isolated, but there is only slight morphological variation between them, and there is too much overlap to separate them conclusively as different species, so their status can only be assessed by molecular research.

Another example of differences between populations in the Mediterranean and the Black Sea is the copepod *Calanus helgolandicus* (Claus, 1863) that has diverged behaviourally and morphologically between the two locations (Fleminger, 1986). Sinclair (1988) discussing fish populations argues that life-cycle selection, the combined sexual selection, behaviour and distributional phenomena, can be an internal driving force for speciation in the neritic environments (causing genetic differences, often without noticeable morphological differences between populations). This is well illustrated in Atlantic herring populations of which there are 27 populations each with a different spawning time and different larval development times, e.g. the duration of the larval phase of autumn-spawning populations exceeds 6 months, whereas for spring-spawning populations it can be as short as 2–3 months.

External forcing can come from the fragmentation of distributional ranges by natural environmental fluctuations. Inlet neritic species do not usually drift out into the oceanic realm because of the hydrographic conditions along the coast. However, neritic species may be found in the neighbouring oceanic waters but generally cannot survive in oceanic conditions, whereas there are regular influxes of oceanic species into

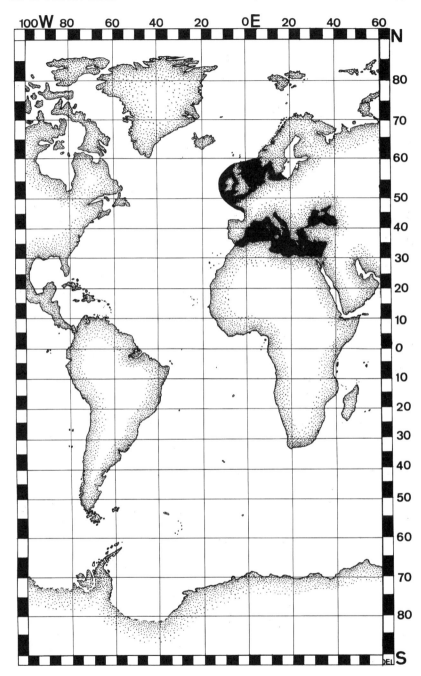

Fig. 4.6. The distribution of *Sagitta setosa*, an example of a neritic distribution with disjunct populations.

the neritic environments (Tokioka, 1979), although even these are not usually found in inlet-waters.

4.4 Relations between biomass and biodiversity

Geographically the biomass maxima of Chaetognatha occur at latitudes >40°, and the mean body sizes also increase towards high latitudes. Bathymetrically, the maxima in both abundance and biomass of Chaetognatha occur in the upper 200 m of the water column and decrease with increasing depth. The species inhabiting the epipelagic zone, such as *Sagitta minima* Grassi, 1881 and *Pterosagitta draco*, are smaller in size than deeper-living species such as *S. lyra* Krohn, 1853 and *S. planctonis* Steinhaus, 1896. Species that live at higher latitudes such as *S. gazellae* Ritter-Zahony, 1909 and *S. maxima* Conant, 1896 have bigger maximum sizes than species living at comparable depths in tropical waters such as *S. serratodentata* and *S. bipunctata* Quoy & Gaimard, 1827.

Although chaetognaths prey mainly on copepods, they will prey on any organism that is appropriately sized, including fish larvae. The estimated daily turnover of the total zooplankton production is about 40–50%, and chaetognath biomass averages about 30–35% of the copepod biomass (Reeve, 1970; Feigenbaum, 1991). This suggests that although chaetognaths are low in species richness they are important functionally as predators, competitors and prey in oceanic pelagic systems. In any given location there seem to be a few very abundant species, a fair number of common species and a large number of 'rare' species (van der Spoel, 1994; Fig. 4.7). This frequency distribution seems to be fractal in nature (van der Spoel, 1994). McGowan (1990) stated that 80–90% of the pelagic species are consistently rare. In chaetognath samples from the North Atlantic between 60–80% of the species were found to be consistently rare, and the dominant species shift with the different depth layers (Pierrot-Bults, 1982). In species-low environments there are more extremes in relative abundances (i.e. greater dominance), while in species-rich environments there is more evenness in abundances between the various species. Off Bermuda (Pierrot-Bults, 1982) four species occurred in the top 10 m with just one contributing 94% of the specimens. At 100–200 m there were eight species, five with relative abundances of 28%, 24%, 20%, 14% and 8%, and the remaining three species sharing 6% of the relative abundance. The same is true for the horizontal variability. In the Southern Ocean the dominant species *S. gazellae* Ritter-Zahony, 1909 contributes 95% of the total

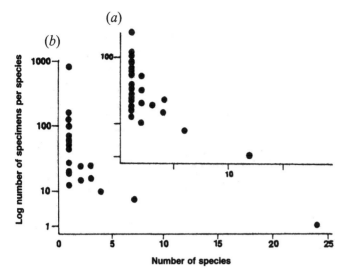

Fig. 4.7. The relationship between abundance and number of species of Myctophidae (a) and Hydromedusae (b) in the North Atlantic (after van der Spoel, 1994).

chaetognath population. So both vertically and horizontally there seems to be an inverse relationship between biomass and biological diversity. Van der Spoel (1994) illustrated the relationship between biomass and diversity at the species level as a function of time and place using Haury et al.'s (1977) use of a Stommel diagram (Fig. 4.8). This clearly indicates the relationship between large biomass variability and low diversity fluctuation and vice versa. In the time/space envelope of seconds to a month and centimetre to kilometre there is large variability, because of currents and waves, seasonal change and migration and seasonal species succession. In the envelope that includes decades and 10 km there is another maximum caused by shifts of major current patterns and ENSO (El Niño Southern Oscillation) type events. These relatively 'unstable' scales in the time/space regime are probably less sensitive to human influence than the areas with lower variability.

As stated above, there are regional differences in biomass (Fig. 4.1). There are also differences in the productivity regimes in the different regions. The polar production cycle is characterised by a single peak in summer, the North Atlantic temperate cycle has peaks in both spring and autumn (blooms), whereas (sub)tropical regimes tend to have relatively continuous production albeit low and showing a slight maximum in winter. The relationship between productivity and species richness might

86   *Biological diversity in oceanic macrozooplankton*

(a)

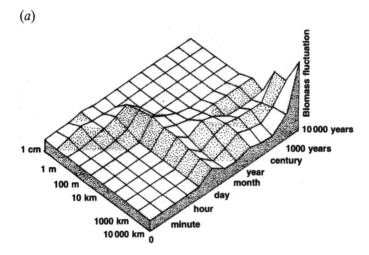

(b)

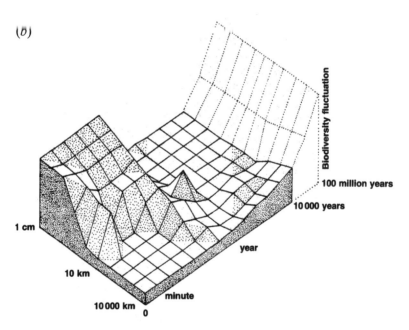

Fig. 4.8. The relation between biomass and biological diversity (after van der Spoel, 1994). (a) Biomass variability simplified after the Stommel diagram by Haury et al. (1977). (b) Diversity variability at the species level. Diversity resulting from evolution takes place at larger time scales and is added in the dotted area; space on y-axis, time on x-axis, variability on z-axis.

not be a direct one but rather reflects regions where, for example, vertical mixing in the water column is relatively high or where there are fewer perturbations. The greater the mixing the greater the supply of nutrients for the primary producers, whereas the lower the perturbation the greater the predictability with more chance for a higher degree of specialisation. Superimposing plots of the known distributions of different taxa shows that each species has a slightly different pattern (van der Spoel & Heyman, 1983; Fig. 4.9), suggesting that environmental tolerances are subtly different between the taxa, so that species tend to separate both horizontally and vertically. Although the general distributions of species overlap, their main centres of abundance are slightly displaced, either horizontally or vertically, as shown by the horizontal distribution of Chaetognatha in the Indian Ocean (David, 1963: Fig. 4.10) and the vertical distribution of Chaetognatha in the Atlantic Ocean (Pierrot-Bults, 1982; Pierrot-Bults & Nair, 1991). Another mode of separation is through a lack of synchronisation in reproduction times between the different chaetognath species (Pierrot-Bults, 1982) suggesting that separate niches do occur in the oceanic environment.

Although the neritic environment has more pronouncedly different habitats, there is segregation in the times of reproduction, between species or between populations as illustrated by Sinclair (1988) for herring populations.

4.5 Conclusions

The number of species present today and their distribution patterns is the result of both their evolutionary history and present-day environmental circumstances. Chaetognatha are considered to be a geologically old marine lineage. Recent RNA research (Telford & Holland, 1993) establishes them as an early offshoot of metazoan evolution. Fossil evidence is scarce but they have been identified from the mid-Cambrian (Wright, 1993). The coherence in zoogeographic distribution patterns of different groups of animals with physico-chemical and climatological properties suggest a strong link between the general oceanic circulation and productivity regimes and the pelagic distributions.

It is necessary to link the present with the past, to be able to make reasonable, accurate extrapolations in order to try to make predictions about the future. Unfortunately phylogenetic analyses of pelagic animals and areo-cladograms are not available yet. Linking the history of the ocean basins and water-masses gives an indication of the age of the

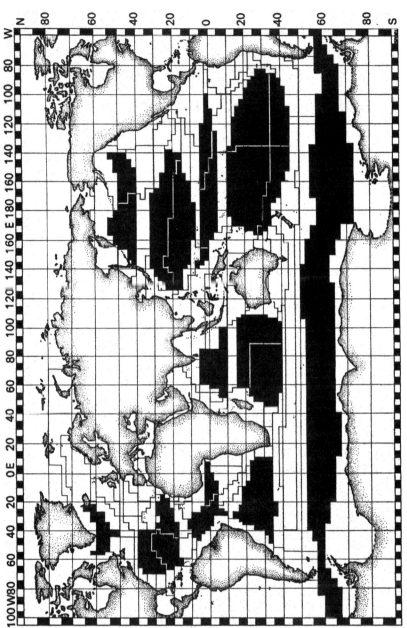

Fig. 4.9. Faunal boundaries of pelagic taxa. Black areas indicate Central Waters and cyclic currents.

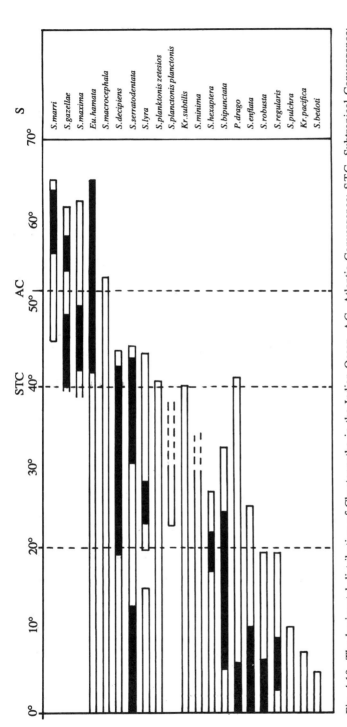

Fig. 4.10. The horizontal distribution of Chaetognatha in the Indian Ocean. AC, Atlantic Convergence; STC, Subtropical Convergence; S, South. Black bars indicate areas of maximum abundance (after David, 1963).

groups. For example the Atlantic is not as old as the Pacific, so for old marine groups such as Chaetognatha the Atlantic has been populated by dispersal from the Pacific. A group that is thought to have been more recently developed in the Atlantic are the Pteropoda (van der Spoel, 1967). This may be the reason why pteropods are more speciose in the Atlantic than in the Pacific, the reverse of the situation for Chaetognatha and Euphausiidae (see Table 4.1). Tokioka (1965) argued that groups that are richer in species in the Indo-Malayan region than in other parts of the world's oceans have their origin in the Pacific and are old marine groups.

The distribution boundaries of individual species show changes in time and space because of the seasonal cycles of populations and the moving environment. Phenomena such as rings and eddies, seasonal shifting of water-masses, ontogenetic, seasonal and vertical migration, sterile expatriates etc give a false picture of species richness if not properly understood. Species that are advected out of their normal range rarely survive in their new environment and so do not increase the species richness of that new environment in the long term. For example the chaetognath *S. tasmanica* Thomson, 1946 is found in cold-core rings in both the North-West Atlantic and the South-East Pacific. When these rings are gradually mixing in more warm water the specimens become smaller and smaller, shrinking from adult sizes of 16 mm to 6 mm in length before they disappear. Although the species is found in subtropical waters, it is unable to persist and so is not really part of the biodiversity of that area.

Species richness is the result of speciation processes in the past. In some groups speciation (and extinction) have resulted in there being many extant taxa in some groups, notably the Crustacea, but very few in others, such as the Chaetognatha. Also within groups some genera have radiated into many species and others have hardly radiated at all. The phylum Chaetognatha is exclusively marine and there are only about 100 species recognised of which about 20 are benthic coastal forms. The monospecific genus *Pterosagitta* is epipelagic and widespread (Fig. 4.4), whereas the genus *Sagitta*, which some authors want to split, currently includes about 50 species. The Pteropoda and the pelagic Cephalopoda belonging to the phylum Mollusca have about 250 and 400 species, respectively (van der Spoel, 1967), and the Crustacea (belonging to the phylum Arthropoda, the most species-rich phylum of all) are also the most numerous in the zooplankton, with about 2200 species of marine pelagic copepods, 400 of pelagic mysids (which are mainly neritic) and 86 of euphausiids (which are mainly oceanic). However, the number of species in the pelagic system is much lower than on the land in the same phylum or in the same group. For example the estimated total number of all copepod species is about 7500 (Pearse & Buchsbaum, 1987) while the estimated number of pelagic copepods is about 2200.

Molecular research is needed to define the amount of genetic infraspecific variation in widespread oceanic species, and to determine whether we deal with cryptic species or a continuum of population differences within one species without clear separations. The mosaic of different populations with genetic differences of a widespread species could be a blueprint for speciation when changes in the environment occur by global warming/cooling or sea-level fluctuations. However, species with restricted distributions and/or restricted variability might go extinct, being unable to adapt when changes in the environment occur. Another question to be answered is whether the amount of genetic variation is comparable in oceanic and neritic species; do neritic species have more infraspecific genetic variation, because they have to cope with a variable environment, than oceanic species, which live in more predictable surroundings? Or do species that are able to live in extreme conditions such as (sub)polar waters show more genetic variation than species inhabiting less stressful conditions? The ability of the present gene pool to (re)create diversity in the future from its present diversity by using now hidden but vital resources in their genetic make-up when environmental changes occur, is an important aspect for nature conservation. Whether it makes much ecological difference if there are a number of closely related species or one polymorphic species within a given range is not yet clear. In other words how much advantage is there in keeping a large flexible gene pool *versus* speciation in separate species that do not interbreed any more?

The importance of the open ocean in global processes such as fixation of atmospheric carbon dioxide makes it important to understand the role of the organisms living there, and the processes that give rise to fluctuations in species distributions and give rise to either speciation or extinction of species. Biodiversity of the open ocean must not be neglected nor its ecological role underestimated.

References

Angel, M.V., 1993. Biodiversity of the pelagic ocean. *Conservation Biology*, **7**, 760–72.
Bieri, R., 1991. Systematics of the Chaetognatha. In *The Biology of Chaetognaths*, ed. Q. Bone, H. Kapp & A.C. Pierrot-Bults, pp. 122–36. Oxford: Oxford University Press.
Bucklin, A., LaJeunesse, T.C., Curry, E., Wallinga, J. & Garrison, K., 1996. Molecular genetic diversity of the copepod, *Nannocalanus minor*: Genetic evidence of species and population structure in the N. Atlantic Ocean. *Journal of Marine Research*, (in press).
David, P.M., 1963. Some aspects of speciation in the Chaetognatha. In *Speciation in the Sea*, ed. M.P. Harding & N. Tebble, pp. 129–43. The Systematics Association Publication no. 5. London: The Systematics Association.
Fagetti, E.G., 1972. Bathymetric distribution of chaetognaths in the southeastern Pacific Ocean. *Marine Biology*, **17**, 7–30.
Feigenbaum, D., 1991. Food and feeding behaviour. In *The Biology of Chaetognaths*, ed. Q. Bone, H. Kapp & A.C. Pierrot-Bults, pp. 45–54. Oxford: Oxford University Press.
Fleminger, A., 1986. The Pleistocene barrier between the Indian and Pacific Oceans and a likely cause for Wallace's line. In *Pelagic Biogeography*, ed. A.C. Pierrot-Bults, S. van der Spoel, R.K. Johnson & B.J. Zahuranec, pp. 84–97. UNESCO Technical Papers in Marine Science, no. 49. Paris: UNESCO.
Fleminger, A. & Hulsemann, K., 1973. Relationship of Indian Ocean epiplankton calanoids to the World Ocean. In *The Biology of the Indian Ocean*, ed. B. Zeitschel & S.A. Gerlach, pp. 339–48. Berlin: Springer-Verlag.
Funch, P. & Kristensen, R.M., 1995. Cycliophora is a new phylum with affinities to Entoprocta and Ectoprocta. *Nature*, **378**, 711–14.
Haury, L.R., McGowan, J.J. & Wiebe, P.H., 1977. Patterns and processes in the time-space scales of plankton distribution. In *Spatial Pattern in Plankton Communities*, ed. J.H. Steele, pp. 227–328. New York: Plenum Press.
May, R.M., 1994. Conceptual aspects of the quantification of the extent of biological diversity. *Philosophical Transactions of the Royal Society of London, Series B*, **345**, 13–20.
McGowan, J.A., 1990. Species dominance-diversity patterns in oceanic communities. In *The Earth in Transition*, ed. G.M. Woodwell, pp. 395–421. New York: Cambridge University Press.
Medlin, L.K., Elwood, H.J., Stickel, S. & Sogin, M.L., 1991. Morphological and genetic variation within the diatom *Skeletonema costatum* (Bacillariophyta): Evidence for a new species *Skeletonema pseudocostatum*. *Journal of Phycology*, **27**, 514–24.
Omori, M., van der Spoel, S. & Norman, C.P., 1994. Impact of human activities on pelagic biogeography. *Progress in Oceanography*, **34**, 211–19.
Pearse, J. & Buchsbaum, A., 1987. *Living Invertebrates*. Oxford: Blackwell Scientific Publications.
Pierrot-Bults, A.C., 1976. Zoogeographic patterns in Chaetognatha and some other planktonic organisms. *Bulletin of the Zoological Museum of the University of Amsterdam*, **5**, 59–72.
Pierrot-Bults, A.C., 1982. Vertical distribution of Chaetognatha in the central Northwest Atlantic near Bermuda. *Biological Oceanography*, **2**, 31–61.
Pierrot-Bults, A.C. & Nair, V.R., 1991. Distribution patterns in Chaetognatha. In

The Biology of Chaetognaths, ed. Q. Bone, H. Kapp & A.C. Pierrot-Bults, pp. 86–116. Oxford: Oxford University Press.
Pierrot-Bults, A.C. & van der Spoel, S., 1979. Speciation in macrozooplankton. In Zoogeography and Diversity in Plankton, ed. S. van der Spoel & A.C. Pierrot-Bults, pp. 144–67. Utrecht: Bunge Scientific Publications.
Reeve, M.R., 1970. The biology of Chaetognatha. I. Quantitative aspects of growth and egg production in *Sagitta hispida*. In *Marine Food Chains*, ed. J.H. Steele, pp. 168–89. Edinburgh: Oliver & Boyd.
Reid, J., Brinton, E., Fleminger, A., Venrick, E.L. & McGowan, J.A., 1978. Ocean circulation and marine life. In *Advances in Oceanography*, ed. H. Charnock & Sir G. Deacon, pp. 65–130. New York: Plenum Press.
Sinclair, M., 1988. *Marine Populations: An Essay on Population Regulation and Speciation*. Seattle, London: University of Washington Press, Washington Sea Grant Program.
Sournia, A., Chretiennot-Dinet, M.-J. & Ricard, M., 1991. Marine phytoplankton: How many species in the world ocean? *Journal of Plankton Research*, **13**, 1093–9.
Telford, M.J. & Holland, P.W.H., 1993. The phylogenetic affinities of the chaetognaths: A molecular analysis. *Molecular and Biological Evolution*, **10**, 660–76.
Tokioka, T., 1965. Supplementary notes on the systematics of Chaetognatha. *Publications of the Seto Marine Biology Laboratory*, **13**, 231–42.
Tokioka, T., 1979. Neritic and oceanic plankton. In *Zoogeography and Diversity of Plankton*, ed. S. van der Spoel & A.C. Pierrot-Bults, pp. 126–43. Utrecht: Bunge Scientific Publications.
van der Spoel, S., 1967. *Euthecosomata, A Group with Remarkable Developmental Stages (Gasteropoda, Pteropoda)*. Gorinchem: Noorduyn & Zn.
van der Spoel, S., 1994. A biosystematic basis for pelagic diversity. *Bijdragen tot de Dierkunde*, **64**, 3–31.
van der Spoel, S. & Heyman, R.P., 1983. *A Comparative Atlas of Zooplankton*. Utrecht: Bunge Scientific Publishers.
Wright, J.C., 1993. Some comments on the inter-phyletic relationships of chaetognaths. In *Proceedings of the 2nd International Workshop of Chaetognatha*, ed. I. Moreno, pp. 51–61. Palma de Mallorca: Universitat de les Illes Balears.

Chapter 5
Large-scale patterns of species diversity in the deep-sea benthos

MICHAEL A. REX, RON J. ETTER and
CAROL T. STUART
*Department of Biology, University of Massachusetts, Boston,
MA 02125, USA*

Abstract

As in other environments, species diversity in the deep sea reflects an integration of ecological and evolutionary processes operating at different spatial and temporal scales. Contemporary deep-sea research has focused primarily on the importance of small-scale phenomena that permit species coexistence. While this work has provided important insights into the mechanisms that regulate local diversity, it is unclear how small-scale events can account for geographic patterns of diversity. A complete understanding of diversity must incorporate the influence of historical, biogeographic and oceanographic processes that are imposed at much larger scales. In this chapter, we review large-scale bathymetric and geographic patterns of species diversity in the deep-sea benthos and discuss how ecological and evolutionary factors might shape these patterns.

In the western North Atlantic, the most thoroughly sampled region of the World Ocean, species diversity is low on the shelf, increases to a maximum at intermediate depths and then decreases in the abyssal plain. Analyses of diversity in other deep basins of the Atlantic indicate that this parabolic trend may not be universal, but good comparative data are extremely limited. The marked variation in diversity at any particular depth indicates that extensive sampling is necessary to accurately assess bathymetric patterns. The causes of these patterns are not well understood, but appear to involve environmental gradients in nutrient flux, biotic interactions and environmental heterogeneity.

There is also inter-regional variation in deep-sea diversity. Bathyal species diversity in the Atlantic shows a latitudinal decline in the northern hemisphere and pronounced basin-to-basin variation in the southern hemisphere. Global-scale patterns of diversity are still poorly characterised, but are likely to result from both ecological and regional–historical processes. Environmental gradients at the surface may be translated to the deep sea through surface–benthic coupling. Local diversity in deep-sea prosobranch gastropods can be predicted by regional diversity and the dispersal potential of the regional species pool suggesting that the evolutionary development of regional diversity and the mechanism of regional enrichment contribute to patterns of diversity on global scales. An important priority for future research is to design sampling programmes on very large scales to explore the regional–historical causes of deep-sea biodiversity.

5.1 Introduction

Changes in the focus and scale of research in deep-sea ecology have followed closely the trends in coastal marine and terrestrial ecology during the last several decades. Emphasis has shifted from an early interest in large-scale patterns of species diversity and their evolutionary causes to more controlled studies on ecological mechanisms of species co-existence at small scales. The use of large trawls in the 1960s to document bathymetric gradients of diversity gave way in the mid-1970s to precision sampling with box-cores and manipulative experiments using submersibles, as deep-sea ecologists grew disenchanted with trying to explain large-scale patterns in an environment where the potential causes on these scales were difficult or impossible to measure. Work on small scales yielded much more accurate estimates of diversity and spatial dispersion (Jumars, 1976), and improved our understanding of how disturbance and patch dynamics might mediate local co-existence (Smith *et al.*, 1986). Recently, as ecologists became more aware that community structure can only be understood fully in a larger biogeographic and historical context (Ricklefs, 1987), attention has returned to the problems of global-scale patterns and generation of the regional biotas that ultimately provide the species participating in local assemblages. The stochastic patch dynamics that seem so pervasively important in maintaining diversity at very small scales of time and space in nature cannot account for geographic variation in diversity.

Sanders (1968) hypothesised that both contemporary ecological factors and historical processes must be responsible for patterns of species diversity in the deep sea, as well as other marine habitats. The biogeographic signature of historical phenomena is usually manifested only at very large scales. In this chapter we review the two best-known examples of large-scale patterns of diversity in the deep-sea benthos, bathymetric and latitudinal gradients. The single most important observation is that both kinds of patterns appear to vary geographically. Bathymetric patterns change within and among deep-sea basins, and latitudinal patterns differ between northern and southern hemispheres. We also discuss recent evidence for the possible roles of ecology and evolution in regulating these patterns. Ecological gradients are probably important at both regional and global scales. Historical processes may come into play at inter-regional and global scales.

5.2 Bathymetric gradients of diversity

More is known about the physical and biological changes along depth gradients than for any other physiographic feature of the deep-sea environment. Increase in depth is accompanied by decreases in mean temperature and temperature variability, increased hydrostatic pressure, progressively less light and nutrient flux, as well as shifts in sediment grain size and composition, and near-bottom current activity (Gage & Tyler, 1991). The most striking biological change with depth is the exponential decline in standing stock (both biomass and abundance) of the benthos that results from decreased rates of nutrient input (Rowe & Pariente, 1992). Community composition changes rapidly with depth; upper bathyal and abyssal assemblages are often completely distinct (Rex, 1977; Carney et al., 1983; Hecker, 1990). Both the rate of faunal turnover and the degree of clinal variation in individual species correspond to the rate of change in depth, being highest in the steep upper bathyal zone and very subtle in the deep abyssal plain (Etter & Rex, 1990). Indirect evidence from the trophic organisation of the community (Rex, 1981), the incidence of sublethal damage to prey from predators (Vale & Rex, 1988) and the size structure of closely related species assemblages (Rex et al., 1988) suggest that the kind and intensity of biotic interactions vary with depth. Life-history characteristics of some taxa also undergo change with depth; for example, the incidence of planktotrophic development among prosobranch gastropods increases steadily from the upper bathyal to the abyss (Rex & Warén, 1982; Etter & Caswell, 1994). Biochemical systems of individual species are known to change with depth (Hochachka & Somero, 1984). Given the dramatic changes in the biotic and physical environment, it is not surprising that species diversity also varies with depth.

The most extensively sampled area of the deep sea is the depth gradient of the western North Atlantic (Etter & Grassle, 1992; Rex et al., 1993). Here, diversity of several macrofaunal groups, and the invertebrate and fish megafauna vary parabolically with depth, being low on the continental shelf, increasing to a maximum at mid-bathyal depths and declining in the abyssal plain (Rex, 1981, 1983). All groups analysed exhibit essentially the same pattern with depth, though peak diversity varies between 2000 and 3000 m.

Because the samples used in early analyses (Rex, 1981, 1983) were collected with various techniques (anchor-dredge, epibenthic sled, trawl), some of which may traverse an extensive (sometimes exceeding 1 km) and discontinuous region of the sea-floor, it remained unclear to what extent the observed patterns reflected alpha or beta diversity. However, an analysis of 186 box-cores (0.25 m^2) collected south-east of Massachusetts as part of the Atlantic Continental Slope and Rise Study provided similar patterns (see Fig. 5.1(a)) for the entire invertebrate macrofaunal assemblage (Etter & Grassle, 1992). The box-core data confirm the pattern identified earlier and indicate that the general parabolic relationship between species diversity and depth in this region of the deep sea is robust to the exact methodology used to sample the community.

Is the pattern found in the western North Atlantic south of New England representative of other geographic regions? Along the east coast of the USA, the best data come from the Atlantic Continental Slope and Rise Study mentioned above. A total of 558 box-cores were analysed and contained 272 009 individuals distributed among 1597 species (data are reported in Blake et al., 1985, 1987; Maciolek et al., 1987a,b). Box-cores were collected from three geographic regions off the east coast of the USA: the North (south-east of Massachusetts), the Mid (off the coast of New Jersey and Delaware) and the South (east of North and South Carolina). The northern samples mentioned above are closest geographically to the samples used by Rex (1981, 1983) and exhibit a similar bathymetric pattern of diversity (Fig. 5.1(a)). The Mid region, several hundred kilometres to the south, was sampled over a much narrower depth range (1500–2500 m) and provides only a limited view of bathymetric variation. However, even over this narrow depth range, diversity varies parabolically (Fig. 5.1(b)). The southern region exhibits considerable variability in diversity at the 600 m and 2000 m stations and there is no consistent pattern with depth (Fig. 5.1(c)). In part, this probably reflects the sampling programme. Samples were collected along four different depth transects separated by tens to hundreds of kilometres

98 *Species diversity patterns in the deep-sea benthos*

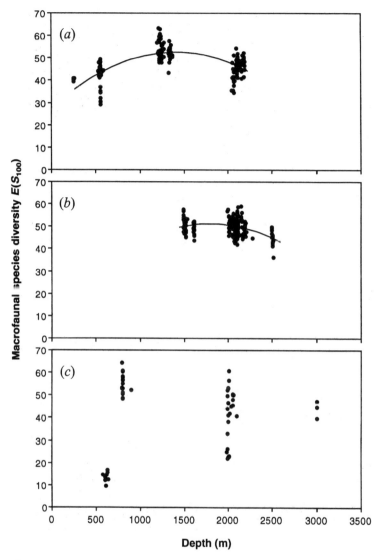

Fig. 5.1. Bathymetric patterns of macrofaunal diversity along the east coast of North America. Samples are from the Atlantic Continental Slope and Rise Study (Blake *et al.*, 1985, 1987; Maciolek, 1987a,b). Three geographic regions were sampled: (*a*) the North (south-east of Massachusetts), (*b*) the Mid (off the coast of New Jersey and Delaware) and (*c*) the South (east of North and South Carolina). Each point represents the macrofaunal diversity for the inner nine subscores (0.09 m^2) of a single 0.25 m^2 box-core. Diversity is given as Hurlbert's (1971) expected number of species normalised to 100 individuals. The regression lines for each region were fitted using standard least-squares techniques and all were significant at $p = 0.0001$. No line is shown on

and were designed to sample major physical and oceanographic features of this region (Blake & Grassle, 1994). For example, some of the samples for the 600 and 2000 m depths were collected from the Charleston Bump, which is impacted by the Gulf Stream, other samples were collected off Cape Lookout where there are no major topographic or hydrographic features and others were collected from a deep-gullied slope off Cape Hatteras, which is influenced by the Western Boundary Undercurrent and experiences downslope transport from the shelf. It is difficult to tease out the bathymetric trends from the variation due to other geographic and hydrographic differences.

Limited data on bathymetric patterns of species diversity from other regions of the World Ocean make it difficult to assess the generality of the pattern found in the western North Atlantic. In the eastern North Atlantic, Paterson & Lambshead (1995) found that polychaete diversity estimated from box-cores varied parabolically with depth between 400 and 2800 m. Isopods from the Norwegian and Greenland Seas, also collected with box-cores, show a simple decline in diversity with depth between 794 and 3709 m (Svavarsson *et al.*, 1990). Other invertebrates (polychaetes and ophiuroids) and foraminiferans from Arctic waters also exhibit a decline in diversity with depth, but the data are very restricted in number of samples or bathymetry (for a review, see Svavarsson *et al.*, 1990). Stuart (1991) compared diversity estimates for gastropod assemblages collected with epibenthic sleds from nine deep-sea basins in the North and South Atlantic to bathymetric gradients in the western North Atlantic (Fig. 5.2). The regression line and confidence limits plotted in Fig. 5.2 represent samples from the western North Atlantic. This is the only area that has been reasonably well sampled and shows a convincing and highly significant ($p = 0.0003$) bathymetric pattern of diversity. As with the box-core data discussed above (Fig. 5.1), epibenthic sled samples show high variation in diversity indicating that depth trends are only distinguishable by using many samples collected over a broad bathymetric range. In the Norwegian Sea (Fig. 5.2), there is a

---

Caption for Fig. 5.1 (*cont.*)
the plot for the South region because the relationship is not significant ($n = 55$, $F = 1.079$, $r^2 = 0.0400$, $p = 0.3475$). A line fit to the data with all three regions combined is also parabolic and significant at $p = 0.0001$ (Etter & Grassle, 1992). The equations for the lines are:
North: $y = 27.645 + 0.035x - 1.251 \times 10^{-5}x^2$ ($n = 186$, $F = 68.194$, $r^2 = 0.427$);
Mid: $y = 10.314 + 0.046x - 1.267 \times 10^{-5}x^2$ ($n = 231$, $F = 21.331$, $r^2 = 0.158$);
Combined: $y = 28.718 + 0.028x - 8.766 \times 10^{-6}x^2$ ($n = 472$, $F = 46.353$, $r^2 = 0.165$).

100   Species diversity patterns in the deep-sea benthos

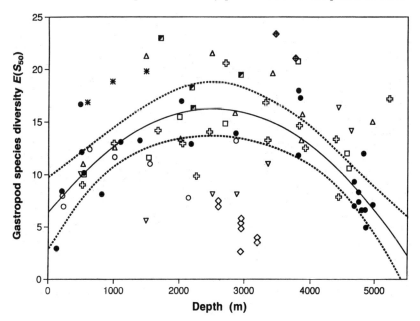

Fig. 5.2. The relationship between gastropod diversity and depth in 85 epibenthic sled samples collected in 10 basins from the Norwegian Sea and Atlantic Ocean. Estimates of diversity were normalised to 50 individuals by using Hurlbert's (1971) expected number of species. The regression line and 95% confidence limits are for samples from the North American Basin.
Equation for the regression line: $y = 6.436 + 0.008x - 1.561 \times 10^{-6}x^2$ ($F = 12.766$, $r^2 = 0.561$, $p = 0.0003$, $n = 23$). ⊕, equator; ◊, Norwegian Sea; ○, Cape Basin; ◼, Gambia Basin; ●, North American Basin; □, Angola Basin; △, Guiana Basin; ▽, West European Basin; ✳, Brazil Basin; ✣, Argentine Basin.

significant ($p = 0.016$) decrease in diversity with depth supporting Svavarsson et al.'s (1990) analysis for isopods, but the depth range available for gastropod data is too narrow to infer an overall bathymetric pattern. The sparse data available for the West European Basin in the eastern North Atlantic fit a second order polynomial ($p = 0.019$) suggesting a parabolic trend, but one that is concave upward with a minimum at intermediate depths – just the opposite of the western North Atlantic. No other basin displays a significant bathymetric diversity gradient; but, again, we caution that the number of samples and depth coverage are probably inadequate to detect large-scale patterns. Samples from other regions that fall outside the confidence limits tend to cluster above or below the confidence limits. Diversity in the Norwegian Sea falls consistently below the 95% confidence intervals for the North American

Basin and some tropical areas are grouped above. Part of the variation in diversity shown in Fig. 5.2 may be due to different depth trends among basins, and some is associated with the latitudinal changes in the level of diversity discussed later.

5.2.1 Processes that shape bathymetric gradients of diversity

Most research on the forces that govern spatial and temporal patterns of diversity in the deep sea has focused on the importance of small-scale patchiness, which presumably permits a large number of similar species to coexist by specialising on different types of patches or different successional stages of a patch. Recent work has identified several types of patches that may be important in maintaining diversity. These include biogenic structures (Jumars, 1975, 1976; Eckman & Thistle, 1991; Levin, 1991), phytodetrital pulses from surface blooms (Billett et al., 1983; Lampitt, 1985; Rice et al., 1986), carcass falls (Smith, 1985, 1986; Smith et al., 1989), bioturbation (Smith, 1986; Smith et al., 1986), megafaunal predation and benthic storms (Hollister & McCave, 1984; Hollister & Nowell, 1991). Small-scale manipulative experiments and monitoring programmes (Smith, 1985; Grassle & Morse-Porteous, 1987; Smith & Hessler, 1987; Kukert & Smith, 1992; Snelgrove et al., 1992) have demonstrated that these patches can persist for 2–5 years, which is much longer than similar patches in shallow-water environments. The experiments also indicated: (a) that opportunists can rapidly respond to disturbed habitats and coexist with more equilibrium-type species during the early stages of succession, (b) that succession is spatially and temporally variable – the species involved in succession can vary as well as the time required to reach background community characteristics, (c) certain species were only found in patches suggesting that disturbance is necessary for persistence, (d) patches increased in diversity sometimes exceeding that of ambient sediments during recovery. These experiments provide some evidence that the mosaic nature of deep-sea environments might permit high local diversity (compare, however, Schaff & Levin, 1994).

It remains unclear how these small-scale effects are integrated into larger bathymetric and geographic patterns. At small scales, stochastic processes that are tied to patch formation and dispersal seem to be important. At larger scales, environmental gradients and the evolutionary history of community development appear to become more important in forming geographic patterns of diversity.

Rex (1981, 1983), applying Huston's (1979) dynamic equilibrium model, suggested that the bathymetric pattern of diversity in the western North Atlantic represents a balance between rates of competitive displacement and the frequency of disturbance from predation or physical perturbation. The rate of competitive displacement is a positive function of population growth, which in turn depends on nutrient flux. Population growth rates determine how quickly communities approach competitive equilibrium, where exclusion by dominant species depresses diversity. Disturbance, due to biotic or physical causes, crops down populations interrupting the approach to competitive equilibrium and permitting more species to coexist. The relative rates of these two processes determine local diversity.

There is some evidence for the action of competition (Rex et al., 1988) and predation (Vale & Rex, 1988) in the deep-sea fauna, but there is only very inferential evidence that the intensities of competition and predation vary bathymetrically in a way that is consistent with predictions of the dynamic equilibrium model (Rex, 1983). Both direct (Turley et al., 1995) and indirect (Rowe & Pariente, 1992) measures of nutrient supply indicate that food availability decreases exponentially with depth. If population growth rates are dependent on food availability (Huston, 1979), then the rate of approach to competitive equilibrium should decline exponentially with depth. Little is known about how disturbance varies with depth. If predation pressure can be inferred from the depth-related patterns of predatory snail diversity and megafaunal diversity (Rex, 1983), then biological disturbance varies parabolically with depth. Combining these two gradients provides predictions that are consistent with the bathymetric patterns observed in the western North Atlantic.

Etter & Grassle (1992) proposed that much of the spatial and temporal variability in species diversity in the deep sea reflects changes in sediment characteristics, specifically particle size diversity. They tested this hypothesis using 558 box-cores collected from bathyal depths (250–3029 m) in the western North Atlantic (Etter & Grassle, 1992). The samples were part of the Atlantic Continental Slope and Rise Study and represent the most extensive quantitative sampling of deep-sea community structure. Macrofaunal species diversity within box-cores is positively related to sediment particle size diversity. More importantly, when sediment diversity is held constant statistically, species diversity is uncorrelated with depth suggesting that the parabolic relationship observed in the western North Atlantic may be largely attributable to changes in sediment characteristics with depth. These findings indicate that sediment diversity may play an important role in determining the number of species within a community, and identify for the first time a direct environmental factor that potentially influences species diversity in the deep sea.

Because the relationship between macrofaunal diversity and particle diversity is consistent across inter-regional, regional and local scales and obviously operates on a scale within box-cores, sediment characteristics offer a powerful unifying mechanism for coupling theories on the forces that shape patterns of diversity at different scales. Biogenic structures, various forms of disturbance and nutrient flux are all thought to create small patches that ultimately permit the remarkably high local diversity in deep-sea communities (e.g. Grassle & Sanders, 1973; Jumars, 1975; Thistle, 1983; Levin, 1991; Kukert & Smith, 1992), but the ecological qualities of patches that permit them to be partitioned remain poorly understood. Some patches are known to provide refugia from predation (Thistle & Eckman, 1990) or particular hydrodynamic regimes (Eckman & Thistle, 1991); how other patches are partitioned remains unclear. Changes in the distribution of particle sizes may be one of the essential features of these patches that allows them to be partitioned. At larger scales, gradients in sediment characteristics may be important in regulating diversity, such as the bathymetric patterns found by Etter & Grassle (1992).

Stevens (1989, 1992) suggested that species diversity gradients in terrestrial and coastal marine communities may be related to patterns in the size of geographic ranges (Rapoport's Rule). At high latitudes (or altitudes) organisms must adapt to broad temporal environmental variation that permits them to occupy broader geographic ranges and decreases the number of species inhabiting a specific area. In contrast, at low latitudes (or altitudes), organisms adapt to much narrower temporal environmental variation and thus experience the spatial environment in a much more fine-grained manner resulting in smaller geographic ranges. The smaller geographic ranges permit more species per unit area. According to Stevens' hypothesis, diversity should be inversely related to geographic range size. Pineda (1993) tested whether bathymetric patterns in deep-sea organisms could be explained by variation in geographic range size. The most diverse communities occur where bathymetric ranges are broadest, just the opposite of theoretical predictions based on Rapoport's rule.

Studies on local co-existence and bathymetric gradients of diversity within deep-sea basins have centred exclusively on the ecological mechanisms of biological interaction, environmental heterogeneity and disturbance. It seems quite reasonable that ecological opportunity, population dynamics and dispersal would largely determine relative levels and patterns of diversity at these spatial scales (centimetres to tens or even hundreds of kilometres). Historical causes, generally involving much longer spans of time and spatial scales large enough to promote speciation, seem less likely to be implicated directly in maintaining geographic trends of diversity at within-basin scales. However, this possibility has not been explored well, and definitely should not be dismissed. Research on the evolutionary diversification of the deep-sea benthos is just beginning. At bathyal depths, gastropods in the western North Atlantic show the strongest intraspecific clinal variation in shell architecture and the highest rate of species turnover (Etter & Rex, 1990). The bathyal region in this basin is also topographically complex (Rex et al., 1988; Mellor & Paull, 1994). The potential for geographic isolation by physical features and adaptive constraints, and the pattern of geographic variation within species suggest that the bathyal region could act as a diversity pump for the deep-sea fauna. Much additional research on geographic distribution, population differentiation, speciation and phylogeny is needed to determine whether the diversity generated in the deep sea is extensively re-shaped on within-basin scales by contemporary ecological factors, or whether biogeographic features, such as the diversity peak at bathyal depths, bear the imprint of past evolutionary–historical processes.

5.3 Latitudinal patterns of species diversity

Latitudinal gradients of species diversity are the most conspicuous and well-known features of global biogeography. The biotas of terrestrial (Stevens, 1989; Currie, 1991), freshwater (France, 1992), coastal marine (Fischer, 1960; Roy et al., 1994) and open-ocean pelagic (Angel, 1993; McGowan & Walker, 1993; Dodge & Marshall, 1994) ecosystems all show general poleward declines in biodiversity, although these patterns often exhibit meso-scale idiosyncrasies and longitudinal shifts associated with the life histories and ecology of different taxa. Recently, Rex et al. (1993) reported latitudinal diversity gradients in the North Atlantic and marked inter-regional variation in the South Atlantic (Figs 5.3 and 5.4) for deep-sea isopods, gastropods and bivalves. These trends were unexpected since it was long assumed that the tremendous depth of overlying water would effectively buffer deep-sea communities from

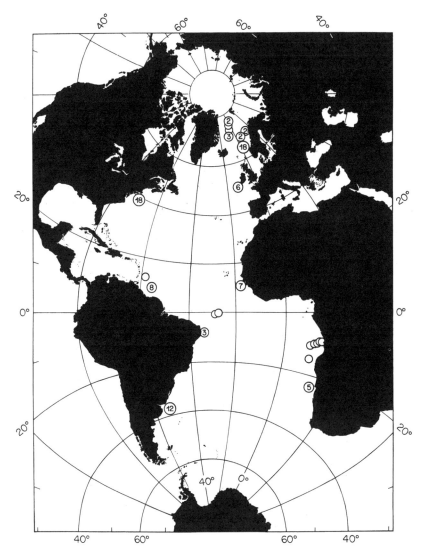

Fig. 5.3. Locations of 97 epibenthic sled samples collected at bathyal depths (500–4000 m) from 10 regions in the Atlantic Ocean and Norwegian Sea. The samples ranged in latitude from 77°N to 37°S. The basins included the Norwegian Sea, West European Basin, North American Basin, Guiana Basin, Gambia Basin, Mid-equatorial Atlantic, Brazil Basin, Angola Basin, Cape Basin and Argentine Basin. Circles represent individual samples. Numbers in circles represent the number of samples taken in that region. Reprinted with permission from *Nature* (Rex *et al.*, 1993, vol. **365**, 636–9), copyright (1993) Macmillan Magazines Limited.

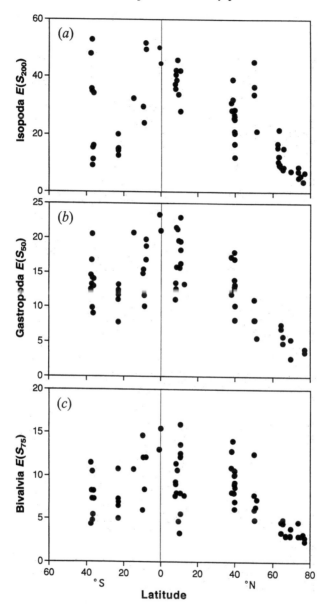

Fig. 5.4. The relationships between species diversity and latitude for deep-sea isopods (a), gastropods (b) and bivalves (c) in the North and South Atlantic. Species diversity was calculated as Hurlbert's (1971) expected number of species $E(S_n)$. For each taxon, samples were normalised to the largest sample size (n) that included the greatest number of samples over a broad latitudinal range. For the North Atlantic, diversities of all three groups are significantly correlated with latitude (isopods, $r = -0.865$, $p = 0.0001$; gastropods, $r = -0.849$, $p = 0.0001$; bivalves, $r = -0.687$, $p = 0.0001$). For the

the environmental gradients that are thought to influence global diversity patterns. Below we discuss latitudinal patterns in the deep North and South Atlantic and their potential causes, particularly regional–historical processes.

5.3.1 The North Atlantic

The North Atlantic and Norwegian Sea have been much more extensively sampled than the South Atlantic. North of the equator, isopods, gastropods, and bivalves all display significant latitudinal gradients of diversity (Fig. 5.4) much like those found in surface systems. The variation in diversity within separate basins shown in Fig. 5.4 represents, in part, the bathymetric changes discussed above. Since diversity can vary both with depth and horizontally, differences in depth coverage among regions potentially can confound latitudinal trends. We attempted to control for this to some extent by restricting the analysis to bathyal depths. Moreover, when the effect of depth is removed by partial regression analysis, the latitudinal gradients remain highly significant (Rex et al., 1993).

Gray (1994, p. 208) claimed, without any actual analysis, that the latitudinal gradient discovered by Rex et al. (1993) in the north 'was largely determined by the low diversity found in the Norwegian Sea as the other data had high variance and showed no distinct pattern.' Figure 5.4 shows that the relationships between diversity and latitude are strongly anchored by fairly consistent low diversities in the Norwegian Sea. In addition to the variation associated with depth, it should be borne in mind that the data in Fig. 5.4 are estimates of diversity for individual samples. Hence, they show more inherent variability than classic examples from terrestrial and coastal marine environments that are based on the number of coexisting species per unit area or in latitudinal bands compiled from extensive long-term biotic surveys. The latter method effectively averages diversity over very large scales of time and space

---

Caption for Fig. 5.4 (cont.)
South Atlantic, the gastropods and bivalves show significant relationships (gastropods, $r = -0.438$, $p = 0.0254$; bivalves, $r = -0.600$, $p = 0.0031$). See Rex et al., 1993, table 1, for regression equations and other statistics. Figure adapted with permission from Nature (Rex et al., 1993, vol. **365**, 636–9), copyright (1993) Macmillan Magazines Limited.

obscuring the kind of local (intersample) variability indicated in Fig. 5.4. We show elsewhere that this approach applied to deep-sea data reveals a typical simple monotonic decline in diversity in the North Atlantic and Norwegian Sea. Also, when data for the Norwegian Sea are removed from the analysis of individual samples (Fig. 5.4), the latitudinal gradients for the most diverse taxa, isopods and gastropods, remain highly significant ($r = -0.607$, $p = 0.0004$, and $r = -0.690$, $p = 0.0001$, respectively). Only the relationship for bivalves, with their comparatively low levels of diversity and high variability at tropical latitudes, becomes insignificant. However, it is quite significant when just two exceptionally low outliers for the Guiana Basin are removed ($r = -0.400$, $p = 0.0157$). So, the contention that significant northern latitudinal gradients are attributable only to low diversity in the Norwegian Sea is incorrect. Of course, the Norwegian Sea is an integral and interesting part of the deep-sea ecosystem north of the equator and should be included in the analysis.

5.3.2 The South Atlantic

The deep sea of the South Atlantic has been sampled over a much smaller latitudinal range (0–37°S), and large-scale biogeographic patterns there are less clear (Fig. 5.4). Gastropods and bivalves show significant latitudinal decreases in diversity over this limited span, but isopods do not. Isopods reach their highest local diversity at temperate latitudes in the Argentine Basin. Tropical and temperate regions of the deep South Atlantic appear to be characterised by strong inter-regional differences in diversity (Fig. 5.4). Whether a general poleward decline in diversity underlies this variation is still uncertain.

Brey *et al.* (1994, p. 297) compared diversities of deep-sea isopods, gastropods and bivalves collected from the Weddell Sea to those published by Rex *et al.* (1993) for the tropical Atlantic (20°N–20°S in Fig. 5.4), and concluded 'that there is no steady latitudinal decrease in deep-sea benthic diversity towards south polar regions'. However, the two studies used fundamentally different sampling methods and analyses of diversity. Consequently, they do not allow a direct comparison of diversity between the tropics and Antarctica as Brey *et al.* (1994) assumed.

Brey *et al.* (1994) combined species lists and relative abundance data from an unspecified number of Agassiz trawl samples that spanned 500–2000 m in depth and 8 deg. of latitude. Estimates of regional-scale diversity based on the combined data were compared to diversity values from individual epibenthic sled samples given in Rex et al. (1993). Since estimates of regional diversity in the Weddell Sea fell within the range of local (sample) diversities reported for the tropical Atlantic, Brey et al. (1994) proposed that polar diversity is similar to tropical diversity. What the comparison actually shows is that the species pool of a very large region in the Weddell Sea corresponds roughly to diversities found in individual samples from the tropics that each cover about 1000 m^2. This is like comparing the flora of Britain to the diversity of a hectare of rain forest in Costa Rica and claiming that, because they were similar, a latitudinal diversity gradient does not exist. The normalising (rarefaction) method used by Brey et al. (1994) to compare Antarctic and tropical diversities at common sample sizes cannot compensate for the much larger geographic scale represented by the Weddell Sea data.

Another problem is that sampling methods used in the two studies collect different components of the benthos. The Agassiz trawl typically has a larger mesh size (10-20 mm) and is deployed primarily to sample the megafauna (Gage & Tyler, 1991). The smaller epibenthic sled with a 1 mm mesh size is designed to sample the macrofauna (Hessler & Sanders, 1967). The Weddell Sea assemblage is dominated by megafaunal taxa (mainly echinoderms, decapods and polynoid worms). This may explain the surprisingly low abundances for macrofaunal groups (peracarids and molluscs) reported by Brey et al. (1994). Half of the individual epibenthic sled samples from the tropics yielded more molluscs or isopods than the entire Weddell Sea survey. Thus, sampling efficiency varies significantly between the two studies for the taxa used to compare species diversity.

Brandt (1995) recently concluded that the analysis of Brey et al. (1994) rejected the notion of latitudinal patterns in the deep sea. The research of Brey et al. (1994) is an important contribution to Antarctic deep-sea ecology, but it allows no critical insight into relative macrofaunal diversity between tropical and southern polar regions. Whether the south polar regions support diversity that is higher, lower or similar to tropical regions remains an open question. The South Atlantic does appear to show strong inter-regional variation. Everything known about large-scale biogeographic patterns in the deep sea is based on remote sampling and this involves a tremendous amount of uncertainty and error. It is impossible to interpret comparisons that are not controlled for sampling gear, bathymetric and horizontal scales of sampling, habitat type, faunal composition and the analytical methods used to measure community structure.

The existing picture of global-scale patterns in the deep-sea benthos is still very fragmentary. The geographic coverage of sampling sites in most regions is sparse, and our understanding is restricted to bathyal depths in several basins of the Atlantic (Fig. 5.3). It would be very interesting to document patterns for other faunal groups, particularly polychaetes which are the most abundant and diverse macrofaunal taxon, and the megafaunal assemblage. Diversity estimates from box-core (Fig. 5.1) and epibenthic sled samples (Figs 5.2 and 5.4) display high variability on small and large scales. This indicates that detecting meaningful intra- and inter-regional patterns of diversity requires a substantial data base and broad geographic coverage. Grassle & Maciolek (1992) have emphasised that the deep-sea fauna is probably grossly undersampled. So, an important assumption is that large-scale patterns based on available samples and selected taxa accurately reflect actual patterns, at least in a relative sense.

5.3.3 Processes that shape latitudinal patterns of diversity

Despite the intensive and long-standing interest in latitudinal diversity gradients as a central problem in evolutionary ecology, their explanation has remained elusive (Rohde, 1992). It is likely that both ecological and historical causes are responsible for large-scale patterns in the deep sea (Sanders, 1968; Rex et al., 1993). Rex et al. (1993) suggested that large-scale environmental gradients at the surface may be translated to great depths through surface–benthic coupling. Except for vent-seep habitats, food in the deep sea is entirely of extrinsic origin. It seems inevitable that large-scale species diversity gradients in the deep-sea benthos must reflect, to some extent, the pattern and rate of surface production, the consequent pattern and rate of descent of organic material and its transformation in the water column. Elsewhere, Etter & Rex (unpublished results) elaborate on the potential importance of environmental gradients on global scales. Below, we look at the potential importance of historical factors, primarily to make the point that causality can change with scale. First we discuss the possible role of regional enrichment in maintaining diversity in the North Atlantic, and then briefly consider the impact of glaciation in the Norwegian Sea.

5.3.4 Regional–historical processes in the North Atlantic

Stuart & Rex (1994) explored the evolutionary basis of large-scale patterns in prosobranch snails of the deep North Atlantic and Norwegian Sea. Prosobranchs are the largest subclass of deep-sea gastropods. They reveal a clear latitudinal diversity gradient from the equator to 77°N (Fig. 5.5). Stuart & Rex (1994) showed that local (sample) diversity is significantly and positively related to regional diversity (Fig. 5.6), suggesting that local diversity might be maintained by the mechanism of regional enrichment (Cornell, 1985a,b, 1993; Ricklefs, 1987, 1989; Ricklefs & Schluter, 1993). In the regional enrichment model, local diversity represents a balance between local extinction and colonisation from the regional species pool. Extinction is attributed to biological interactions such as competition and predation, and physical sources of disturbance that act to depress diversity locally. The size of the regional species pool is a function of the evolutionary history of speciation and adaptive radiation – this is the assumed historical connection. Local communities are relatively unstructured and open to colonisation from the regional species pool. Under these circumstances, local diversity should reflect the size of the regional pool in a relative sense. A larger

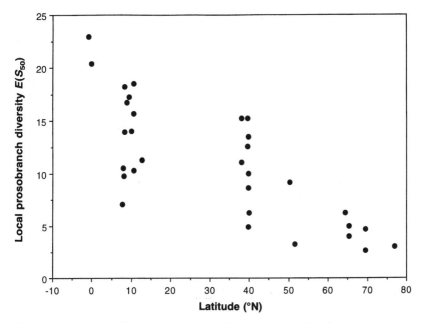

Fig. 5.5. The relationship between species diversity of prosobranch gastropods and latitude for six regions in the North Atlantic and Norwegian Sea. Diversity was estimated by using Hurlbert's (1971) expected number of species $E(S_n)$ with $n = 50$. Figure adapted from Stuart & Rex (1994) in *Reproduction, Larval Biology, and Recruitment of the Deep-Sea Benthos*, ed. C.M. Young & K.J. Eckelbarger, copyright © 1994 by Columbia University Press. Reprinted with the permission of the publisher. Regression equation: $y = 16.601 - 0.176x$ ($F = 41.828$, $r^2 = 0.590$, $p = 0.0001$, $n = 31$).

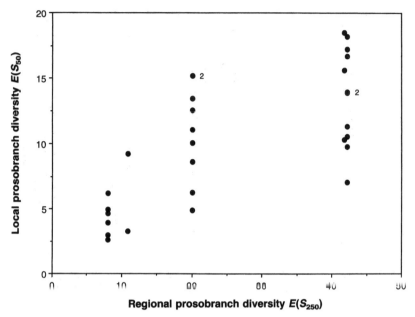

Fig. 5.6. The relationship between local and regional prosobranch gastropod diversity in the North Atlantic and Norwegian Sea. Local diversity was calculated by normalising each of the individual samples to $n = 50$. For regional diversity, species abundance data were combined for each basin and normalised to $n = 250$. Figure adapted from Stuart & Rex (1994) in *Reproduction, Larval Biology, and Recruitment of the Deep-Sea Benthos*, ed. C.M. Young & K.J. Eckelbarger, copyright © 1994 by Columbia University Press. Reprinted with the permission of the publisher.
Regression equation: $y = 4.117 + 0.237x$ ($F = 25.332$, $r^2 = 0.484$, $p = 0.0001$, $n = 29$).

regional pool of colonists will, on average, support higher local diversities. The general predictions of the regional enrichment model appear to be borne out in a wide range of ecosystems and habitats (Ricklefs & Schluter, 1993), suggesting that dispersal and the historical–evolutionary development of biotas on regional scales can affect large-scale patterns of local diversity.

Most tests of the regional enrichment model have relied on the relationship between local and regional diversity (as in Fig. 5.6). A reasonably linear monotonic relationship is consistent with the predictions of the model and its underlying assumption that natural communities tend to be open to colonisation and unsaturated with species. If, however, local communities reach a saturated level of diversity that is strictly enforced by local ecological circumstances, then local and regional diversity should remain statistically independent. A growing body of evidence indicates that many communities do behave as unsaturated assemblages of species (Cornell, 1993; Hugueny & Paugy, 1995).

An equally important and seldom tested part of the model concerns dispersal among the constituent local communities that make up the regional species pool. Recruitment dynamics and dispersal are often recognised as important factors influencing diversity (e.g. Roughgarden *et al.*, 1988; Underwood & Fairweather, 1989; Etter & Caswell, 1994), but these phenomena are very difficult to measure even in environments that are much more accessible than the deep ocean. Prosobranchs are particularly well suited for assessing the contribution of dispersal because their mode of development can be inferred from larval shell morphology (Jablonski & Lutz, 1983; Bouchet & Warén, 1994; and see Rex and Warén, 1982 and Stuart & Rex, 1994 for exceptions). In most deep-sea taxa, species either have planktotrophic development involving a swimming–feeding dispersal phase, or non-planktotrophic development in which the young hatch crawling and experience very limited dispersal. The proportion of species with planktotrophic development can serve as a rough estimate of the dispersal potential of the regional species pool (Josefson, 1985; Stuart & Rex, 1994).

We tested the regional enrichment model for prosobranchs in the North Atlantic and Norwegian Sea by performing a multiple regression analysis with local (sample) diversity as the response variable, and regional diversity and the proportion of dispersing species as explanatory variables. Depth was also included as an independent variable to control for the potential confounding influence of bathymetric coverage discussed earlier. According to the model, local diversity is a positive function of the regional species pool and the dispersal potential of the regional fauna. Table 5.1 gives the ANOVA for the multiple regression analysis. Regional diversity is overwhelmingly the most significant predictor of local diversity, explaining nearly half the variance, and the regional proportion of species with planktotrophic development enters the equation as a significant and positive subordinate variable. An analysis of residuals showed that the percentage of species with planktotrophic development behaves in a consistent way: areas where regional diversity underpredicts local diversity have a high proportion of dispersing species and *vice versa*, as the model predicts (Stuart & Rex, 1994). In general, the analysis corroborates the theory of regional enrichment including the feature that local and regional diversity are coupled by dispersal. Insofar as the size of the regional species pool reflects the history of

Table 5.1. *Multiple regression analysis for predicting local diversity of prosobranch gastropods in the North Atlantic and Norwegian Sea*

Independent variables	Contribution to r^2	t
x_1 regional diversity $E(S_{250})$	0.484	5.629***
x_2 regional % planktotrophic species	0.111	2.990**
x_3 depth	0.092	3.142**

Regression equation:
$y = -5.144 + 0.223x_1 + 0.187x_2 + 0.002x_3$ ($n = 29$, $F = 18.332$, $r^2 = 0.687$, $p < 0.001$).
***, $p < 0.001$
**, $p < 0.01$
Local diversity is the response (dependent) variable. Regional diversity and the regional percentage of species with planktotrophic development are the explanatory variables. Depth was also included in the analysis to correct for potential bias resulting from bathymetric variation in diversity. The results show that regional diversity and the dispersal potential of the regional species pool both contribute significantly to explaining the variance in local diversity. Table reproduced from Stuart & Rex (1994) in *Reproduction, Larval Biology, and Recruitment of the Deep-Sea Benthos*, ed. C. M. Young & K. J. Eckelbarger, copyright © 1994 by Columbia University Press. Reprinted with the permission of the publisher.

evolutionary diversification, this suggests a role for historical processes, at least on inter-regional scales.

There are many difficulties with interpreting the role of history in community development from the relationship between local and regional diversity. Excellent critical discussions of regional enrichment studies as well as recommendations for complementary analyses and future research priorities can be found in Ricklefs (1989), Ricklefs & Schluter (1993) and Cornell (1993). Perhaps the most important general caveat is that a statistically significant relationship between local and regional diversity is only very indirect evidence for how historical processes determine diversity. The causal direction of the relationship between local and regional diversity is hard to discern. Does regional diversity augment local diversity through enrichment, or does it merely represent the accumulation of local diversities? Or as Cornell (1993) put it, is the dependent variable local or regional diversity?

Additional uncertainties surround our particular analysis of regional enrichment in deep-sea prosobranchs. It is unclear which species in the regional pool are adaptively competent to participate in local species assemblages throughout the bathyal realm. As indicated earlier, bathyal depth gradients are accompanied by strong environmental gradients and by pronounced faunal replacement. Another difficulty is that the actual pattern of larval dispersal, as well as the temporal and spatial scales of dispersal are unknown. The proportion of planktotrophic species is only a rough estimate of the dispersal potential of the regional species pool. Recently, it has been shown that some archaeogastropods have a non-feeding pelagic dispersal phase that cannot be detected by larval shell morphology (Hadfield & Strathmann, 1990). If this mode of development occurs in some deep-sea archaeogastropods, then the dispersal potential of the regional pool may be underestimated.

5.3.5 *Quaternary glaciation and the Norwegian Sea*

A specific historical event, Quaternary glaciation, has been invoked to explain depressed diversity in the Norwegian Sea (e.g. Dahl, 1972; Svavarsson *et al.*, 1993). According to this view, the Norwegian Sea has low diversity because there has been insufficient time for recolonisation and extensive evolutionary radiation since glaciation occurred. Even more recently (6000–8000 years before present), massive sediment flows (Bugge *et al.*, 1988), may have had a catastrophic effect on the deep benthos of the Norwegian Sea. Low overall diversity in the Norwegian Sea could also reflect the more pulsed nutrient input typical of high latitudes (Yoder *et al.*, 1993). Another possibility is that diversity is maintained at a low level by biogeographic constraints. The Norwegian Sea is a relatively small, shallow and partially isolated basin that may impose higher extinction rates and offer less ecological opportunity than the larger and more confluent basins to the south in the Atlantic. Low diversity may be enforced by a peninsular or stepping-stone colonisation effect (Sepkoski & Rex, 1974), since the Norwegian Sea is essentially the terminus of a long colonisation route from the south. While the relative impact of ecological and evolutionary processes on the deep fauna of the Norwegian Sea is unclear, there is good reason on taxonomic grounds to think that historical phenomena are part of the explanation (e.g. Svavarsson *et al.*, 1993). A systematic and phylogenetic comparison of Norwegian Sea and Atlantic deep-sea taxa offers a very exciting opportunity to explore historical effects in a precise way.

5.4 Summary and conclusion

Compared to terrestrial and coastal marine ecosystems, our knowledge of large-scale patterns of diversity in the deep sea is still restricted to a few major taxa and based on very limited sampling and geographic coverage. It is clear that local diversity varies on large scales, but the generality of the few well-documented patterns is uncertain. Much more extensive sampling and taxonomic research are needed to describe geographic patterns of diversity well enough to infer their causes. If what we have learned so far about all aspects of deep-sea ecology is any indication, we should keep open minds and expect a variety of patterns to emerge in different taxa and in different areas of the World Ocean.

The last quarter century has seen remarkable strides in deep-sea ecology. We learned unexpectedly that the deep sea is one of Earth's great reservoirs of biodiversity, and have begun to assess geographic patterns of diversity on different scales. As in surface ecosystems (Levin, 1992), it seems likely that both ecological and evolutionary processes contribute to shaping patterns of biodiversity in the deep sea and that their effects are scale dependent. So far, most attention has been devoted to studying ecological influences on community structure. The next great challenge is to explore the evolutionary processes of speciation and adaptive radiation that generate the deep-sea fauna, and determine the relationship between these historical phenomena and contemporary geographic patterns of local species diversity. This will require measuring geographic variation and distributions of individual species (e.g. Etter & Rex, 1990), and phylogenetic analyses of higher taxa on an ocean-wide basis. We need to envision how deep-sea sampling programmes can be designed on larger scales specifically to address evolutionary questions. An important priority in planning how to approach evolutionary problems is to develop the first global data base of deep-sea sampling programmes and make existing collections available for systematic and biogeographic research.

Acknowledgements

We thank A. Rex for reading the manuscript. The gastropod assemblages analysed in the paper were collected by vessels of the Woods Hole Oceanographic Institution and were made available to us by H. Sanders. The box-core data plotted in Fig. 5.1 are from the Atlantic Continental Slope and Rise Study (Blake *et al.*, 1985, 1987; Maciolek *et al.*, 1987a,b;

Etter & Grassle, 1992; Grassle & Maciolek, 1992; Blake & Grassle, 1994). The research of M.A.R. and R.J.E. is supported by grants from the National Science Foundation (OCE-9301687 and OCE-9402855).

References

Angel, M.V., 1993. Biodiversity of the pelagic ocean. *Conservation Biology*, **7**, 760–72.
Billett, D.S.M., Lampitt, R.S., Rice, A.L. & Mantoura, R.F.C., 1983. Seasonal sedimentation of phytoplankton to the deep-sea benthos. *Nature*, **302**, 520–2.
Blake, J.A. & Grassle, J.F., 1994. Benthic community structure on the U.S. South Atlantic slope off the Carolinas: Spatial heterogeneity in a current-dominated system. *Deep-Sea Research II*, **41**, 835–74.
Blake, J.A., Hecker, B., Grassle, J.F., Maciolek-Blake, N., Brown, B., Curran, M., Dade, B., Freitas, S. & Ruff, R.E., 1985. Study of Biological Processes on the U.S. South Atlantic Slope and Rise. Phase 1. Benthic Characterization Study. Final report prepared for the U.S. Department of Interior, Minerals Management Service, Washington DC, pp. 142 and Appendices A–D.
Blake, J.A., Hecker, B., Grassle, J.F., Brown, B., Wade, M., Boehm, P., Baptiste, E., Hilbig, B., Maciolek, N., Petrecca, R., Ruff, R.E., Starczak, V. & Watling, L.E., 1987. Study of Biological Processes on the U.S. South Atlantic Slope and Rise. Phase 2. Final report prepared for the U.S. Department of Interior, Minerals Management Service, Washington, DC, pp. 414 and Appendices A–M.
Bouchet, P. & Warén, A., 1994. Ontogenetic migration and dispersal of deep-sea gastropod larvae. In *Reproduction, Larval Biology, and Recruitment of the Deep-sea Benthos*, ed. C.M. Young & K.J. Eckelbarger, pp. 98–117. New York: Columbia University Press.
Brandt, A., 1995. Peracarid fauna (Crustacea, Malacostraca) of the Northeast Water Polynya off Greenland: Documenting close benthic-pelagic coupling in the Westwind Trough. *Marine Ecology Progress Series*, **121**, 39–51.
Brey, T., Klages, M., Dahm, C., Gorny, M., Gutt, J., Hahn, S., Stiller, M., Arntz, W.E., Wägele, J.-W. & Zimmermann, A., 1994. Antarctic benthic diversity. *Nature*, **368**, 297.
Bugge, T., Belderson, R.H. & Kenyon, N.H., 1988. The Storegga Slide. *Philosophical Transactions of the Royal Society of London, Series A*, **325**, 357–88.
Carney, R.S., Haedrich, R.L. & Rowe, G.T., 1983. Zonation of fauna in the deep sea. In *Deep-Sea Biology*, ed. G.T. Rowe, pp. 371–98. New York: John Wiley.
Cornell, H.V., 1985a. Local and regional richness of cynipine gall wasps on California oaks. *Ecology*, **66**, 1247–60.
Cornell, H.V., 1985b. Species assemblages of cynipid gall wasps are not saturated. *American Naturalist*, **126**, 565–9.
Cornell, H.V., 1993. Unsaturated patterns in species assemblages: The role of regional processes in setting local species richness. In *Species Diversity in Ecological Communities: Historical and Geographical Perspectives*, ed. R.E. Ricklefs & D. Schluter, pp. 243–52. Chicago: University of Chicago Press.
Currie, D.J., 1991. Energy and large-scale patterns of animal- and plant-species richness. *American Naturalist*, **137**, 27–49.

Dahl, E., 1972. The Norwegian Sea deep water fauna and its derivation. *Ambio Special Report*, **2**, 19–24.

Dodge, J.D. & Marshall, H.G., 1994. Biogeographic analysis of the armored planktonic dinoflagellate *Ceratium* in the North Atlantic and adjacent seas. *Journal of Phycology*, **30**, 905–22.

Eckman, J.E. & Thistle, D., 1991. Effects of flow about a biologically produced structure on harpacticoid copepods in San Diego Trough. *Deep-Sea Research*, **38**, 1397–416.

Etter, R.J. & Caswell, H., 1994. The advantages of dispersal in a patchy environment: Effects of disturbance in a cellular automaton model. In *Reproduction, Larval Biology, and Recruitment of the Deep-sea Benthos*, ed. C.M. Young & K.J. Eckelbarger, pp. 284–305. New York: Columbia University Press.

Etter, R.J. & Grassle, J.F., 1992. Patterns of species diversity in the deep sea as a function of sediment particle size diversity. *Nature*, **360**, 576–8.

Etter, R.J. & Rex, M.A., 1990. Population differentiation decreases with depth in deep-sea gastropods. *Deep-Sea Research*, **37**, 1251–61.

Fischer, A.G., 1960. Latitudinal variations in organic diversity. *Evolution*, **14**, 64–81.

France, R., 1992. The North American latitudinal gradient in species richness and geographical range of freshwater crayfish and amphipods. *American Naturalist*, **139**, 342–54.

Gage, J.D. & Tyler, P.A., 1991. *Deep-sea Biology: A Natural History of Organisms at the Deep-sea Floor*. Cambridge: Cambridge University Press.

Grassle, J.F. & Maciolek, N.J., 1992. Deep-sea species richness: Regional and local diversity estimates from quantitative bottom samples. *American Naturalist*, **139**, 313–41.

Grassle, J.F. & Morse-Porteous, L.S., 1987. Macrofaunal colonization of disturbed deep-sea environments and the structure of deep-sea benthic communities. *Deep-Sea Research*, **34**, 1911–50.

Grassle, J.F. & Sanders, H.L., 1973. Life histories and the role of disturbance. *Deep-Sea Research*, **20**, 643–59.

Gray, J.S., 1994. Is deep-sea species diversity really so high? Species diversity of the Norwegian continental shelf. *Marine Ecology Progress Series*, **112**, 205–9.

Hadfield, M.G. & Strathmann, M.F., 1990. Heterostrophic shells and pelagic development in trochoideans: Implications for classification, phylogeny and palaeoecology. *Journal of Molluscan Studies*, **56**, 239–56.

Hecker, B., 1990. Variation in megafaunal assemblages on the continental margin south of New England. *Deep-Sea Research*, **37**, 37–57.

Hessler, R.R. & Sanders, H.L., 1967. Faunal diversity in the deep sea. *Deep-Sea Research*, **14**, 65–78.

Hochachka, P.W. & Somero, G.N., 1984. *Biochemical Adaptation*. Princeton, NJ: Princeton University Press.

Hollister, C.D. & McCave, I.N., 1984. Sedimentation under deep-sea storms. *Nature*, **309**, 220–5.

Hollister, C.D. & Nowell, A.R.M., 1991. HEBBLE epilogue. *Marine Geology*, **99**, 445–60.

Hugueny, B. & Paugy, D., 1995. Unsaturated fish communities in African rivers. *American Naturalist*, **146**, 162–9.

Hurlbert, S.H., 1971. The nonconcept of species diversity: A critique and alternative parameters. *Ecology*, **52**, 577–86.

Huston, M., 1979. A general hypothesis of species diversity. *American Naturalist*, **113**, 81–101.
Jablonski, D. & Lutz, R.A., 1983. Larval ecology of marine benthic invertebrates: Paleobiological implications. *Biological Reviews*, **58**, 21–89.
Josefson, A.B., 1985. Distribution of diversity and functional groups of marine benthic infauna in the Skagerrak (eastern North Sea) – can larval availability affect diversity? *Sarsia*, **70**, 229–49.
Jumars, P.A., 1975. Environmental grain and polychaete species' diversity in a bathyal benthic community. *Marine Biology*, **30**, 253–66.
Jumars, P.A., 1976. Deep-sea species diversity: Does it have a characteristic scale? *Journal of Marine Research*, **34**, 217–46.
Kukert, H. & Smith, C.R., 1992. Disturbance, colonization and succession in a deep-sea sediment community: Artificial-mound experiments. *Deep-Sea Research*, **39**, 1349–71.
Lampitt, R.S., 1985. Evidence for the seasonal deposition of detritus to the deep-sea floor and its subsequent resuspension. *Deep-Sea Research*, **32**, 885–97.
Levin, L.A., 1991. Interactions between metazoans and large, agglutinating protozoans: Implications for the community structure of deep-sea benthos. *American Zoologist*, **31**, 886–900.
Levin, S.A., 1992. The problem of pattern and scale in ecology. *Ecology*, **73**, 1943–67.
Maciolek, N., Grassle, J.F., Hecker, B., Boehm, P.D., Brown, B., Dade, B., Steinhauser, W.G., Baptiste, E., Ruff, R.E. & Petrecca, R., 1987a. Study of Biological Processes on the U.S. Mid-Atlantic Slope and Rise. Final report prepared for the U.S. Department of Interior, Minerals Management Service, Washington, DC, pp. 310 and Appendices A–M.
Maciolek, N., Grassle, J.F., Hecker, B., Brown, B., Blake, J.A., Boehm, P.D., Petrecca, R., Duffy, S., Baptiste, E. & Ruff, R.E., 1987b. Study of Biological Processes on the U.S. North Atlantic Slope and Rise. Final report prepared for the U.S. Department of Interior, Minerals Management Service, Washington, DC, pp. 362 and Appendices A–L.
McGowan, J.A. & Walker, P.W., 1993. Pelagic diversity patterns. In *Species Diversity in Ecological Communities: Historical and Geographical Perspectives*, ed. R.E. Ricklefs & D. Schluter, pp. 203–14. Chicago: University of Chicago Press.
Mellor, C.A. & Paull, C.K., 1994. Seam Beam bathymetry of the Manteo 467 Lease Block off Cape Hatteras, North Carolina. *Deep-Sea Research II*, **41**, 711–18.
Paterson, G.L.J. & Lambshead, P.J.D., 1995. Bathymetric patterns of polychaete diversity in the Rockall Trough, northeast Atlantic. *Deep-Sea Research I*, **42**, 1199–214.
Pineda, J., 1993. Boundary effects on the vertical ranges of deep-sea benthic species. *Deep-Sea Research I*, **40**, 2179–192.
Rex, M.A., 1977. Zonation in deep-sea gastropods: The importance of biological interactions to rates of zonation. In *Biology of Benthic Organisms*, ed. B.F. Keegan, P.O. Ceidigh & P.J.S. Boaden, pp. 521–30. New York: Pergamon Press.
Rex, M.A., 1981. Community structure in the deep-sea benthos. *Annual Review of Ecology and Systematics*, **12**, 331–53.
Rex, M.A., 1983. Geographic patterns of species diversity in the deep-sea benthos. In *The Sea*, ed. G.T. Rowe, vol. **8**, pp. 453–72. New York: John Wiley.

Rex, M.A. & Warén, A., 1982. Planktotrophic development in deep-sea prosobranch snails from the western North Atlantic. *Deep-Sea Research*, **29**, 171–84.

Rex. M.A., Watts, M.C., Etter, R.J. & O'Neill, S., 1988. Character variation in a complex of rissoid gastropods from the upper continental slope of the western North Atlantic. *Malacologia*, **29**, 325–39.

Rex, M.A., Stuart, C.T., Hessler, R.R., Allen, J.A., Sanders, H.L. & Wilson, G.D.F., 1993. Global-scale latitudinal patterns of species diversity in the deep-sea benthos. *Nature*, **365**, 636–9.

Rice, A.L., Billett, D.S.M., Fry, J., John, A.W.G., Lampitt, R.S., Mantoura, R.F.C. & Morris, R.J., 1986. Seasonal deposition of phytodetritus to the deep-sea floor. *Proceedings of the Royal Society of Edinburgh*, **88B**, 265–79.

Ricklefs, R.E., 1987. Community diversity: Relative roles of local and regional processes. *Science*, **235**, 167–71.

Ricklefs, R.E., 1989. Speciation and diversity: The integration of local and regional processes. In *Speciation and its Consequences*, ed. D. Otte & J.A. Endler, pp. 599–622. Sunderland, MA: Sinauer Associates.

Ricklefs, R.E. & Schluter, D., 1993. Species diversity: Regional and historical influences. In *Species Diversity in Ecological Communities: Historical and Geographical Perspectives*, ed. R.E. Ricklefs & D. Schluter, pp. 350–63. Chicago: University of Chicago Press.

Rohde, K., 1992. Latitudinal gradients in species diversity: The search for the primary cause. *Oikos*, **65**, 514–27.

Roughgarden, J., Gaines, S. & Possingham, H., 1988. Recruitment dynamics in complex life cycles. *Science*, **241**, 1460–6.

Rowe, G.T. & Pariente, V. (eds), 1992. *Deep-sea Food Chains and the Global Carbon Cycle*. Dordrecht: Kluwer Academic Publishers.

Roy, K., Jablonski, D. & Valentine, J.W., 1994. Eastern Pacific molluscan provinces and latitudinal diversity gradient: No evidence for 'Rapoport's rule.' *Proceedings of the National Academy of Sciences of the USA*, **91**, 8871–4.

Sanders, H.L., 1968. Marine benthic diversity: A comparative study. *American Naturalist*, **102**, 243–82.

Schaff, T.R. & Levin, L.A., 1994. Spatial heterogeneity of benthos associated with biogenic structures on the North Carolina continental slope. *Deep-Sea Research II*, **41**, 901–18.

Sepkoski, J.J. Jr & Rex, M.A., 1974. Distribution of freshwater mussels: Coastal rivers as biogeographic islands. *Systematic Zoology*, **23**, 165–88.

Smith, C.R., 1985. Food for the deep sea: Utilization dispersal and flux of nekton falls at the Santa Catalina Basin floor. *Deep-Sea Research*, **32**, 417–42.

Smith, C.R., 1986. Nekton falls, low-intensity disturbance and community structure of infaunal benthos in the deep sea. *Journal of Marine Research*, **44**, 567–600.

Smith, C.R. & Hessler, R., 1987. Colonization and succession in deep-sea ecosystems. *Trends in Ecology and Evolution*, **2**, 359–63.

Smith, C.R., Jumars, P.A. & DeMaster, D.J., 1986. *In situ* studies of megafaunal mounds indicate rapid sediment turnover and community response at the deep-sea floor. *Nature*, **323**, 251–3.

Smith, C.R., Kukert, H., Wheatcroft, R.A., Jumars, P.A. & Deming, J.W., 1989. Vent fauna on whale remains. *Nature*, **341**, 27–8.

Snelgrove, P.V.R., Grassle, J.F. & Petrecca, R.F., 1992. The role of food patches in maintaining high deep-sea diversity: Field experiments with hydrodynamically unbiased colonization trays. *Limnology and Oceanography*, **37**, 1543–50.

Stevens, G.C., 1989. The latitudinal gradient in geographical range: How so many species coexist in the tropics. *American Naturalist*, **133**, 240–56.
Stevens, G.C., 1992. The elevational gradient in altitudinal range: An extension of Rapoport's latitudinal rule to altitude. *American Naturalist*, **140**, 893–911.
Stuart, C.T., 1991. Regional and Global Diversity Patterns of Deep-sea Gastropods in the Atlantic Ocean. PhD Thesis, University of Massachusetts/Amherst.
Stuart, C.T. & Rex, M.A., 1994. The relationship between developmental pattern and species diversity in deep-sea prosobranch snails. In *Reproduction, Larval Biology, and Recruitment of the Deep-sea Benthos*, ed. C.M. Young & K.J. Eckelbarger, pp. 118–36. New York: Columbia University Press.
Svavarsson, J., Brattegard, T. & Strömberg, J.-O., 1990. Distribution and diversity patterns of asellote isopods (Crustacea) in the deep Norwegian and Greenland Seas. *Progress in Oceanography*, **24**, 297–310.
Svavarsson, J., Strömberg, J.-O. & Brattegard, T., 1993. The deep-sea asellote (Isopoda, Crustacea) fauna of the Northern Seas: Species composition, distributional patterns and origins. *Journal of Biogeography*, **20**, 537–55.
Thistle, D., 1983. The stability-time hypothesis as a predictor of diversity in deep-sea soft-bottom communities: A test. *Deep-Sea Research*, **30**, 267–77.
Thistle, D. & Eckman, J.E., 1990. The effect of a biologically produced structure on the benthic copepods of a deep-sea site. *Deep-Sea Research*, **37**, 541–54.
Turley, C.M., Lochte, K. & Lampitt, R.S., 1995. Transformations of biogenic particles during sedimentation in the northeastern Atlantic. *Philosophical Transactions of the Royal Society of London*, **348**, 179–89.
Underwood, A.J. & Fairweather, P.G., 1989. Supply-side ecology and benthic marine assemblages. *Trends in Ecology and Evolution*, **4**, 16–20.
Vale, F.K. & Rex, M.A., 1988. Repaired shell damage in deep-sea prosobranch gastropods from the western North Atlantic. *Malacologia*, **28**, 65–79.
Yoder, J.A., McClain, C.R., Feldman, G.C. & Esaias, W.E., 1993. Annual cycles of phytoplankton chlorophyll concentrations in the Global Ocean: A satellite view. *Global Biogeochemical Cycles*, **7**, 181–93.

Chapter 6
Diversity, latitude and time: Patterns in the shallow sea

ANDREW CLARKE and J. ALISTAIR CRAME
British Antarctic Survey, High Cross, Madingley Road, Cambridge, CB3 0ET, UK

Abstract

The latitudinal cline in diversity is a well established, though poorly understood, feature of terrestrial communities and although it is often assumed that a similar cline is to be found in the sea, the evidence for this is still equivocal. Attempts to develop an overall picture of diversity in the sea are made difficult by the small number of key studies, the varied sampling protocols employed in these studies, the different measures of diversity utilised and the varying levels of taxonomic resolution. Nevertheless, there is clear evidence for greater species richness in the tropics for several shallow-water taxa that depend on calcareous skeletons, including gastropod and bivalve molluscs, foraminifera and hermatypic corals. There is also an increasing body of evidence for a latitudinal diversity cline in the deep sea, at least in the northern hemisphere. However, any cline in species richness in the northern hemisphere will, in part, be a necessary consequence of the depauperate Arctic fauna and the intense species richness of the Indo-West Pacific. The Southern Ocean is now known to support a rich and diverse shallow-water benthic fauna, the striking contrast to the Arctic being the result of a very different tectonic and evolutionary history. Whereas the Arctic marine system is very young and still being colonised, the Southern Ocean marine fauna has essentially evolved *in situ*, and has been influenced by a dynamic glacial and climatic history. Latitudinal range shifts induced by these cycles have resulted in a rich fauna (the taxonomic diversity pump), and several taxa have subsequently colonised lower latitudes. Evidence for a global latitudinal diversity cline in non-calcareous marine taxa remains equivocal, and there is as yet no evidence for a cline in within-habitat diversity for shallow-water soft-bottom infaunal communities. These results suggest that patterns are very different in the northern and southern hemispheres, and that the assumption of a simple and universal latitudinal diversity cline in the sea is, on present evidence, probably unwarranted. An improved understanding of the role of spatial heterogeneity in regulating diversity in the sea is urgently required.

6.1 Introduction

The study of biological diversity is at an interesting stage. Following an intense phase of development in the 1960s and early 1970s, when the causes of biological diversity were the subject of much discussion and many of the techniques for measuring and expressing diversity were developed, studies of ecological diversity entered a period of relative decline. Taxonomic work was viewed as of low priority by funding agencies, and ecologists were finding explanations for the observed patterns of diversity rather elusive. The development of new taxonomic and analytical techniques, together with an increasing awareness of the rapid loss of species, have stimulated a resurgence of interest in questions of biological diversity.

Many of our views of diversity in the sea have their roots in results from the far better known terrestrial ecosystem. Here, we review the present understanding of diversity in the shallow marine ecosystem, contrasting the patterns observed with those long known from the land and, where relevant, the deep sea. We start by considering the overall number of species in the sea and then examine critically the evidence for a latitudinal cline in marine diversity. We then look in detail at the Southern Ocean, where recent work has changed our views of polar marine diversity. We examine the role of time (evolutionary and tectonic history) in regulating marine diversity, contrasting the patterns observed in the northern and southern polar oceans, and finish by re-examining the case for a latitudinal cline in shallow water marine diversity.

6.2 Global marine diversity: How many species in the sea?

Despite the increasing attention being given to biological diversity, ecologists are still in the frustrating position of being unable to give even approximate answers to quite fundamental questions, such as how many species exist at present. Estimates of the number of species currently known and described vary from roughly 1.4–1.8 million (Stork, 1988;

124   *Diversity, latitude and time: Patterns in shallow seas*

Minelli, 1993; May, 1994), although the validity of adding together protist, plant and animal 'species' is perhaps questionable. Estimates of the total number of species on the globe are far more difficult to make (see discussion in Chapter 1).

Current knowledge suggests that most biological diversity is to be found on the land, but this may simply indicate our ignorance of the sea. May (1994) has pointed out that most species described to date are either insects (56%) or plants (14%). Among animals, however, most phyla are marine and benthic. Whilst tropical rain forest is probably the most species-rich environment on the planet, the most phyletically diverse habitats are probably coral reefs.

Difficulties of sampling means that our knowledge of marine diversity remains well behind that of the terrestrial system. Indeed we are still discovering (or recognising) marine taxa representing major clades, such as the enigmatic medusa-like echinoderm *Xyloplax*, or the poorly understood *Trichoplax*. Long known but only recently adequately described, this latter organism may represent a whole new phylum (Placozoa).

Estimates of the total number of marine species are few. The most recent thorough estimate of the number of benthic marine macroinvertebrate species (Grassle & Maciolek, 1992) has prompted revisions both downwards (May, 1992) and upwards (Poore & Wilson, 1993) (see Pimm *et al.*, 1995, and discussion in Chapter 1). These differences of opinion serve to emphasise just how little we know of the species to be found in some marine habitats. Even among taxa believed to be well known, such as molluscs or crustaceans, estimates of the total number of species can vary significantly (see discussion in Minelli, 1993).

Nevertheless, some clear broad-scale patterns have emerged from studies of marine diversity and biogeography. Of these, one of the most distinct is the latitudinal cline in species diversity.

6.3 Latitudinal clines in diversity

The latitudinal cline in the diversity of terrestrial organisms is a long-established feature of biogeography. Since it is only relatively recently (since the 1950s) that evidence for a similar cline has been recognised in the sea, an outline of terrestrial patterns of diversity forms an essential introduction to any consideration of such patterns in the sea.

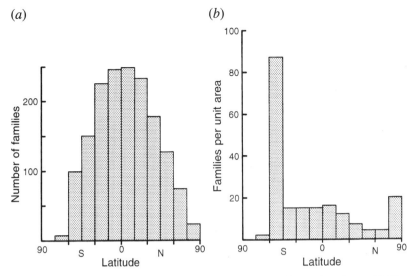

Fig. 6.1. Latitudinal variation in the number of families of vascular plants. (*a*) The number of families pooled by 10° bands of latitude. (*b*) The same data corrected for land area in each latitudinal band. The single strikingly high value is the result of the high family diversity in the vascular flora of the semi-arid fynbos of South Africa. Redrawn from Woodward (1987).

6.3.1 Latitudinal diversity clines on the land

The number of species in many terrestrial taxa decreases from the tropics to the poles. Although there are clearly local variations (for example tropical deserts harbour fewer species than do rain forests), this pattern is found in many terrestrial plant and animal taxa when viewed on a broad scale. A good example is that of vascular plants, which show an almost symmetrical distribution in family level diversity about the equator (Fig. 6.1(*a*)). There is, as yet, no consensus on the explanation for this deceptively simple pattern. Thus Rosenzweig & Abramsky (1993) regard this problem as having been resolved in the mid-1970s, with the explanation being essentially a combined effect of the species/ area relationship and the role of allopatric speciation (Terborgh, 1973; Rosenzweig, 1975, 1977), whereas Ricklefs (cited in Lewin, 1989) has called the latitudinal cline in diversity 'the major unexplained feature of natural history'.

A recurring theme in explanations of the latitudinal diversity cline on land has been the relationship between species richness and area. Clear, bell-shaped diversity clines centred on the equator can be produced by

126 *Diversity, latitude and time: Patterns in shallow seas*

very simple models assuming a random association between latitude and the size and placement of species ranges (Colwell & Hurtt, 1994). This result is a simple consequence of geometry but if the data for families of vascular plants are corrected on a broad scale for actual land area, then a more or less uniform pattern is obtained (Fig. 6.1(*b*)). The obvious exception is caused by the unusually rich floras of dry Mediterranean-type communities such as the semi-arid fynbos community of southern Africa, and this suggests that there is more to the relationship than simply area (O'Brien, 1993; Huston, 1994). Simple correction for available land area (as against total surface area of the globe between different latitudes as used in the Colwell & Hurtt model) is only valid, however, if the relationship between diversity and area is linear. Numerous studies have shown that this is not so; the relationship between species richness and area is usually best described by an exponential relationship with the value of the exponential parameter generally lying in the range 0.2–0.4 (Connor & McCoy, 1979).

Correction for land area does not remove the cline in plant diversity at the species level (Reid & Miller, 1989) and recent theoretical studies have revealed subtle scale effects on the relationship between species richness and area (Palmer & White, 1994). Although this relationship remains a favoured underlying explanation for the latitudinal diversity cline on the land, at least 14 other different explanations have been suggested (Pianka, 1966; Stevens, 1989; Pagel *et al.*, 1991; Huston, 1994).

6.3.2 Latitudinal diversity clines in the sea

The sea constitutes a vastly more extensive habitat than does the land, and incorporates a third dimension (depth), which is of far greater magnitude and ecological significance than on land. Oceanic systems also differ fundamentally from terrestrial systems in the pattern of spatial and temporal variability, which is of major significance to an understanding of physical and biological processes in the two environments. We should not expect *a priori* that explanations for the patterns of diversity we see on land will also be valid in the sea.

The enormous size of the ocean means that we need to consider separately patterns of diversity in the benthos of the shallow seas (essentially the continental shelves), the benthos of the deep sea, and the plankton of the water column.

The first suggestion for a latitudinal cline in the species richness of

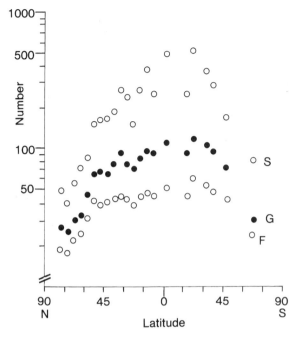

Fig. 6.2. Latitudinal variation in the number of species (S), genera (G) and families (F) of bivalve molluscs. All data pooled by 5° latitude classes. Redrawn from Stehli et al. (1967).

shallow-water, hard-bottom communities came from Thorson (1952, 1957). Subsequently a number of studies showed distinct maxima in tropical regions for species richness in bivalve and gastropod molluscs, and foraminifera (Fischer, 1960; Stehli et al., 1967, 1972; Flessa & Jablonski, 1995; Fig. 6.2), and a recent detailed analysis of the ranges of 2838 eastern Pacific molluscs has provided a clear gradient in species richness (Roy et al., 1994). Taylor & Taylor (1977) demonstrated a gradient in the diversity of North Atlantic predatory prosobranch gastropods that was, however, complicated by a pronounced step at about 40°N.

Thorson (1957) commented that he could find no evidence of a latitudinal cline in the species richness of shallow-water soft-bottom communities, and more recent studies in shallow waters have failed to provide evidence for such a cline (Richardson & Hedgpeth, 1977; Kendall & Aschan, 1993). By contrast, work on the diversity of deep-sea communities has tended to indicate the presence of a distinct latitudinal cline in diversity. The classic early work from the Woods Hole Oceanographic Institution that first established the high diversity of the deep sea also suggested a latitudinal cline in diversity in the shallow-water benthos (Sanders, 1968). Further work on continental slope and deep-sea soft-bottom communities has produced evidence both for and against a latitudinal cline in diversity (Grassle & Maciolek, 1992; Rex et al., 1993; and see Chapter 5).

The factors that regulate the distribution of oceanic (planktonic) species are quite different from those that govern the distribution of plankton to water-masses, and subsequently the relationship to the major gyral structure of the oceans was clarified (McGowan, 1974). On a broad scale, many oceanic taxa show a latitudinal cline in diversity with maximum species richness in the tropics. Present knowledge of the distribution of oceanic plankton has been summarised recently by Angel (1994; and see Chapter 3), and water column species will not be considered further here.

Studies of marine diversity have tended to concentrate on the northern hemisphere, where the recognition of a latitudinal cline in marine diversity was influenced critically by two key factors: the intense species-richness of some shallow tropical seas, especially in the Indo-West Pacific, and the depauperate Arctic fauna. The Southern Ocean, in contrast to the Arctic basin, has a rich and diverse fauna. This fauna was well described by the pioneering oceanographic expeditions, but has again been receiving taxonomic attention in recent years. We will therefore take our improving knowledge of the shallow-water marine fauna around Antarctica as a starting point for a critical examination of the evidence for the existence of a global latitudinal diversity cline in the sea.

6.4 Marine diversity in Antarctica

Although the richness of the Southern Ocean benthic fauna has been known since the pioneering work of the scientists aboard HMS *Challenger*, the polar regions are still traditionally regarded as areas hostile to life and occupied by sparse and depauperate faunas somehow hanging on in the face of extreme physical conditions. On land this is to a large extent true, and the Arctic marine system is also genuinely low in species richness (Curtis, 1975; Dayton, 1990; Dunton, 1992). The marine fauna of the Antarctic continental shelves is, by contrast, generally rich in species (Dell, 1972; Arntz et al., 1994, in press). Unfortunately little is known of the deep-sea regions of the Southern Ocean (Dell, 1972; Arntz et al., 1994, 1997). In this review we will

Table 6.1. *Numbers of species in selected Antarctic benthic taxa compiled from the literature*

	Antarctica	Weddell Sea	Admiralty Bay
Molluscs			
Gastropods	604†	145‡	35‡
Bivalves	166†	43	27
Echinoderms			
Asteriods	104	50	15
Crinoids	27	6	1
Echinoids	25	ND	4
Ophiuroids	87	43	15
Holothurians	103	35	3
Crustaceans			
Decapods	4	4	2
Amphipods	520		
Gammaridea	459	174	31
Polychaetes	562	225	ca 100

Data are for the Southern Ocean south of the Polar Front (Antarctic Convergence).
†South of 50°S.
‡Prosobranch gastropods only.
Principal sources are cited by Arntz et al. (1997); additional material from Arnaud et al. (1986), de Broyer & Jazdzewski (1993) and Siciński (1993). ND, no data.

therefore concentrate on patterns of diversity in the faunas of the Antarctic continental shelf, and compare these with selected sites elsewhere.

Although the fauna of the Southern Ocean is reasonably well documented, some taxa are better known than others. Winston (1992) has estimated that for many taxa in the Southern Ocean between 70 and 95% of species are known; for others she estimates that only 20–40% may have been described. In at least two of these groups, nemerteans and bryozoans, recent taxonomic work is resulting in rapid improvements in our knowledge.

In Table 6.1 we list the species numbers known for selected major taxa, and compare these with species numbers for other areas. The totals for some taxa and sites differ from those published previously (e.g. by Knox & Lowry, 1977) because of new biogeographical or taxonomic work. The data reflect two features of Antarctic marine diversity that have long been recognised: that the Antarctic has a higher diversity than the Arctic, and that the Antarctic marine fauna is characterised by a mixture of relatively species-poor (for example bivalve and gastropod molluscs) and species-rich (for example polychaetes, bryozoans, amphipods and sponges) taxa.

These topics have been reviewed extensively in the past (Dell, 1972; Knox & Lowry, 1977; White, 1984; Dayton, 1990; Arntz et al., 1997). We will therefore concentrate on only three aspects, namely the nature of the data, geographic variation around Antarctica and a comparison of the two polar regions, before moving to a discussion of the role of evolutionary history in influencing the patterns we see today.

6.4.1 Antarctic marine diversity: What is being measured?

Whittaker (1960, 1972) first advanced the concept of partitioning regional (gamma) diversity into local (alpha) and turnover (beta) components. Ecologists have tended to concentrate on local (or within-habitat) alpha diversity, whereas biogeographers have concentrated on regional faunal lists (which approximate to gamma diversity). Relatively little attention has been directed at turnover, which suffers from a plethora of mathematical formulations (see Magurran, 1988). Furthermore, ecologists currently have no firm theoretical basis for the relationship between alpha and gamma diversity, or for those processes underlying beta diversity (Cornell & Lawton, 1992; Harrison et al., 1992), despite their importance in disentangling the relative roles of local and regional processes in determining species richness.

The data in Table 6.1 are simple species lists rather than true diversity values. Although a species list can be taken as an approximation to a regional species richness value, comparison with other areas requires some degree of comparability in geographic area and sampling intensity. Thus, although much of the Antarctic benthos is believed to have a circumpolar distribution, the data for the well-sampled high Antarctic Weddell Sea shows much lower overall species richness than the Southern Ocean as a whole (Table 6.1). Where areas have been sampled quantitatively, species/abundance plots indicate for some taxa that although the majority of species have probably been recorded, representation of rare species is not complete. Examples from Signy Island in the maritime Antarctic are given in Fig. 6.3.

The problems of comparability of sampling technique in studies of marine diversity have long been recognised (Abele & Walters, 1979a,b; Clarke, 1992; Arntz et al., 1997). For Antarctic ecologists these problems are exacerbated by two particular features. The first is that many polar

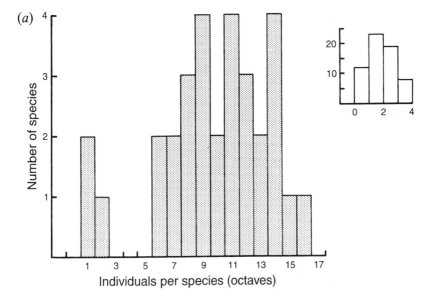

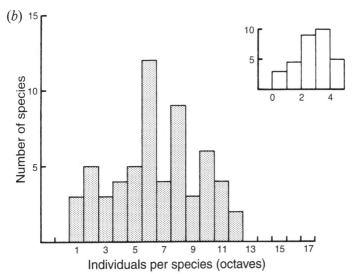

Fig. 6.3. Species/abundance (Preston, 1962) plots for amphipods and prosobranch gastropods at Signy Island, Antarctica. Units for the abscissa are $\log 2(n)$ where n is the number of individuals per species (octaves: Preston, 1962). Inset histograms show the same data but with abscissa as $\log 10(n)$. Note differing ordinate scales for the two plots. (a) Prosobranch gastropods; data from a collection of 138 650 individuals of 31 species collected by Picken (1980). (b) Amphipods; data from a collection of 34 718 amphipods of 62 species, analysed by Thurston (1972). The truncated shape of the distributions indicates that some rare species were not sampled.

132 *Diversity, latitude and time: Patterns in shallow seas*

marine invertebrates are small and hence may be missed by conventional sampling gear where the mesh is too large. The second is that the recruitment of juvenile stages is often very seasonal; diversity statistics that include a substantial evenness component may thus be distorted by the presence of recent recruits. In fact this is not currently a problem for there has been to date only a single study of within-habitat (alpha) diversity for any Antarctic marine community (Richardson, 1976; Richardson & Hedgpeth, 1977).

6.4.2 Patterns of diversity around Antarctica

Detailed examination of patterns of diversity within the Southern Ocean are hampered by the poor state of current knowledge. Relatively few sites have been studied in any detail, and in most cases all that exists are species lists of varying degrees of completeness.

The data collated in Table 6.1 show the expected trend that as the size of area being covered increases, so the species list increases. Although many Antarctic species are known to be circumpolar, there are clear biogeographic patterns superimposed on this general distribution; many of these patterns reflect the history of faunal invasion (for example along the Scotia arc: Hedgpeth, 1969, 1971; Knox & Lowry, 1977). A generally agreed broad-scale distribution of benthic biogeographic provinces around Antarctica is shown in Fig. 6.4.

As would be expected, where species lists are available for large areas such as the Weddell Sea, detailed analyses indicate the presence of distinct assemblages within the overall area (Fig. 6.5). Generally these assemblages have been determined from non-selective techniques such as bottom trawls, which may mix representatives from different small-scale assemblages thereby obscuring important ecological detail relating to fine-scale heterogeneity. Photographic techniques, though possessing their own obvious limitations for small organisms, are now providing much valuable ecological data and hold considerable promise for the future (Barthel & Gutt, 1992; Arntz *et al.*, 1997).

6.4.3 A comparison of marine benthic diversity in the Arctic and the Antarctic

Even broad regional comparisons suffer from difficulties of definition. For example, what is the most sensible definition of the Arctic for biogeographic purposes? Unlike the Southern Ocean, which has a defined

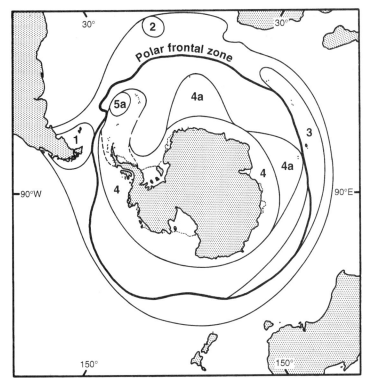

Fig. 6.4. Major biogeographic provinces for Southern Ocean benthos. (Redrawn from Hedgpeth, 1969).

northern limit of clear biogeographical validity in the Polar Frontal Zone (or Antarctic Convergence), there is no simple oceanographic definition of the Arctic. A working definition is usually taken to be the mean southerly extent of surface Arctic water (which like all oceanographic features is variable from year to year: see discussion in Dunton, 1992). Whatever definition is used, however, it is an unavoidable conclusion that the Arctic is relatively species-poor when compared with the Antarctic (Knox & Lowry, 1977; Dayton, 1990; Dunton, 1992). This difference is particularly marked for polychaetes, amphipods and molluscs.

The area of the Arctic Ocean is less than half that of the Southern Ocean (Dayton, 1990) but since a relatively small fraction of the waters around Antarctica is comprised of continental shelf, a simple species/area relationship cannot be invoked to explain the differences in overall species richness. Of the two most frequently discussed explanations for the relatively low diversity of the Arctic, one is an equilibrium expla-

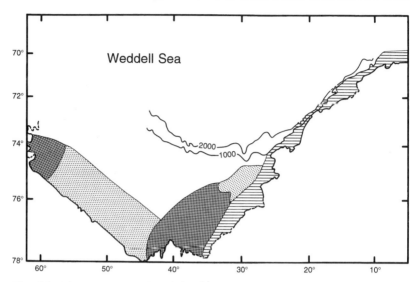

Fig. 6.5. Distribution of different benthic community assemblages in the eastern Weddell Sea. Communities defined by cluster analysis of 430 benthic taxa obtained by trawl and grab samples. ☰, eastern shelf community; ▦, southern shelf community; ▦, southern trench community. Depth in metres. (Redrawn from Voß, 1988).

nation (the harshness of the environment) and the other a non-equilibrium explanation (that the Arctic marine ecosystem is young and not yet fully occupied).

About the only similarities of the Arctic and Antarctic marine environments are that they are both very cold and ice-covered for at least part of the year. The Arctic is characterised by a broad and relatively shallow shelf, substantial fluvial sediment input and significant disturbance from both iceberg ploughing and marine mammal foraging activity (Dayton, 1990). Associated with these factors is a relative lack of any benthic communities exhibiting substantial three-dimensional (tiering) structure. The combination of a lack of habitat structure and extreme physical disturbance could mean that present diversity is in equilibrium for the particular circumstances of the Arctic, but low in comparison with other areas of the globe.

The Arctic is, however, a very young area (Dunton, 1992) and it is likely that with only narrow connections to lower latitudes, colonisation is still taking place. The present situation is thus almost certainly non-equilibrium. Indeed Vermeij (1991) has shown in a detailed faunal analysis that the North Pacific is the centre of boreal molluscan diversity, and has demonstrated convincingly that the Arctic basin is still undergoing colonisation. Interestingly, whilst the Southern Ocean molluscan fauna is dominated by epifaunal gastropods, that of the Arctic is characterised by infaunal bivalves such as *Astarte* and *Modiolus*.

In contrast, the available evidence suggests that the benthic communities of the Antarctic have experienced a long period of evolution *in situ*, albeit with periodic glacial advances restricting available shelf areas even further than at present (reviewed by Clarke & Crame, 1989). Shallow-water communities suffer frequent disturbance from ice-scour and, as in the Arctic, intertidal communities are highly depauperate. Deeper-water assemblages, particularly those on the continental shelf, are relatively protected from ice impact and rich, highly tiered, benthic communities have developed (Voß, 1988; Arntz *et al.*, 1997).

It is not possible to do more than speculate whether the benthic communities of Antarctica are at equilibrium, but the weight of evidence suggests not. Although there has been a long period of evolution *in situ*, the onset of the present oceanographic regime has resulted in an isolation of the Southern Ocean from long-distance dispersal. The communities have also been subjected to complete eradication in some areas at some times and certain taxa are now very poorly represented despite being present formerly (for example decapod crustaceans and many groups of fish). These factors all suggest an unsaturated, non-equilibrium benthic community in Antarctica, and this in turn points to the importance of processes operating over both ecological and evolutionary time in regulating the level of diversity we see today.

6.5 Patterns in time: The history of the Southern Ocean benthic fauna

The landmass that now forms the continent of Antarctica is formed of two unequally sized fragments that both once formed part of the large continental landmass of Gondwana. In the early Cretaceous (about 120 million years before present (my BP)) the southern hemisphere was dominated by Gondwana, with the Tethys Sea to the north and the proto-Pacific to the south. By the early Cenozoic (60 my BP) Gondwana had fragmented significantly and South America, Africa and India had all moved well way from Antarctica/Australia. Many of the details of the break-up of Gondwana are still unknown. It is not clear, for example, precisely when the separation of western Antarctica and South America was complete, but the best estimate for the opening of the Drake Passage is probably 25–30 my BP (Barker & Burrell, 1977). The final deep-water separation of Australia and Antarctica probably occurred shortly afterwards, allowing establishment of the circum-Antarctic current and the basic features of the oceanographic regime we observe today.

The climatic history of the fragments of Gondwana that now form Antarctica has been complex. Superimposed upon an overall decline from warm mid-Cretaceous waters of perhaps 20°C to the present glaciation, have been several episodes of rapid cooling as well as long periods of slow warming, and possibly occasional rapid climatic shifts (Clarke & Crame, 1989, 1992). The end result today is a steep meridional temperature cline from tropics to poles, a cline that dominates our thinking about physical controls on biological distribution but which is distinctly atypical of the world in the past. Conservative estimates suggest that for over 90% of the Cenozoic, tropical/polar temperature gradients in the sea were less than today, and for perhaps 75% of the Cenozoic substantially so (Crame, 1993).

The glacial history of Antarctica is also complex, but is less well understood than the thermal history. Although it was long felt that the major ice sheets of Antarctica did not form until the middle Miocene (14 my BP), more recent data from sediment cores has suggested the presence of an extensive and dynamic ice cover since at least 37 my BP (Webb, 1990). The important point is that both climate and ice cover have been variable throughout the Cenozoic. Not only have there been at least ten major cooling or warming episodes during the Cenozoic, but superimposed on these has been a whole series of smaller-scale events. In the last 1.5 my alone there may have been as many as 50 climatic cycles (Crame, 1993).

These repeated climatic cycles are likely to have resulted in repeated latitudinal range shifts of many marine taxa living in the shallow waters around Antarctica. Such shifts, resulting in fragmentation of distributions and consequent allopatric speciation, are the basis of so-called taxonomic diversity pumps (Valentine, 1984: Fig. 6.6). Coupled with periodic extensions of the ice cover resulting in a reduced area of continental shelf, and retreats exposing extensive areas of new shallow marine habitat, these climatic cycles have clearly been instrumental in influencing the composition of the fauna that exists today. Where glacial extensions have eradicated the previous flora and fauna, recolonisation will necessarily have occurred either from refugia (perhaps in deeper water) or possibly even from outside the Southern Ocean. Such historical factors may explain the present distinct cline in macroalgal diversity along the Antarctic peninsula (Moe & DeLaca, 1976; Clayton, 1994).

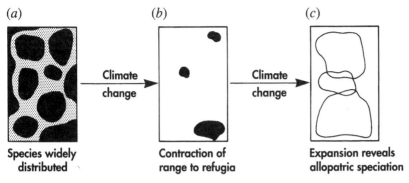

Fig. 6.6. Conceptual model of the climate diversity pump. Initially a species is widely distributed over an extensive area (*a*), with the population exhibiting classic metapopulation structure comprising some areas of high abundance and successful reproduction (represented by the black areas) and sink areas of unsuccessful reproduction (stippled areas). Following climatic change the range contracts to a small number of refugia, such as areas protected from an expanded continental ice shelf (*b*). Following further climate change, such as a contraction of the ice sheet, the refuge populations can once more expand, coming into contact (*c*). Under these circumstances, genetic differences accumulated during the allopatric contraction to refugia may be sufficient to prevent interbreeding and hence result in speciation. Reproduced, with permission, from Clarke (1996).

The contrast in the diversity of the Arctic and Antarctic shallow-water marine faunas is thus largely explained by their vastly different evolutionary histories. The Arctic is a young system still undergoing colonisation, mostly from the North Pacific (Vermeij, 1991; Dunton, 1992). The continental shelf fauna of Antarctica has been present in some areas of Gondwana since the Palaeozoic and has evolved *in situ*, tracking climate change (Lipps & Hickman, 1982; Clarke & Crame, 1989, 1992).

6.5.1 Evolution in the polar regions

The polar regions have traditionally been viewed as the recipients of taxa that have evolved elsewhere rather than centres of evolutionary novelty in themselves (see review by Crame, 1992). It is now recognised that some taxa may be diversifying at present in the Southern Ocean (Arnaud & Bandel, 1976) and two groups that exhibit the dynamic nature of this speciation are serolid isopods and notothenioid fish.

Phylogenetic analysis by Brandt (1991) has shown that the origin of the marine isopod family Serolidae can be placed some time after the separation of Africa from the rest of Gondwana, which occurred at about 80–90 my BP, but before the separation of Australia at about 55 my BP.

This is indicated by the presence in South America of the sister taxon Bathynataliidae, and in Australia of the genus *Serolina*. The main radiation of serolid isopods occurred in the Southern Ocean after 55 my BP (the early Eocene), and a number of taxa have subsequently penetrated into the northern hemisphere. These include *Cristaserolis* and *Leptoserolis* along the continental shelf to the east coast of the Americas, and *Atlantoserolis* and *Glabroserolis* in the Atlantic deep sea. Serolid isopods are thus an example of a taxon that evolved at high latitude and that has spread northwards out of the Southern Ocean during the Cenozoic.

In comparison with shallow seas of similar extent, the Southern Ocean fish fauna is very poor in species. The fauna is also unusual in being dominated by a single group, the perciform suborder Notothenioidea (Eastman, 1993). Although there is as yet no convincing fossil that might indicate the age of this particular group, fossil remains from the early Cenozoic suggest a typical diverse shallow-water fish fauna (Eastman, 1993). At some time this fauna died out to be replaced by the notothenioid radiation. The cause of the extinction of the previous fauna is not known but it is likely to be associated with the onset of glaciation (Clarke & Crame, 1989; Eastman & Grande, 1989). The deepening of the continental shelf by the weight of the ice-cap and the eradication of many habitats typical of near-shore fish (such as estuaries, tidal flats and rivers) may have reduced the available niches for fish (Eastman, 1993). It is even possible that a major extension of the ice may have eradicated all continental shelf habitat, although invertebrate data suggest this did not happen (Clarke & Crame, 1989). Whatever the cause, the notothenioids have subsequently undergone a major radiation, which has involved the evolution of neutral buoyancy to overcome the lack of a swimbladder in the original benthic stock (Eastman, 1993). The resultant radiation may have been a relatively recent event (Bargelloni *et al.*, 1994) and the present fauna represents an example of a marine species flock akin to the radiation of sculpins in Lake Baikal (Eastman, 1993).

These two examples emphasise the dynamic nature of evolutionary processes at high latitudes (see also Chapter 11) and the importance of habitat diversity in regulating overall regional (gamma) diversity. This sets the scene for a re-examination of the evidence for the existence of a latitudinal diversity cline in the shallow sea.

6.6 The latitudinal diversity gradient revisited

6.6.1 Patterns of regional (gamma) diversity in the shallow sea

The critical evidence for the existence of a latitudinal diversity cline in the shallow seas came from the early studies of Thorson (1952, 1957), and the later studies of molluscan and foraminiferal distribution by Fischer (1960), and Stehli *et al.* (1967, 1972). The pattern shown so clearly by molluscs (Fig. 6.2) may, however, not be a general one, for investigations of other taxa have failed to find such clear patterns. For example the number of species in bryozoan assemblages shows no clear latitudinal trend in the North Atlantic (Fig. 6.7). Unfortunately the molluscan data (Fig. 6.2) and bryozoan data (Fig. 6.7) are not strictly comparable, exemplifying once again the problems that beset global-scale studies of diversity. The molluscan data pool all known species within a given range of latitudes, regardless of habitat or location, and thus approximate to values of gamma diversity. The bryozoan data are plotted as individual species numbers for different assemblages. These data show very clearly the heterogeneity characteristic of natural assemblages, but are essentially point measures of local area (alpha) diversity.

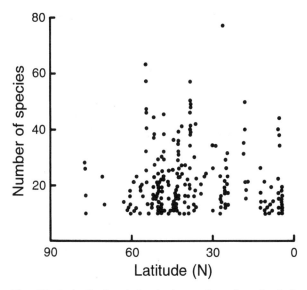

Fig. 6.7. Latitudinal variation in the number of species in bryozoan assemblages in the North Atlantic. Each data point represents a separate assemblage. Redrawn from Lidgard (1990).

140 *Diversity, latitude and time: Patterns in shallow seas*

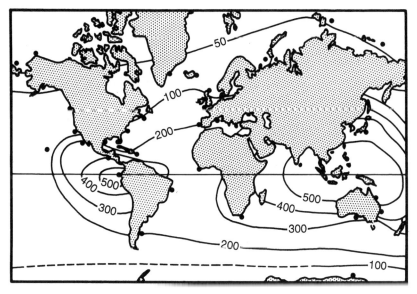

Fig. 6.8. Species-richness of bivalve molluscs, with contours. Redrawn from Stehli et al. (1967).

Without knowledge of the degree of overlap in species between the different bryozoan assemblages at any one latitude (which could be regarded as a form of beta diversity), we cannot compute regional (gamma) bryozoan diversity figures for different latitude classes to compare with the molluscan data.

The evidence for a general latitudinal diversity cline in the sea is greatly influenced by three key features: the selective nature of the data (with data lacking for many important groups), the very different biogeographic patterns in the northern and southern hemispheres (especially the vast differences in the species richness of the two polar regions), and the distribution of the overall pattern by the intense species-richness of the Indo-West Pacific. The last point shows up clearly in contoured plots of molluscan diversity (Fig. 6.8) but could also be shown for other shallow-water calcareous taxa. The global pattern for molluscs is not so much a latitudinal diversity cline, as a centre of species richness in the Indo-West Pacific (with a smaller centre in the eastern Pacific). An equatorial longitudinal diversity cline centred about 120°W would show a similar pattern to the latitudinal cline shown in Fig. 6.2.

Nevertheless, even allowing for the effect of the highly diverse areas in the Pacific there remains a cline in bivalve diversity (in terms of species, genus and family richness) from the tropics towards both polar regions in the Atlantic (Flessa & Jablonski, 1995; Fig. 6.8). Although recent taxonomic work (Mühlenhardt-Siegel, 1989; Hain, 1990; Dell, 1990), has increased the number of bivalve species known in Antarctica from the 86 quoted by Stehli *et al.* (1967), this does not alter the qualitative picture.

Unfortunately data are too few and scarce for other taxa to undertake similar analyses. All that we can say is that several taxa are known to be diverse in the Southern Ocean (for example polychaetes, bryozoans, ascidians, sponges, amphipods and isopods) and the pattern of a cline in diversity with latitude may not be a general one. Clarke (1992) speculated that the increasing thermodynamic cost of producing a carbonate skeleton as seawater temperature decreases may mean that the cline is limited to those taxa requiring skeletal carbonate. Rarely, however, do simple ecological patterns have simple explanations, and factors such as predation and reproductive ecology are undoubtedly also involved (Clarke, 1991, 1994).

6.6.2 Is there a latitudinal cline in alpha diversity?

Global comparisons of regional (gamma) diversity have provided only equivocal evidence for a universal latitudinal cline in shallow-water marine diversity. An alternative approach is to look for a cline in within-habitat diversity. Such studies are again hampered by the small size of the data base, but at least one study (that of Kendall & Aschan, 1993) has undertaken a tropical/temperate/polar comparison with rigorously comparable sampling and analytical techniques. This revealed no significant difference in the structure and diversity (as indicated by k-dominance plots of species abundance) for the macrobenthic infauna of soft sediments from sites in Svalbard (78°N), the North Sea (55°N) and Java (7°S).

Early studies by Thorson (1952, 1957) had drawn attention to the difference in latitudinal clines between epifauna and infauna. The only comparison of infaunal diversity involving an Antarctic site is that at Arthur Harbour, Anvers Island (Richardson, 1976; Richardson & Hedgpeth, 1977). The diversity at this site, expressed both as Shannon–Wiener and Margalef indices, was fully comparable with a range of other, lower latitude, soft-bottom communities. This comparison is beset with the traditional difficulties of comparability of sampling and analytical protocols (see discussion by Arntz *et al.*, 1997) and is also affected by the impact of seasonality of recruitment. Nevertheless we are left with the result that no study of shallow-water soft-bottom infauna communities has yet provided convincing evidence of a latitudinal cline in alpha diversity.

6.6.3 Latitudinal diversity clines in the deep sea

Studies of alpha diversity from the continental slope or the deep sea have, however, tended to suggest a latitudinal cline in diversity (e.g. see Sanders, 1968; Grassle & Maciolek, 1992; Rex *et al.*, 1993). These are described elsewhere in this book, and so will not be discussed further here. It is at present an intriguing but unresolved question whether this contrast reflects differences in sampling and analysis (Abele & Walters, 1979a,b), or a real difference between shallow waters and the deep sea.

Acknowledgements

We thank Jenny Twelves for much hard bibliographic work on the Southern Ocean marine fauna, and also Professor Wolf Arntz for providing us with a preprint of his extensive manuscript on Antarctic marine diversity.

References

Abele, L.G. & Walters, K., 1979a. The stability-time hypothesis: A re-evaluation of the data. *American Naturalist*, **114**, 559–68.
Abele, L.G. & Walters, K., 1979b. Marine benthic diversity: A critique and alternative explanation. *Journal of Biogeography*, **6**, 115–26.
Angel, M.V., 1994. Spatial distribution of marine organisms: patterns and processes. In *Large-scale Ecology and Conservation Biology*, ed. P.J. Edwards, R.M. May & N.R. Webb, pp. 59–109. Oxford: Blackwell Scientific Publications.
Arnaud, P.M. & Bandel, K., 1976. Comments on six species of marine Antarctic Littorinacea (Mollusca, Gastropoda). *Tethys*, **8**, 213–30.
Arnaud, P.M., Jazdzewski, K., Presler, P. & Sicinski, J., 1986. Preliminary survey of benthic invertebrates collected by Polish Antarctic expeditions in Admiralty Bay (King George Island, South Shetland, Antarctica). *Polish Polar Research*, **7**, 7–24.
Arntz, W.E., Brey, T. & Gallardo, V.A., 1994. Antarctic zoobenthos. *Oceanography and Marine Biology: An Annual Review*, **32**, 241–304.
Arntz, W.E., Gutt, J. & Klages, M., 1997. Antarctic marine biodiversity. In *Antarctic Communities: Species, Structure and Survival*, ed. B. Battaglia, J. Valencia & D.W.H. Walton, pp. 3–14. Cambridge: Cambridge University Press. In press.

Bargelloni, L., Ritchie, P.A., Patarnello, T., Battaglia, B., Lambert, D.M. & Meyer, A., 1994. Molecular evolution at subzero temperatures: Mitochondrial and nuclear phylogenies of fishes from Antarctica (suborder Notothenioidei), and the evolution of antifreeze glycopeptides. *Molecular Biology and Evolution*, **11**, 854–63.

Barker, P.F. & Burrell, J, 1977. The opening of Drake Passage. *Marine Ecology*, **25**, 15–34.

Barthel, D. & Gutt, J., 1992. Sponge associations in the eastern Weddell Sea. *Antarctic Science*, **4**, 137–50.

Brandt, A., 1991. Zur Besiedlungsgeschichte des antarktischen Schelfes am Beispiel der Isopoda (Crustaces, Malacostraca). *Berichte fur Polarforschung*, **98**, 1–120.

de Broyer, C. & Jazdzewski, K., 1993. Contribution to the marine biodiversity inventory: A checklist of the Amphipoda (Crustacea) of the Southern Ocean. *Studiedocumenten van het Koninklijk Belgisch Instituut voor Natuurwetenschappen*, **73**, 1–154.

Clarke, A., 1991. What is cold adaptation and how should we measure it? *American Zoologist*, **31**, 81–92.

Clarke, A., 1992. Is there a latitudinal diversity cline in the sea? *Trends in Ecology and Evolution*, **7**, 286–7.

Clarke, A., 1994. Temperature and extinction in the sea: A physiologist's view. *Paleobiology*, **19**, 499–518.

Clarke, A., 1996. The influence of climate change on the distribution and evolution of organisms. In *Animals and Temperature*, ed. I.A. Johnston & A.F. Bennett, Cambridge: Cambridge University Press.

Clarke, A. & Crame, J.A., 1989. The origin of the Southern Ocean marine fauna. In *Origins and Evolution of the Antarctic Biota*, ed. J.A. Crame, pp. 253–68. Geological Society Special Publication no. 47. London: The Geological Society.

Clarke, A. & Crame, J.A., 1992. The Southern Ocean benthic fauna and climate change: An historical perspective. *Philosophical Transactions of the Royal Society, Series B*, **338**, 299–309.

Clayton, M.N., 1994. Evolution of the Antarctic marine benthic algal flora. *Journal of Phycology*, **30**, 897–904.

Colwell, R.K. & Hurtt, G.C., 1994. Nonbiological gradients in species richness and a spurious Rapoport effect. *American Naturalist*, **144**, 570–95.

Connor, E.F. & McCoy, E.D., 1979. The statistics and biology of the species–area relationship. *American Naturalist*, **113**, 791–833.

Cornell, H.V. & Lawton, J.H., 1992. Species interactions, local and regional processes, and limits to the richness of ecological communities: A theoretical perspective. *Journal of Animal Ecology*, **61**, 1–12.

Crame, J.A., 1992. Evolutionary history of the polar regions. *Historical Biology*, **6**, 37–60.

Crame, J.A., 1993. Latitudinal range fluctuations in the marine realm through geological time. *Trends in Ecology and Evolution*, **8**, 162–6.

Curtis, M.A., 1975. The marine benthos of Arctic and subarctic continental shelves. *Polar Record*, **17**, 595–626.

Dayton, K., 1990. Polar Benthos. In *Polar Oceanography, Part B: Chemistry, Biology and Geology*, ed. W.O. Smith, pp. 631–85. San Diego: Academic Press.

Dell, R.K., 1990. Antarctic Mollusca: With special reference to the fauna of the Ross Sea. *Bulletin of the Royal Society of New England*, **27**, 1–310.

Dunton, K., 1992. Arctic biogeography: The paradox of the marine benthic fauna and flora. *Trends in Ecology and Evolution*, **7**, 183–9.

Eastman, J.T., 1993. *Antarctic Fish Biology: Evolution in a Unique Environment*. San Diego: Academic Press.

Eastman, J.T. & Grande, L., 1989. Evolution of the Antarctic fish fauna with emphasis on the recent notothenioids. In *Origins and Evolution of the Antarctic Biota*, ed. J.A. Crame, pp. 241–52. Geological Society Special Publication no. 47. London: The Geological Society.

Fischer, A.G., 1960. Latitudinal variations in organic diversity. *Evolution*, **14**, 64–81.

Flessa, K.W. & Jablonski, D., 1995. Biogeography of Recent marine bivalve molluscs and its implications for paleobiogeography and the geography of extinction: A progress report. *Historical Biology*, **10**, 25–47.

Grassle, J.F. & Maciolek, N.J., 1992. Deep-sea species richness: Regional and local diversity estimates from quantitative bottom samples. *American Naturalist*, **139**, 313–41.

Hain, S., 1990. Die beschatten benthischen Mollesken (Gastropoda und Bivalvia) des Weddellmeeres, Antarktis. *Berichte fur Polarforschung*, **70**, 1–181.

Harrison, S., Ross, S.J. & Lawton, J.H., 1992. Beta diversity on taxonomic gradients in Britain. *Journal of Animal Ecology*, **61**, 151–8.

Hedgpeth, J.W., 1969. Marine biogeography of the Antarctic regions. In *Antarctic Ecology*, vol. **1**, ed. M.W. Holdgate, pp. 97–104. London: Academic Press.

Hedgpeth, J.W., 1971. Perspectives of benthic ecology in Antarctic. In *Research in the Antarctic*, ed. Quam Lo, pp. 93–136. Washington, DC: American Association for the Advancement of Science.

Huston, M., 1994. *Biological Diversity: Coexistence of Species*. Cambridge: Cambridge University Press.

Kendall, M.A. & Aschan, M., 1993. Latitudinal gradients in the structure of macrobenthic communities: A comparison of Arctic, temperate and tropical sites. *Journal of Experimental Marine Biology and Ecology*, **172**, 157–69.

Knox, G.A. & Lowry, J.K., 1977. A comparison between the benthos of the Southern Ocean and the North Polar Ocean with special reference to the Amphipoda and the Polychaeta. In *Polar Oceans*, ed. M.J. Dunbar, pp. 423–62. Calgary: Arctic Institute of North America.

Lewin, R., 1989. Biologists disagree over bold signature of nature. *Science*, **244**, 527–8.

Lidgard, S., 1990. Growth in encrusting cheilostone bryozoans. II. Circum-Atlantic distribution patterns. *Paleobiology*, **16**, 304–21.

Lipps, J.H. & Hickman, C.S., 1982. Origin, age and evolution of Antarctic and deep-sea faunas. In *Environment of the Deep Sea*, ed. W.G. Ernst & J.G. Morris, pp. 324–56. Englewood Cliffs, NJ: Prentice-Hall.

Magurran, A.E., 1988. *Ecological Diversity and its Measurement*. Princeton, NJ: Princeton University Press.

May, R.M., 1992. Bottoms up for the oceans. *Nature*, **357**, 278–9.

May, R.M., 1994. Biological diversity: Differences between land and sea. *Philosophical Transactions of the Royal Society, Series B*, **343**, 105–11.

McGowan, J.A., 1974. The nature of oceanic ecosystems. In *The Biology of the Oceanic Pacific*, ed. C. Miller, pp. 9–28. Corvallis: Oregon State University Press.

Minelli, A., 1993. *Biological Systematics. The State of the Art*. London: Chapman & Hall.

Moe, R.L. & DeLaca, T.E., 1976. Occurrence of macroscopic algae along the Antarctic Peninsula. *Antarctic Journal of the United States*, **11**, 20–4.

Mühlenhardt-Siegel, U., 1989. Antarktische Bivalvia der Reisen des FS 'Polarstern' und des FFS 'Walter Herwig' aus den Jahren 1984 bis 1986. *Mitteilungen aus den Hamburgischen Zoologischen Museum und Institut*, **86**, 153–78.
O'Brien, E.M., 1993. Climatic gradients in woody plant species richness: Towards an explanation based on an analysis of southern Africa's woody flora. *Journal of Biogeography*, **20**, 181–98.
Pagel, M.D., May, R.M. & Collie, A.R., 1991. Ecological aspects of the geographical distribution and diversity of mammalian species. *American Naturalist*, **137**, 791–815.
Palmer, M.W. & White, P.S., 1994. Scale dependence and the species–area relationship. *American Naturalist*, **144**, 717–40.
Pianka, E.R., 1966. Latitudinal gradients in species diversity: A review of concepts. *American Naturalist*, **100**, 65–75.
Picken, G.B., 1980. The Nearshore Prosobranch Gastropod Epifauna of Signy Island, South Orkney Islands. PhD thesis. University of Aberdeen.
Pimm, S.L., Russell, G.J., Gittleman, J.L. & Brooks, T.M., 1995. The future of biodiversity. *Science*, **269**, 347–50.
Poore, G.C.B. & Wilson, G.D.F., 1993. Marine species richness. *Nature*, **361**, 597–8.
Preston, F.W., 1962. The canonical distribution of commonness and rarity. *Ecology*, **43**, 185–212.
Reid, W.V. & Miller, K.R., 1989. *Keep Options Alive: The Scientific Basis for Conserving Biodiversity*. Washington, DC: World Resources Institute.
Rex, M.A., Stuart, C.T., Hessler, R.R., Allen, J.A., Sanders, H.A. & Wilson, G.D.F., 1993. Global-scale latitudinal patterns of species diversity in the deep-sea benthos. *Nature*, **365**, 636–9.
Richardson, M.D., 1976. The Classification and Structure of Macrobenthic Assemblages of Arthur Harbour, Anvers Island, Antarctica. PhD thesis. Oregon State University.
Richardson, M.D. & Hedgpeth, J.W., 1977. Antarctic soft-bottom, macrobenthic community adaptations to a cold, stable, highly productive, glacially affected environment. In *Adaptations within Antarctic Ecosystems*, ed. G.A. Llano, pp. 181–95. Washington, DC: The Smithsonian Institution.
Rosenzweig, M.L., 1975. On continental steady states of species diversity. In *Ecology and Evolution of Communities*, ed. M.L. Cody & J.M. Diamond, pp. 121–40. Cambridge, MA: The Belknap Press.
Rosenzweig, M.L., 1977. Geographical speciation: On range size and the probability of isolate formation. In *Proceedings of the Washington State University Conference on Biomathematics and Biostatistics*, ed. D. Wollkind, pp. 172–94. Washington, DC: Pullman.
Rosenzweig, M.L. & Abramsky, Z., 1993. How are diversity and productivity related? In *Species Diversity in Ecological Communities*, ed. R.E. Ricklefs & D. Schluter, pp. 52–65. Chicago: University of Chicago Press.
Roy, K., Jablonski, D. & Valentine, J.W., 1994. Eastern Pacific molluscan provinces and latitudinal diversity gradient: No evidence for 'Rapoport's Rule'. *Proceedings of the National Academy of Sciences of the USA*, **91**, 8871–4.
Sanders, H.L., 1968. Marine benthic diversity: A comparative study. *American Naturalist*, **102**, 243–82.
Siciński, J., 1993. Polychaeta. In *The Maritime Antarctic Coastal Ecosystem of Admiralty Bay*, ed. S. Rakusa-Suszczewski, pp. 101–7. Warsaw: Polish Academy of Sciences.

Stehli, F.G., McAlester, A.L. & Helsley, C.E., 1967. Taxonomic diversity of recent bivalves and some implications for geology. *Geological Society of America Bulletin*, **78**, 455–66.

Stehli, F.G., Douglas, R. & Kafescegliou, I., 1972. Models for the evolution of planktonic foraminifera. In *Models for the Evolution of Planktonic Foraminifera*, ed. T.J.M. Schopf, pp. 116–28. San Francisco: Freeman.

Stevens, G.C., 1989. The latitudinal gradient in geographical range: How so many species coexist in the tropics. *American Naturalist*, **133**, 240–56.

Stork, N.E., 1988. Insect diversity: Facts, fiction and speculation. *Biological Journal of the Linnean Society*, **35**, 321–37.

Taylor, J.D. & Taylor, C.N., 1977. Latitudinal distribution of predatory gastropods on the eastern Atlantic shelf. *Journal of Biogeography*, **4**, 73–81.

Terbogh, J., 1973. On the notion of favourableness in plant ecology. *American Naturalist*, **107**, 481–501.

Thorson, G., 1952. Zur jetzeigen Lage der marinen Bodentier-Ökologie. *Zoologischer Anzeiger (supplement)*, **16**, 276–327.

Thorson, G., 1957. Bottom communities (sublittoral or shallow shelf). In *Treatise on Marine Ecology and Paleoecology*, ed. J.W. Hedgpeth, pp. 461–534. New York: Geological Society of America.

Thurston, M.H., 1972. *The Crustacea Amphipoda of Signy Island, South Orkney Islands*. British Antarctic Survey Scientific Reports no. 71. London: BAS.

Valentine, J.W., 1984. Neogene marine climate trends: Implications for biogeography and evolution of the shallow-sea biota. *Geology*, **12**, 647–59.

Vermeij, G.J., 1991. Anatomy of an invasion: The trans-Arctic interchange. *Paleobiology*, **17**, 281–307.

Voß, J., 1988. Zoogeographie und Gemeinschaftsanalyse des Makrozoobenthos des Weddellmeeres (Antarktis). *Berichte fur Polarforschung*, **45**, 1–145.

Webb, P.-N., 1990. The Cenozoic history of Antarctica and its global impact. *Antarctic Science*, **2**, 3–21.

White, M.G., 1984. Antarctic benthos. In *Antarctic Ecology*, vol. II, ed. R.M. Laws, pp. 421–61. London: Academic Press.

Whittaker, R.H., 1960. Vegetation of the Siskiyou Mountains, Oregon and California. *Ecological Monographs*, **30**, 279–338.

Whittaker, R.H., 1972. Evolution and measurement of species diversity. *Taxon*, **21**, 213–51.

Winston, J.E., 1992. Systematics and marine conservation. In *Systematics, Ecology and the Biodiversity Crisis*, ed. N. Eldredge, pp. 144–68. New York: Columbia University Press.

Woodward, F.I., 1987. *Climate and Plant Distribution*. Cambridge: Cambridge University Press.

Note added in proof:

Since this text was written in 1993, several major reviews of diversity have appeared. These include extensive discussions of large-scale patterns of marine diversity. Rosenzweig (1995) has a predominantly terrestrial emphasis, but includes discussion of marine diversity, as does Huston (1994: see main reference list), who favours a different explanation for the underlying patterns, and Gaston (1996). The reanalysis by Roy et al. (1994) of molluscan distribution along the Pacific coast of North America, which provided no support for Rapoport's Rule, has been joined by other marine examples where the Rapoport effect was not detected (Rohde et al.,

1993; Rohde, 1996). Finally, Palumbi (1996) has brought molecular evidence to bear on the causes of the particularly high marine diversity of the Indo-West Pacific, emphasising the importance of recent changes in sea level to speciation in this area.

Gaston, K.J. (ed.), 1996. *Biodiversity. A Biology of Numbers and Differences.* Oxford: Blackwell Scientific Publications.
Palumbi, S.R., 1996. What can molecular genetics contribute to marine biogeography? An urchin's tale. *Journal of Experimental Marine Biology and Ecology,* **203**, 75–92.
Rohde, K., 1996. Rapoport's Rule is a local phenomenon and cannot explain latitudinal gradients in species diversity. *Biodiversity Letters,* **3**, 10–13.
Rohde, K., Heap, M. & Heap, D., 1993. Rapoport's Rule does not apply to marine teleosts and cannot explain latitudinal gradients in species richness. *American Naturalist,* **142**, 1–16.
Rosenzweig, M.L., 1995. *Species Diversity in Space and Time.* Cambridge: Cambridge University Press.
Roy, K., Jablonski, D. & Valentine, J.W., 1994. Eastern Pacific molluscan assemblages and latitudinal diversity gradient: No evidence for 'Rapoport's Rule'. *Proceedings of the National Academy of Sciences of the USA,* **91**, 8871–4.

Chapter 7
High benthic species diversity in deep-sea sediments: The importance of hydrodynamics

JOHN D. GAGE
Scottish Association for Marine Science (formerly Scottish Marine Biological Association), Dunstaffnage Marine Laboratory, P.O. Box 3, Oban, Argyll, PA34 4AD, UK

Abstract

The surprisingly high species diversity found among the mostly small-bodied invertebrates inhabiting the deep-sea sediment is thought to be maintained by biologically generated habitat heterogeneity and patchy food resources acting in concert with biological disturbance at the metre to centimetre scale or less (Grassle & Maciolek, 1992). Disturbance is thought to thin existing populations, and is followed by recolonisation of the disturbed patch, to that the populations never develop to a full competitive equilibrium before the next disturbance event.

This review focuses on the possible importance of larger, >kilometre scale disturbance, generated by currents at the deep-sea floor. Data on species richness and abundance from quantitative benthic samples from five areas presenting varying space–time scales in hydrodynamic disturbance are reviewed, in order to show that such diffuse disturbance of the sediment community resulting from episodes of high bed flow in the supposedly tranquil physical conditions at the deep-sea bed may be an important structuring agent for the deep-sea benthic community. The first area is located at the High Energy Benthic Boundary Layer (HEBBLE) study site on the Nova Scotian continental rise in the North West Atlantic, where the sediment floor periodically experiences strong flow and massive re-suspension during vorticity-driven 'benthic storms' and hence where eddy kinetic energy, K_E, is very high in the benthic mixed layer. The second site is in the southern Rockall Trough (North-East Atlantic) depth where K_E is also high and where the sediment may experience seasonally varying low-frequency episodes of strong, sediment-suspending bed flow. The third and fourth areas contrast a mid-basin site on the Tagus Abyssal Plain (where K_E is minimal) with a station on the adjacent Portuguese continental margin in the lower end of the Setubal Canyon. Here, high frequencies and levels of tidal/inertial flow are inferred from ripple bedforms visible in sea-bed photographs. The fifth area is located beneath the subtropical gyre in the central north Pacific where abyssal eddy kinetic energy is also extremely weak.

7.1 Introduction

Despite more than 150 years of intensive work, marine biologists have described only about 160 000 species in the oceans (Thorson, 1971; Barnes, 1983) compared to the total of around 1.8 million recorded on earth, with possibly more than 1 million described insects alone on land (Stork, 1988). What does this mean about the processes of marine biodiversity?

Arguments based on the supposed greater environmental heterogeneity and structural complexity on land, on the greater age and phyletic diversity of marine environments, and on differences in patterns in herbivory and species–size relationships, have been put forward (Fenchel, 1993; May, 1994). What is not in doubt is that marine biology has been very much a fringe science, not in the sense that it occupies a marginal territory of science but that its land-based viewpoint has not extended very far beyond the landward margins of its environment. Even there, the concerns of amateur natural historians and conservation agencies alike have been concentrated on either swimming or crawling animals, or biota attached to hard rocky surfaces that are large enough to see. This certainly neglects the equally interesting suite of generally small-sized organism forms inhabiting the sediments that make up most of all species described from the coastal seas, the latter total making up probably less than 15% or so of all recorded species on Earth. Around half or more of these animals are burrowing, worm-like species belonging to a variety of phyla, but mostly to the Annelida, class Polychaeta (bristle worms) and the Nematoda (thread worms). Other phyla that are well represented in the sediment include various small, peracarid crustaceans, such as isopods, amphipods, tanaids and copepods, and molluscs, especially small bivalves and gastropods.

Yet coastal sediments represent only a small proportion of the total marine sediment habitat. The remainder lies in the deep ocean. Owing to the immense practical difficulties of studying this huge ecosystem, the fauna inhabiting the drape of sediment over the deep-sea floor is still very poorly known. Although generally smaller in individual size, the deep-sea fauna is made up of similar sorts of animals to those found in coastal sediments. Many deep-sea species belong to the same families or even genera as those in shallow water. Because these organisms exist at very low densities, the vast tracts of deep-sea sediment resemble a desert in terms of density of biomass. Yet, surprisingly, species diversity of the deep-sea macrobenthos (that includes the polychaete worms) is now thought, from detailed analysis of samples covering about 500 m^2 of deep-sea bed, to rival that of the tropical rain forest and coral reefs (Grassle et al., 1990; Grassle & Maciolek, 1992). Knowledge of the even smaller-bodied animals of the deep-sea meiobenthos (largely made up of nematode worms) is even less well-developed, being based on samples covering less than 1 m^2, but it also suggests very high species diversity (Lambshead et al., 1995). In making comparisons it is important to realise that studies in the rain forest, where most of the estimated total insect species diversity (ranging from about 1 million upwards to 200 million!) can be found, have been going on for a longer period of time, and are easier than studying the deep-sea floor.

Marine biodiversity studies need to address three related topics. The first is essentially descriptive, involving taxonomic and biogeographical studies of what species are there and where they can be found. The second addresses the geological-scale, evolutionary processes and patterns determining how these species have evolved and are distributed over the oceans (see Chapter 5); while the third is concerned with the ecological processes thought to be implicated in allowing this biodiversity to co-exist in time. I shall in this chapter address the last, and certainly the most contentious, of these three aspects. However, the size of the regional or global pool of species, and determining the large-scale patterns described by Rex et al. (Chapter 5), cannot be completely dissociated from how, at the local scale, this incredibly rich biodiversity is maintained in what appears to be a highly resource-limited environment. Yet we do not possess any certainty in explanations for what seems an astonishingly high and unexpected co-occurrence of species in seemingly the most uniform of habitats. Perhaps this is because the habitat, lacking the three-dimensional complexity of the rain forest, seems the last place to find such an exuberant expression of biodiversity.

This species richness is not very obvious. The low densities of the animals living in benthic sediments mean that large quantities of mud need to be washed through a fine-meshed sieve in order to catch more than very small numbers of the most abundant of these species. We should remember, too, how little of this habitat our present knowledge is based on. The approximately 500 m^2 of sea-bed that has so far been sampled by quantitative cores comes from a total of 270 million km² of the deep ocean floor. Even if the area of the sampling paths of deep-sea trawl and sled samples are added, this figure remains a vanishingly small proportion of the total environment. In a coastal area one might aim to sample about one millionth of the bottom in order to characterise the benthic community. To equal this effort in the deep ocean one would have to sample 500 m² every day for at least a millennium!

7.2 The role of disturbance and habitat heterogeneity in benthic community structure

Howard Sanders during the 1960s at the Woods Hole Oceanographic Institution in the United States of America was largely responsible for the discovery of high species diversity amongst the small-bodied fauna inhabiting deep-sea sediments. He concluded that, because environmental conditions in the deep sea appeared to be so constant, niche fragmentation has occurred on a lavish scale with a large number of highly specialised species being able to co-exist in these physically undemanding, albeit resource-poor, conditions (Sanders, 1968). Although the deep-sea bed may appear superficially homogeneous, the low sedimentation rates mean biologically generated habitat heterogeneity, such as bioturbatory mounds and burrows, is more persistent and therefore may contribute more to niche diversification at the deep-sea floor than in shallow, more energetic, sediment environments (Jumars, 1975, 1976; Jumars & Gallagher, 1982; Jumars & Eckman, 1983). Diversity in the granulometric structure of the sediment itself may also play a part. Etter & Grassle (1992) showed that macrofaunal species diversity is positively associated with sediment particle diversity in the North-East Atlantic.

Until comparatively recently all explanations of biodiversity in ecological time have made the critical assumption that nature as we see it represents a dynamic equilibrium in competitive effects that follow a predictable, successional sequence after colonisation towards a stable 'climax' community. The assumption that species richness is regulated entirely by competition derives from the dogma that as a result of competitive exclusion there cannot be any equilibrium state between two species sharing the same resource. However, the implicit assumption of steady state has been questioned in other explanations for species-rich habitats, including the deep sea.

Dayton & Hessler (1972) suggested that the high diversities of small-sized species in the sediment might be maintained by the activity of large benthopelagic and megabenthic 'croppers', such as elasipod holothurians. This predatory disturbance acts by thinning their prey populations so that prey species are less able to displace other small species through competition for resources.

The idea of the importance of disturbance frequency was first developed by Connell (1978) in his 'intermediate disturbance hypothesis'. This was based on observations that the frequencies of natural disturbance in both terrestrial and aquatic habitats may be faster than the rates of recovery from perturbations. At one extreme, disturbances are so frequent or severe that populations never develop any equilibrium, while at the opposite extreme disturbances are either so weak or infrequent that competitive exclusion occurs. Highest species richness thus occurs at intermediate levels or frequencies of disturbance. In contrast to Dayton & Hessler's (1972) view of disturbance, Grassle & Sanders (1973) considered disturbances in the deep sea to be rare and small-scale and to be caused by a variety of phenomena, including patchiness in organic resources such as sunken wood or carcasses of large animals, with highly localised species successions.

Experiments and observations on the deep-sea bed have supported the idea of small-scale patchiness and processes in maintaining species richness. The patchiness includes small-scale, transient biogenic structures, such as burrows, empty burrows and sediment mounds (Jumars, 1976; Aller & Aller, 1986; Smith *et al.*, 1986) and patch food supply (Grassle, 1977; Desbruyères *et al.*, 1980, 1985; Smith, 1985, 1986; Kukert & Smith, 1992; Rice & Lambshead, 1994). Food supply patches may arise, for example, from local concentrations of phytodetrital flocs and the organic remains of zooplankton and floating macrophytes from the surface (e.g. Levin & Smith, 1984; Grassle & Morse-Porteous, 1987; Rice & Lambshead, 1994). For the marine benthic community in general, these small-scale processes are perceived as creating a random spatiotemporal mosaic of patches at differing stages of recovery from previous localised disturbance events (Johnson, 1970, 1973; Osman & Whitlach, 1978; Thistle, 1981; Botton, 1984; Levin, 1984). Observed species diversity will then be an expression of a dynamic balance between the frequency of disturbance providing opportunities for re-colonisation and the rate of competitive exclusion in 'mature' patches where the populations exist at carrying capacity. Populations in patches of different successional age will then be characterised by differing demographies resulting from the contrasting potential population growth rates of opportunists versus 'climax' species (Zajac & Whitlach, 1991). It still remains unclear whether in marine soft bottoms colonising fauna are responding to the absence of competitors or merely to the release of resources by the disturbances (Thistle, 1981).

In the deep sea, input of small patches of ephemeral food resources along with microhabitat heterogeneity and patch-creating disturbance from grazing and bioturbation by large megafauna are thought to be important (Smith & Hessler, 1987; Rice & Lambshead, 1994). Such patchiness is generally thought to be the chief agent in generating local variation in the development of populations (Grassle, 1989; Grassle & Maciolek, 1992). This effect acts in tandem with random, low-intensity recruitment in a system with few barriers to dispersal of propagules from a huge, widely distributed species pool. Clearly not only the frequency but also the scale and intensity of disturbance will be important to such patch dynamics in generating high species richness. The evidence cited above suggests that disturbance operates as part of a suite of essentially small-scale processes in maintaining high species richness. Indeed, some workers have suggested that it is only processes resulting in resource or habitat partitioning at scales of less than 0.01 m^2 that provide the environmental grain recognisable by individual animals (Jumars, 1976; Jumars & Eckman, 1983). The importance of larger scale heterogeneity as revealed by side-scan sonar and swath bathymetry and caused by structures such as turbidites, levees and channels in the sea-bed, remains unknown. Equally, periodic effects such as hydrodynamic disturbance affecting spatial scales from tens of metres to tens of kilometres, have been considered unimportant compared to the fine-scale patch dynamics operating at the metre scale or less. However, theoretically at least such broad-scale sources of habitat heterogeneity and diffuse disturbance may also be important in regulating species diversity (Petraitis *et al.*, 1989). The following account addresses hydrodynamic conditions at the deep-sea bed and attempts to assess its rôle in structuring the sediment-dwelling community.

7.3 Hydrodynamic regime in the deep sea

Until comparatively recently the deep-sea bed has been thought to differ from most terrestrial and shallow aquatic habitats in its constant, almost chemostat-like environment. Recent discoveries have challenged this view. Although at abyssal depths temperature shows annual variations of no more than tenths of a degree Celsius, with constancy in lack of light and in chemical conditions, the full space–time range and variability in hydrodynamic benthic energy is only beginning to be understood as long time-series observations of near-bed currents are made using deep-sea current-meter moorings. This review therefore attempts to help towards a better appreciation of such conditions in relation to the regulation of deep-sea benthic species richness.

At the shortest time-scale of variability, semi-diurnal internal tides have been measured far out into the ocean basins and usually as a sharply defined spike in the energy spectra of current-meter records. At least in the North-East Atlantic tidal velocities may range up to 10 cm s^{-1}, with rapid changes in flow direction and velocity (Heezen & Hollister, 1971), and will act to enhance or diminish other currents. On the continental slope, intense currents may result from internal tides determined by the slope gradient and water column stability (Pingree & New, 1989). Internal tides may be focused as vigorous axial bed flows in submarine canyons and valleys, and may attain speeds sufficient to transport sand-sized particles (Shepard et al., 1979). Leaving aside the relatively few and isolated occurrences of massive sediment slides that have apparently occurred within the past few hundred years, canyons are thought largely to reflect a relict landscape compared to the regime of active down-slope erosion during the lowered sea levels of glacial times (Robb et al., 1981; Prior et al., 1984). However, turbidity flows of up to 3 km hr^{-1} are thought to occur with a frequency well within the generation times of larger deep-sea sediment-dwelling metazoans in canyon settings where rivers introduce large quantities of sediment near the canyon head (Shepard et al., 1979).

Other sources of sediment disturbance at frequencies of months to years lie in storm-driven motions, down-slope 'cascading' of cooled dense water over the shelf edge, and internal waves that may suspend sediment causing benthic nepheloid layers (BNLs) and spreading turbid plumes across isopycnal surfaces (e.g. Dickson & McCave, 1986; Gardner, 1989). Periodic sediment erosion at a more rapid frequency on the continental slope may be caused by contour-following geostrophic currents and internal tides (Huthnance, 1981). Sloping and irregular topography may be directly responsible for flow separation and eddy shedding from strong surface flow or slope currents, which may generate vorticity-driven topographic waves and intense near-bottom currents (Booth & Ellett, 1983; Csanady et al., 1988; Pingree & LeCann, 1992). Overall, such variable and generally higher frequency motions result in a bottom mixed layer, or 'benthic boundary layer' (BBL), much thinner (approximately 5 m) and more dynamic than in the abyss (Dickson,

1983; Thorpe et al., 1990). For example, long-term current-meter time-series show higher maximum near-bed flow vclocitics at about 2000 m depth on the Goban Spur and Meriadzek Terrace than on the adjacent Porcupine Abyssal Plain (Vangriesheim, 1985).

Periods of very high bottom-current activity, however, may be most dramatic on the continental rise under western boundary currents. Here, the strongest of these persistent, low-frequency events, termed 'benthic storms', are characterised by periods of daily averaged flow >15 cm s^{-1} maintained for two or more days (Kerr, 1980). One area that was intensively investigated during the early 1980s during the High Energy Benthic Boundary Experiment (HEBBLE) lies at 4820 m depth on the Nova Scotia Rise in the North-West Atlantic. Here, the top few millimetres of sediment may be completely stripped off as a layer and suspended when the normally equator-ward flow is reversed with intermittently strong flow peaking at velocities greater than 40 cm s^{-1} (Hollister & McCave, 1984; Hollister et al., 1984; Weatherly & Kelley, 1985; Gross et al., 1988). Such periods may last from a few days to several weeks. Large quantities of eroded sediment may be transported as nepheloids and then re-deposited during quiescent periods (Eittreim et al., 1976; Lonsdale & Hollister, 1979). Sediment drifts that characterise the west and east side of the Rockall Bank in the North-East Atlantic (Johnson & Schneider, 1969) are probably formed in this way.

These energetic conditions at the bottom are thought to be caused by propagated vorticity from a variety of forcing agents including atmospheric storm-driven motions (Gardner & Sullivan, 1981), the dynamics of the upper water column structure associated with areas of strong surface flow (Weatherly & Kelley, 1985), propagated disturbances associated with seasonal periodicity in such western boundary currents flowing intermittently over sills (e.g. Dickson et al., 1986) and bottom-trapped topographic waves (e.g. Grant et al., 1985). Such propagated eddies, with scales up to 50–200 km, may contain up to 100 times the energy of the background thermohaline flow and result in periodic reversals and changes in flow rate, being analogous to the cyclones and anticyclones giving 'weather' in the atmosphere. Although smaller in scale than the latter, they may be more persistent. Benthic storms may also occur at some distance from the continental slope/rise and can be more frequent and last longer than on the slope (Richardson et al., 1993).

Hydrodynamic benthic energy is thought to be much lower in the abyss than on such areas near the continental margin, with flow generally seldom exceeding 2–4 cm s^{-1} (Elliott & Thorpe, 1983). Forcing by wind stress in conjunction with stratification appears much more important in the eastern North Atlantic than in the west, where mesoscale activity is dominated by the Gulf Stream and its rings (Mercier & de Verdière, 1985). Hence, in the eastern basin, benthic storms generated by propagated eddy vorticity can probably be expected to occur even far out on the abyssal plain, and their frequency will thus tend to reflect the seasonal cycle in wind stress at temperate latitudes. For example, in the North-East Atlantic storm belt deep-reaching eddies are generated principally during the winter and early spring and decay to a minimum in late summer (Dickson *et al.*, 1982) and show a marked northwards gradient in intensity and frequency (Dickson, 1983; Dickson *et al.*, 1985). Peak bottom flow speeds in the range of 27–39 cm s^{-1} occur at the mouth of the Rockall Trough (Dickson & Kidd, 1987). Maximum bottom current velocities up to 27 cm s^{-1} have been recorded in the western Porcupine Abyssal Plain (Klein, 1987). Here, benthic flow seems to be enhanced by the undulating topography of abyssal hills where eddy energy may be up to an order of magnitude greater than on smoother abyssal plains (e.g. Klein & Mittelstaedt, 1992), probably as a consequence of topography trapping free propagating eddies (Saunders, 1988).

Elsewhere, benthic storms as periods of much enhanced flow above 20 cm s^{-1} for periods of several days, with peaks more than 40 cm s^{-1} at 10 metres above bottom (mab) and intense nepheloids, have been reported in the Argentine Basin (Richardson *et al.*, 1993), with less intense events under the East Australia current (Mulchern *et al.*, 1986) and in the north-eastern tropical Pacific (Kontar & Sokov, 1994). These, and many other long current-meter time-series taken at depth (summarised in Dickson, 1983), provide ground truth to predictions of the extensive deep-ocean areas subject to benthic storm activity. Such predictions are derived from maxima in eddy kinetic energy, K_E, at the surface (Wyrtki *et al.*, 1976; Hollister & McCave, 1984; Schmitz, 1984, 1988; Weatherly, 1984) and from the global occurrence of sediment bed forms indicating benthic storm activity (Hollister & Nowell, 1991). Hence, as pointed out by Smith (1994), sizeable areas of the muddy abyss may sustain physical disturbance at time-scales relevant to the generation times of benthic metazoans.

7.4 Interactions between hydrodynamics, deep-sea sediment stability and the benthic community

In shallow coastal and intertidal marine sediment, the benthic community may be moulded by bed flow in transporting food and larvae, and through sediment disturbance acting in concert with larval supply and biogenic effects such as bioturbation. Biogenic structure may affect bed roughness and through mucus binding, re-working and pelletisation in turn affect the stability and erodability of the substratum. Even moderate differences in current energy may be reflected in community composition and abundance (see reviews by Hall, 1994; Snelgrove & Butman, 1994 and references therein).

In the deep sea, BNLs derived from re-suspended sediment particles have been found to be of nearly ubiquitous occurrence in ocean basins (Hollister & McCave, 1984; McCave, 1986). Recent observations have shown how re-suspension and lateral advection of fine-grained sediment particles causing benthic turbid events and nepheloid layers are linked to energetic bottom flow driven by deep-reaching mesoscale eddies (Nyffeler & Godet, 1986; Isley et al., 1990). When eddy energy is transmitted to the benthic boundary layer, the resulting increasing bottom friction increases its height. Particles released at the bed are then more easily released as nepheloids into the deep ocean layers above through local detachments, pycnocline breakdown or breaking of internal waves at the upper BBL boundary (Richards, 1990). This results in nepheloids extending well above the zone of turbulent mixing (McCave, 1986).

Re-suspension of sediment depends on the critical erosion velocity, which is more a function of aggregate strength than of the classical bulk parameters such as yield strength or particle size (McCave, 1986). For example, mucus binding by metazoans, diatoms and bacteria may, as in shallow-water sediment, dramatically increase sediment stability (Eckman et al., 1981; Jumars & Nowell, 1984a). Overall, these effects may be less important than in shallow-water sediments because of the reduced faunal densities. Rapidly increasing bed roughness after benthic storms as a consequence of bioturbation and faecal pellet production is thought to reduce the threshold of sediment re-suspension (Nowell et al., 1981; Wheatcroft et al., 1989). Deposition, and subsequent re-suspension, of phytodetritus in the Porcupine Abyssal Plain (see Lochte & Turley, 1988; Riemann, 1989; Thiel et al., 1989) seems to generate particle-rich nepheloid layers extending hundreds of metres above bottom (Auffret et al., 1994). This is very likely because re-suspension of newly deposited phytodetrital material occurs at much lower bed flows (7 cm s^{-1} at 1 mab in the Porcupine Seabight; see Lampitt, 1985) than the erosion thresholds of a 'normal' foraminiferan–coccolithophore ooze (Southard et al., 1971; Miller et al., 1977). How does this unexpected hydrodynamic instability of the surface sediment impact the deep-sea benthic community?

Disturbance effects on the fauna are usefully considered in relation to either bedload movement, where particles are displaced within a few grain diameters of the sediment surface, and suspended load, where particles are transported with and at the same velocity as the flow supported by turbulence (Hall, 1994). In the deep sea the sediment-dwelling macrofauna would seem to be particularly vulnerable to disturbance from sediment erosion because of their generally smaller size and sometimes delicate structure (Gage & Tyler, 1991). The fragile structure of some taxa such as xenophyophores would appear to render them especially vulnerable and because of this may limit their distribution to relatively quiescent conditions. At the HEBBLE site even meiofaunal copepods are probably swept from the sediment during benthic storms (Thistle, 1988; Thistle et al., 1991), while the benthic community in general is conspicuously lacking in larger 'epifaunal' taxa living on the sediment surface (Thistle & Wilson, 1987). However, this may result more from sediment smothering than high bedload transport. Although relevant data are not yet available, possession of shell or heavy exoskeleton may make bedload transport more likely than re-suspension. However, the impact of such transports on individuals or populations remains largely speculative.

Hydrodynamics by controlling sediment properties may affect deep-sea community structure more indirectly in a number of ways, including sediment particle composition (Etter & Grassle, 1992). Most deep-sea benthic metazoans are deposit feeders that may partition the sediment with respect to size in order selectively to ingest the more labile components of the sediment (Jumars et al., 1990). Moderate flow conditions may be exploited by deposit feeders trapping material transported by currents (Jumars & Nowell, 1984b) and may, by transporting phytodetritus, enhance deposit feeding (Aller & Aller, 1986). On the other hand, the meiobenthic populations that rapidly colonise phytodetritus (see Gooday & Turley, 1990) may be subject to a much greater frequency of hydrodynamic disturbance than the sediment-dwelling fauna beneath. More energetic, sediment-eroding flow may, by altering the particle size and organic content, have a temporarily negative effect on deposit feeding during storms while enhancing deposit feeding overall due to microbial stirring and deposition of fresh organic detritus during transitional and depositional periods (Aller, 1989).

Another possible effect is on population dispersal and reproduction through transport of gametes, larvae and probably post-larval and later stages. Reproductive periodicities may benefit from adaptive coupling to episodes of high bed flow. Indeed, reproduction in benthic echinoderms in the Rockall Trough coincides with a seasonal maximum in K_E (Tyler et al., 1982). There are also indications from a bathymetric transect from the lower continental rise to the upper continental slope in the Rockall Trough that disturbance, in the form of high-energy bottom currents, is important in structuring the polychaete community (Paterson & Lambshead, 1995).

At the HEBBLE site bioturbation appears to be intense during the quieter periods between benthic storms (Yingst & Aller, 1982; DeMaster et al., 1985), quickly generating both micro-environmental heterogeneity (e.g. Aller & Aller, 1986) and probably local disturbance by the creation of mounds, pits, burrows, faecal casts and more complex structures (compare Smith et al., 1986). The disruption and smoothing effect of strong flow on bed roughness associated with this complex biogenic micro-landscape will generate broad-scale, diffuse disturbance to the sediment community. Furthermore, by destroying the patch-creating heterogeneity created by bioturbation, high flow may exclude a range of species normally partitioning spatial or other resources associated with it (e.g. Jumars, 1976; Thistle, 1979; Jumars & Gallagher, 1982). For example, the diffuse effect of large-scale cyclic sediment resuspension and deposition will remove patch structure associated with recently deposited food resources, such as phytodetritus, accumulating in relict burrows (Aller & Aller, 1986) and biogenic predation refuges, such as those associated elsewhere with cirratulid polychaete 'mudballs' (Thistle & Eckman, 1988).

Overall, losses of species through loss of microhabitat or even physical damage to individuals may be balanced by the stimulus to population growth through enhanced feeding and recruitment at high energy deep-sea settings (Thistle et al., 1985, 1991). However, we can at present only guess at the effect in terms of species excluded through unsuited morphology or life-style. However, as we shall see in the case studies below, high energy conditions seem to be associated with reduced species richness and evenness in species abundances, even if biomass is higher than would be expected for the depths concerned.

7.5 Four case studies relating hydrodynamic conditions at the deep-sea floor to benthic community structure

Studies on faunal composition in the deep-sea sediment are technically demanding and expensive to obtain, and until recently none of these had been undertaken with the necessary precision at sites subject to high energy bottom currents to allow any kind of quantitative understanding of their significance with respect to the benthic community. Data considered here are derived from quantitative samples taken using box corers (see Hessler & Jumars, 1974 and Gage & Tyler, 1991 for a description) obtained from five deep-sea areas subject to differing hydrodynamic regimes. These were selected to illuminate the significance of broad-scale hydrodynamic disturbance on deep-sea benthic community structure. Faunal abundance and statistics of species diversity are summarised in Table 7.1. The latter are calculated from species abundances as: (a) rarefaction curves of expected species diversity using the expression of Hurlbert (1971); (b) the Shannon-Wiener diversity statistic, H', where $H' = -\Sigma p_i \ln p_i$, where p_i is the proportion of individuals found in the ith species, and evenness, E, where $E = H'/H_{max}$, where $H_{max} = H'/\ln S$, i.e. when all species are equally abundant; (c) rank abundance plots in order to show the degree of dominance of the community by common species; (d) a simple index of the proportional importance, d, of the most common species as $d = n_{max}/n$, where n_{max} is the abundance of the most common species and n is the total abundances in the sample (Berger & Parker, 1970). Magurran (1988) provides further explanation and discussion of these statistics.

7.5.1 The HEBBLE area

Some details of the benthic environment at this site are given above. Bottom currents up to 73 cm s^{-1} have been measured at this site (Richardson *et al.*, 1981). Periods of high bed flow erode superficial sediment taking with it organic particles, micro-organisms, larvae and juveniles of macrofauna, and even small isopods and meiofauna such as harpacticoid copepods (Thistle, 1988). During transitional periods of intermediate current velocities, there is a vertical and horizontal influx of fresh organic particulates, with an increased microbial abundance and activity (Jumars & Nowell, 1984a). During quiescent periods there is rapid sediment deposition (Grant *et al.*, 1985; McCave, 1986), which may cause some disturbance by burying organisms or filling in burrows

Table 7.1. *Abundance and diversity statistics for the polychaete and bivalve components of the sample data sets considered*

Site Taxon	Abundance		Diversity			
	Mean (n 0.25 m^{-2})	n	S	H'	J'	d
HEBBLE						
Polychaetes	405.8	400	27	1.24	0.38	0.74
Bivalves	58.2	23	6	1.54	0.86	0.39
Rockall Trough						
Polychaetes	276.4	1160	66	3.14	0.75	0.22
Bivalves	43.4	239	19	2.34	0.79	0.24
Setubal Canyon						
Polychaetes	560.1	2151	71	2.60	0.61	0.31
Bivalves	66.2	219	21	2.11	0.69	0.25
Tagus Abyssal Plain						
Polychaetes	60.1	165	54	3.50	0.88	0.11
Bivalves	20.6	44	13	2.20	0.86	0.30
Central North Pacific						
Polychaetes	15.6	184	54	3.15	0.79	0.18

Data for Central North Pacific derived from Hessler & Jumars (1974); data for HEBBLE polychaete and bivalve densities from Thistle *et al.* (1985). See the text for explanation of diversity statistics used.
n 0.25 m^{-2}, number of individuals per 0.25 m^2; n, number of samples; S, number of species; H', Shannon–Wiener diversity index; J', evenness; d, Berger–Parker index of dominance.

(Aller, 1989). Although isopod and harpacticoid populations increased during these periods, abundances of other benthic metazoans appeared to be constant, although considerably higher than might be expected at the depth of the study site (4820 m) elsewhere in the deep sea. This was thought to be because of the current-driven flux of organic-rich sediment through the area combined with higher microbial activity and biomass as a result of sediment stirring (Thistle *et al.*, 1985, 1991). Although there is no evidence that larger components of the macrofauna are displaced from the sediment during the most severe benthic storms, the size structure of the bivalve and polychaete populations was unusual in being dominated by subadults. This was interpreted by Thistle *et al.* (1985) as reflecting high rates of recruitment, possibly as a result of enhanced spatial and trophic resources released after storms. In general Thistle et al. (1985) remarked that the benthic community structure resembled that seen in sediment recolonisation studies on the deep-sea floor (Grassle, 1977; Grassle & Morse-Porteous, 1987; Desbruyères et al., 1980).

Species abundance data for the polychaete and bivalve macrofauna in box-core samples from HEBBLE have been used to generate the expected species rarefactions and rank abundance plots presented in Figs 7.5–7.8, below. These permit comparisons with data on species diversity from the other deep-sea sites considered below.

7.5.2 The Rockall Trough

Quantitative data on macrofaunal community composition are available from box-core samples from the Scottish Marine Biological Association (SMBA) Permanent Station at 2900 m depth in the southern Rockall Trough, a marginal deep-sea basin off the west of Scotland and Ireland in the northern part of the North-East Atlantic. Bottom photographs at the SMBA Permanent Station show evidence of sediment erosion around rocks and projecting biogenic tube structures against an otherwise smooth bed (see Gage, 1977) while at other times the bottom has been obscured by a bottom nepheloid (P.R.O. Barnett, personal communication). High benthic kinetic energy and vigorous, periodically sediment eroding, hydrodynamic conditions may be inferred from long current-meter time-series in deeper water at the mouth of the Trough (Dickson, 1983; Dickson et al., 1985; Dickson & Kidd, 1987) and from a position (Station 'M') in 2200 m a little further north of the permanent station. At the latter mooring, several low-frequency events during which speeds occasionally exceeded 50–60 cm s^{-1} at 464 mab (MacDougall & Edelsten, 1987) were recorded. Although flow speed within the BBL will be somewhat lower, it is believed that these periods of high kinetic energy result from downward propagated vortices generated by storm-driven motions at the surface (Dickson et al., 1982). Such conditions seem to be reflected in scour crescents and leeside drifts in the sediment around isolated pebbles against a smoothed and often current-lineated bed and smoothing and high near-bottom turbidity recorded in sea-bed photographs (Lonsdale & Hollister, 1979) in the general area of the Permanent Station. From these bedforms the latter authors estimate currents flow up to about 20 cm s^{-1}.

Figures 7.1–7.3 show sea-bed photographs illustrating the markedly

Fig. 7.1. View of about 20 m² of the sea-bed taken by the multi-instrumented towed package 'Deep Tow' at about 1000 m depth on the upper continental slope in the southern Rockall Trough west of Ireland. The crests of the ripples are about 10–40 cm apart, their sharp crests and symmetric form indicating frequent energetic flow running normal to the line of the ripples. The lack of suspended material in the water indicates currents are intermittent. The small, round objects lying within ripple troughs are probably regular sea urchins, while the large fish is swimming in a roughly south-westerly direction. Photograph courtesy of Dr Peter Lonsdale, Scripps Institution of Oceanography, La Jolla, CA.

differing hydrodynamic regime operating at different depth levels on the Rockall Trough margin. That shown in Fig. 7.3 was located relatively close to the SMBA Permanent Station. Faunal abundance and diversity statistics for macrofaunal apolychaete and bivalve species from the Permanent Station box-cores (Gage, 1977) are presented in Table 7.1 and Figs 7.5–7.8, below. Abundances are relatively high and similar to comparable depths elsewhere (see Thistle *et al.*, 1985, 1991). Despite these results, the two mid-basin abyssal sites, on the Tagus Abyssal Plain and in the Central North Pacific (see below), appear to have both higher species diversity and greater evenness in species abundances.

Fig. 7.2. A heavily bioturbated, muddy sea-bed at about 2000 m depth along the same photo-transect as that of Fig. 7.1. A wealth of pits and mounds and other biogenic traces are visible, along with larger megafauna, such as the large, sausage-shaped holothurian in bottom left, and numerous small, white sea urchins. Bottom current energy would appear to be low at this position. Photograph courtesy of Dr Peter Lonsdale, Scripps Institution of Oceanography, La Jolla, CA.

7.5.3 Tagus Abyssal Plain and the Setubal Canyon

This contrasts a station at 5035 m depth in the middle of the Tagus Abyssal Plain (TAP) with a station about 100 nautical miles (nm) away at 3400 m depth within the topographically complex, canyoned margin of the western coast of Portugal. Although long-term current-meter time-series data are not available from the centre of TAP, data from nearby position NEADS-2½ indicates a minimum in abyssal eddy energy (Dickson, 1983). This indicates that maximum flow is unlikely to approach that measured even a little further north in the northern Iberian Basin and beyond (where a sharp increase in eddy energy is apparent), and is certainly much below values in the Rockall Trough or the HEBBLE area in the North-West Atlantic. Sea-bed photographs (e.g. Thiel, 1983) indicate a highly bioturbated bottom with numerous traces of burrowed megafauna and no detectable indications of current-generated

Fig. 7.3. Muddy sea-bed at 2950 m depth at the base of the continental slope off western Ireland (same transect as Figs 7.1 and 7.2). The mounds and circular patterns are evidence of burrowing megafauna, while the somewhat smoothed sediment and slightly turbid water indicates more energetic bottom current (probably sufficient to re-suspend fine surficial sediment) than that in Fig. 7.2. Photograph courtesy of Dr Peter Lonsdale, Scripps Institution of Oceanography, La Jolla, CA.

sediment structure. Such persistent biogenic traces also suggest a relatively tranquil hydrodynamic environment.

In contrast, bottom photographs (Fig. 7.4) from the other site in the lower Setubal Canyon (SC) suggests a very different hydrodynamic regime. The sharp-edged crests in the large ripple bedforms, which are orientated across the axis of the canyon, indicate a vigorous, probably tidally driven, bottom current regime. Photographs of the bottom in SC at the site of sampling provide good evidence of a vigorous hydrodynamic regime at the valley bed. The size and spacing of the linear ripple bedforms indicate bottom currents capable of transporting sand-sized particles, while the sharp-edged crests signify recent formation. The mud cloud thrown up by bottom contacts of the camera trigger weight suggest somewhat tranquil conditions prevailing at the time of sampling. Hence it seems likely that bottom-current energy varies periodically.

166   Deep-sea benthic sediment diversity

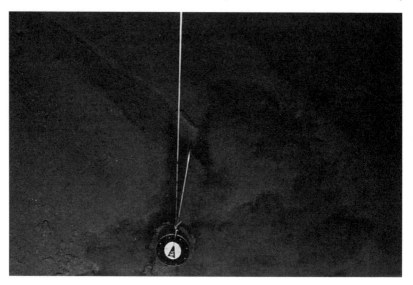

Fig. 7.4. Rippled muddy sea bottom in the floor of the lower Setubal Canyon at the site of the quantitative sampling. The view is much closer than that of Figs 7.1–7.3, and shows the sharp-crested ripples are about 30 cm apart, and therefore comparable to those in Fig. 7.1. The compass shown is attached to the camera's flash trigger.

Extensive records of bottom currents summarised in Shepard et al. (1979) from numerous submarine canyons show that currents flow with a tidal period alternatively up and down the axial floor of these valleys, and frequently attain speeds sufficient to transport sand-sized particles. Furthermore, turbidity flows may be relatively common events where rivers introduce large quantities of sediment to the sea near the canyon head. The frequency of such events may well lie within the life-span of individual benthic metazoans. Comparable current speed data are not, to my knowledge, available for any of the Portuguese canyons (which are among the longest and deepest of such features in the World Ocean). However, the topographic similarity of the SC, as shown from swath bathymetry (see Mougenot, 1988), to canyons of comparable type elsewhere, suggests that hydrodynamic conditions are likely to be similar also.

Comparisons of faunal abundances between HEBBLE sites, the Rockall Trough and similar depths elsewhere (Thistle et al., 1985, 1991) show about tenfold differences in metazoan faunal densities. This must reflect substantial differences in food available to the bottom fauna. Faunal densities from TAP correspond well to abundances recovered from box-cores from other abyssal plain sites at this depth summarised in Thistle *et al.* (1985, 1991). However, macrofaunal abundances for polychaetes, isopods and tanaids at 3400 m depth in SC (see Gage *et al.*, 1995) well exceed the range recorded for this depth, with only the bivalves providing a value falling within the previously observed range. As at other high-energy deep-sea sites, a regime of strong flow seems to provide enhanced nutritional conditions for the benthos (Gage, 1977; Thistle *et al.*, 1985, 1991). The frequent presence of sea-grass fragments in the Canyon cores (Gage *et al.*, 1995) indicates that this high abundance of macrobenthos may also reflect enhanced detrital input from shallow water that is channelled downslope along the canyon bed.

The faunal abundance and diversity data summarised in Table 7.1 and Figs 7.5–7.8 show that despite the much higher densities of benthic fauna at SC, species diversity of polychaetes is much higher at TAP on the adjacent abyssal plain. The difference is less obvious for bivalves. These differences in species richness should be much less affected by possible differences in the regional species pool as these two sites are less than 100 nm apart. Differences in evenness in polychaete species abundances (see Table 7.1) are also very obvious, with the most common polychaete and bivalve species at SC being much more numerically dominant than at TAP (Figs 7.7 and 7.8).

7.5.4 *The Central North Pacific*

The macrobenthic community at this abyssal site in the Central North Pacific (CNP) under the subtropical gyre north of Hawaii was investigated by Hessler & Jumars (1974). Box-core replicates were taken from a red clay sediment with manganese nodules in soundings of 5500–5800 m. Long-term current-meter time-series from this area indicate eddy kinetic energies equivalent to the lowest values from the deep interior of the Atlantic (Taft *et al.*, 1981; Dickson, 1983). Hydrodynamic benthic energy at this site may therefore be presumed to be very low, and comparable to that on the TAP. As at TAP, the low faunal abundances indicate that benthic biomass is very low. It is possible to read polychaete species abundances from figure 7 in Hessler & Jumars (1974). Closely similar rarefactions of polychaete expected species diversity are shown in a comparison of CNP samples with those from TAP (see Fig. 7.5).

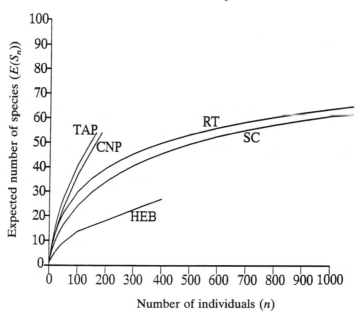

Fig. 7.5. 'Rarefaction' curves of the expected number of polychaete species in progressively larger subsamples of the total samples obtained from the five different deep-sea sites considered. Those for both the Rockall Trough permanent station (RT) and the Setubal Canyon (SC) do not show the final part of the curve up to the full sample size in order to allow the smaller total samples from HEBBLE (HEB), the Tagus Abyssal Plain (TAP) and the Central North Pacific (CNP) to be more easily compared.

7.6 Is there a pattern relating species diversity and evenness to high-energy deep-sea environments?

Diversity of the macrobenthic community shows a negative relationship with increasing bottom current energy for the widely spaced ocean-basin sites considered here. These results are summarised in Figs 7.5 and 7.6 where diversity is expressed as the expected number of species, $E(S_n)$. However, the relationship for polychaetes (Fig. 7.5) seems much clearer than that for the bivalves (Fig. 7.6), with the slope of the curve for the TAP differing only slightly from that for the Rockall Trough permanent station. Evenness in polychaete species abundances (Fig. 7.7) shows a clear trend ranging from the highly uneven distribution at the HEBBLE site, where the most abundant species contributes 74% of total abundance, to TAP where the most common species contributes only 11% of the total. Among the bivalves the trend again is less marked (Table 7.1); the

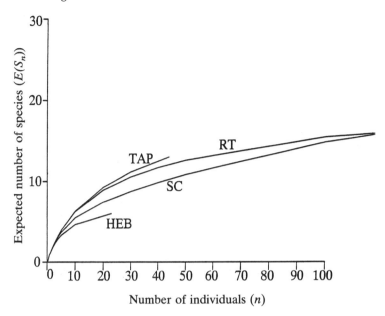

Fig. 7.6. Rarefaction curves for the bivalves in the samples. Other details as given in the legend for Fig. 7.5.

differing response between these two taxa possibly reflecting the protection and greater ability to withstand physical disturbance provided by the bivalve shell compared to the soft, unprotected morphology of polychaete worms. Differences in life-style, particularly burrowing depth, may also be important. The slightly higher species diversities found at the SC site compared to the Rockall Trough permanent station, despite the latter station being 500 m shallower, are more surprising since available evidence from sediment bedforms (Fig. 7.4) suggests that tidal flow conditions are more vigorous than at the Rockall Trough permanent station. An ecological explanation may be that the community may find it easier to approach steady state by adapting to high, but regular frequency of tidal flow than to less frequent and irregular flow events. In evolutionary time an explanation based on differing regional species pools may be important.

7.7 Summary and conclusions

Although data available are still sparse, they indicate that hydrodynamic effects may play a part in determining local species richness among the animals inhabiting the deep-sea sediment. The comparisons above

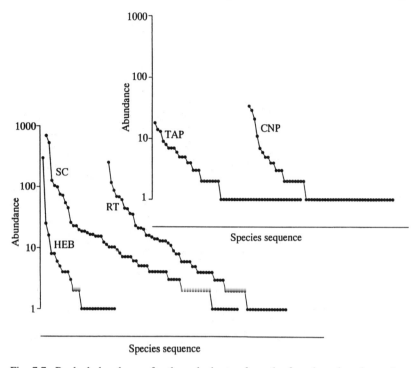

Fig. 7.7. Ranked abundances for the polychaetes from the five sites plotted on a log scale. Each filled circle represents the abundance of each component species. Labelling as given in the legend for Fig. 7.5.

indicate that at least for basin areas eddy kinetic energy will be informative regarding community structure, particularly species diversity and evenness in the macrofauna. However, data from more sites associated with good records of abyssal kinetic energy will be necessary before such associations can be sufficiently refined for predictive purposes.

This is not to say that other determinants of species diversity, particularly the biogenic processes creating small-scale habitat heterogeneity, or the historical processes determining the size of the regional species pool, are not important. Hydrodynamic forcing may, however, be seen as a potent ecological factor that should be considered as part of the suite of contemporary processes responsible for the amazing expression of biodiversity associated with deep-sea sediments.

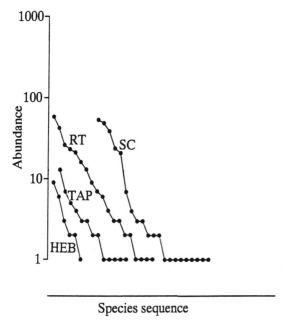

Fig. 7.8. Ranked abundances for the bivalves from the five sites plotted on a log scale. Each filled circle represents the abundance of each component species. Labelling as given in the legend for Fig. 7.5.

Acknowledgements

Polychaete and bivalve species abundances from the HEBBLE site have been made available by courtesy of Dr David Thistle at the State University of Florida, Tallahassee. I am also grateful to Dr G.L.J. Paterson, Natural History Museum, London, for revising polychaete identifications originally undertaken by myself with Dr K. Fauchald, Smithsonian Institute, Washington, for the Rockall Trough samples used in this paper. I am also grateful to Dr Paterson, and Mr Peter Lamont of SAMS, for providing all other species abundance data based on polychaete identifications, except those for the Central North Pacific samples, used here. This chapter is based partly on studies supported by the E.C. MAST programme for whose generous assistance I am very grateful.

References

Aller, J.Y., 1989. Quantifying sediment disturbance by bottom currents and its effects on benthic communities in a deep-sea western boundary current. *Deep-Sea Research*, **36**, 901–34.

Aller, J.Y. & Aller, R.C., 1986. Evidence for localized enhancement of biological activity associated with tubes and burrow structures in deep-sea sediments at the HEBBLE site, western North Atlantic. *Deep-Sea Research*, **33**, 755–90.

Auffret, G., Khripounoff, A. & Vangriesheim, A., 1994. Rapid post-bloom resuspension in the northeastern Atlantic. *Deep-Sea Research I*, **41**, 925–39.

Barnes, R.F.K., 1983. *A Synoptic Classification of Living Organisms.* Oxford: Blackwell Scientific Publications.

Berger, W.H. & Parker, F.L., 1970. Diversity of Foraminifera in deep-sea sediments. *Science*, **168**, 1345–7.

Booth, D.A. & Ellett, D.A., 1983. The Scottish continental slope current. *Continental Shelf Research*, **2**, 127–46.

Botton, M.L., 1984. The importance of predation by horseshoe crabs, *Limulus polyphemus*, to an intertidal and mudflat community. *Journal of Marine Research*, **42**, 139–61.

Connell, J.H., 1978. Diversity in tropical rain forests and coral reefs. *Science*, **199**, 1302–9.

Csanady, G.T., Churchill, J.H. & Butman, B., 1988. Near-bottom currents over the continental slope in the mid-Atlantic Bight. *Continental Shelf Research*, **8**, 653–71.

Dayton, P.K. & Hessler, R.R., 1972. Role of biological disturbance in maintaining diversity in the deep sea. *Deep-Sea Research*, **19**, 199–208.

DeMaster, D.J., McKee, B.A., Nittrouler, C.A., Brewster, D.C. & Biscaye, P.E., 1985. Rates of sediment reworking at the HEBBLE site based on profiles of naturally-occurring ^{234}Th and ^{210}Pb. *Marine Geology*, **66**, 133–48.

Desbruyères, D., Crassous, P., Bevas, J.Y. & Khripounoff, A., 1980. Un cas de colonisation rapide d'une sédiment profond. *Oceanologica Acta*, **3**, 285–91.

Desbruyères, D., Deming, J., Dinet, A. & Khripounoff, A., 1985. Réactions de l'écosystème benthique profond aux perturbations: nouveaux résultats expérimentaux. In *Peuplements Profonds de Golfe de Gascoigne*, ed. L. Laubier and C. Monniot, pp. 121–42. Brest: Institut Français de Recherche pour l'Exploitation de la Mer.

Dickson, R.R., 1983. Global summaries and intercomparisons: Flow statistics from long-term current meter moorings. In *Eddies in Marine Science*, ed. A.R. Robinson, pp. 278–353. Berlin: Springer-Verlag.

Dickson, R.R. & Kidd, R.B., 1987. Deep circulation in the Southern Rockall Trough – the oceanographic setting of site 610. *Initial Reports of the Deep Sea Drilling Programme*, **9A**, 1061–74.

Dickson, R.R. & McCave, I.N., 1986. Nepheloid layers on the continental slope west of the Porcupine Bank. *Deep-Sea Research*, **33**, 791–818.

Dickson, R.R., Gould, W.J., Gurbutt, P.A. & Killworth, P.D., 1982. A seasonal signal in the ocean currents to abyssal depths. *Nature*, **295**, 193–8.

Dickson, R.R., Gould, W.J., Müller, T.J. & Maillard, C., 1985. Estimates of the mean circulation in the deep (>2000 m) layer of the eastern North Atlantic. *Progress in Oceanography*, **14**, 103–27.

Dickson, R.R., Gould, W.J., Griffiths, C., Medler, K.J. & Gmitrowicz, E.M., 1986.

Seasonality in currents of the Rockall Channel. *Proceedings of the Royal Society of Edinburgh*, **88B**, 103–25.

Eckman, J.A., Nowell, A.R.M. & Jumars, P.A., 1981. Sediment destabilization by animal tubes. *Journal of Marine Research*, **39**, 361–74.

Eittreim, S.L., Thorndike, E.M. & Sullivan, L., 1976. Turbidity in the Atlantic Ocean. *Deep-Sea Research*, **23**, 1115–27.

Elliott, A.J. & Thorpe, S.A., 1983. Benthic observations on the Madiera abyssal plain. *Oceanologica Acta*, **6**, 463–6.

Etter, R.J. & Grassle, J.F., 1992. Patterns of species diversity in the deep sea as a function of sediment particle size diversity. *Nature*, **360**, 576–8.

Fenchel, T., 1993. There are more small than large species? *Oikos*, **68**, 375–8.

Gage, J.D., 1977. Structure of the abyssal macrobenthic community in the Rockall Trough. In *Biology of Benthic Organisms*, ed. B.F. Keegan, P.O. O'Ceidigh & P.J.S. Boaden, pp. 247–60. Oxford: Pergamon Press.

Gage, J.D. & Tyler, P.A., 1991. *Deep-sea Biology: A Natural History of Organisms at the Deep-sea Floor*. Cambridge: Cambridge University Press.

Gage, J.D., Lamont, P.A. & Tyler, P.A., 1995. Deep-sea macrobenthic community structure at contrasting sites off Portugal, preliminary results. 1. Introduction and diversity comparisons. *Internationale Revue der gesamten Hydrobiologie*, **80**, 235–50.

Gardner, W.D., 1989. Baltimore canyon as a modern conduit of sediment to the deep sea. *Deep-Sea Research*, **36**, 323–58.

Gardner, W.D. & Sullivan, L.G., 1981. Benthic storms: Temporal variability in a deep-ocean nepheloid layer. *Science*, **213**, 329–31.

Gooday, A.J. & Turley, C.M., 1990. Responses by benthic organisms to inputs of organic material to the ocean floor: A review. *Philosophical Transactions of the Royal Society of London, Series A*, **331**, 119–38.

Grant, W.D., Williams, A.J. & Gross, T.F., 1985. A description of the bottom boundary layer at the HEBBLE site: Low-frequency forcing, bottom stress and temperature structure. *Marine Geology*, **66**, 217–41.

Grassle, J.F., 1977. Slow recolonization of deep-sea sediment. *Nature*, **265**: 618–19.

Grassle, J.F., 1989. Species diversity in deep-sea communities. *Trends in Ecology and Evolution*, **4**, 12–15.

Grassle, J.F. & Maciolek, N.J., 1992. Deep-sea species richness: Regional and local diversity estimates from quantitative bottom samples. *American Naturalist*, **139**, 313–41.

Grassle, J.F. & Morse-Porteous, L.S., 1987. Macrofaunal colonization of disturbed deep-sea environments and the structure of deep-sea benthic communities. *Deep-Sea Research*, **34**, 1911–50.

Grassle, J.F. & Sanders, H.L., 1973. Life histories and the role of disturbance. *Deep-Sea Research*, **20**, 643–59.

Grassle, J.F., Maciolek, N.J. & Blake, J.A., 1990. Are deep-sea communities resilient? In *The Earth in Transition: Patterns and Processes of Biotic Impoverishment*, ed. G.M. Woodwell, pp. 385–93. New York: Cambridge University Press.

Gross, T.F., Williams, A.J. & Nowell, A.R.M., 1988. A deep-sea sediment transport storm. *Nature*, **331**, 518–21.

Hall, S.J., 1994. Physical disturbance and marine benthic communities: Life in unconsolidated sediments. *Oceanography and Marine Biology: An Annual Review*, **32**, 179–239.

Heezen, B.C. & Hollister, C.D., 1971. *The Face of the Deep*. New York: Oxford University Press.
Hessler, R.R. & Jumars, P.A., 1974. Abyssal community analysis from replicate box cores in the central North Pacific. *Deep-Sea Research*, **21**, 185–209.
Hollister, C.D. & McCave, I.N., 1984. Sedimentation under deep-sea storms. *Nature*, **309**, 220–5.
Hollister, C.D. & Nowell, A.R.M., 1991. HEBBLE epilogue. *Marine Geology*, **99**, 445–60.
Hollister, C.D., Nowell, A.R.M. & Jumars, P.A., 1984. The dynamic abyss. *Scientific American*, **250**, 42–53.
Hurlbert, S.H., 1971. The nonconcept of species diversity: A critique and alternative parameters. *Ecology*, **52**, 577–86.
Huthnance, J.M., 1981. Waves and currents near the continental shelf edge. *Progress in Oceanography*, **10**, 193–226.
Isley, A.E., Pillsbury, R.D. & Laine, E.P., 1990. The genesis and character of benthic turbid events, Northern Hatteras Abyssal Plain. *Deep-Sea Research*, **37**, 1099–119.
Johnson, G.L. & Schneider, E.D., 1969. Depositional ridges in the North Atlantic. *Earth and Planetary Science Letters*, **6**, 416–22.
Johnson, R.G., 1970. Variations in diversity within benthic marine communities. *American Naturalist*, **104**, 285–300.
Johnson, R.G., 1973. Conceptual models of benthic communities. In *Models in Paleobiology*, ed. T.J.M. Schopf, pp. 148–59. San Francisco: Freeman Cooper and Co.
Jumars, P.A., 1975. Environmental grain and polychaete species' diversity in a bathyal community. *Marine Biology*, **30**, 253–66.
Jumars, P.A., 1976. Deep-sea species diversity: Does it have a characteristic scale? *Journal of Marine Research*, **34**, 217–46.
Jumars, P.A. & Eckman, J.E., 1983. Spatial structure within deep-sea benthic communities. In *The Sea*, ed. G.T. Rowe, vol. **8**, pp. 399–451. New York: Wiley-Interscience.
Jumars, P.A. & Gallagher, E.D., 1982. Deep-sea community structure: Three plays on the benthic proscenium. In *The Environment of the Deep Sea*, ed. W.G. Ernst & J.G. Morin, pp. 217–55. Englewood Cliffs, NJ: Prentice-Hall.
Jumars, P.A. & Nowell, A.R.M., 1984a. Effects of benthos on sediment transport: Difficulties with functional grouping. *Continental Shelf Research*, **3**, 115–30.
Jumars, P.A. & Nowell, A.R.M., 1984b. Fluid and sediment dynamic effects on marine benthic community structure. *American Zoologist*, **24**, 45–55.
Jumars, P.A., Mayer, L.M., Deming, J.W., Baross, J.A. & Wheatcroft, R.A., 1990. Deep-sea deposit-feeding strategies suggested by environmental and feeding constraints. *Philosophical Transactions of the Royal Society of London, Series A*, **331**, 85–101.
Kerr, R.A., 1980. A new kind of storm beneath the sea. *Science*, **208**, 484–6.
Klein, H., 1987. Benthic storms, vortices, and particle dispersion in the deep West European Basin. *Deutsche Hydrographische Zeitschrift*, **40**, 87–102.
Klein, H. & Mittelstaedt, E., 1992. Currents and dispersion in the abyssal Northeast Atlantic. Results from the NOAMP field program. *Deep-Sea Research*, **39**, 1727–45.
Kontar, E.A. & Sokov, A.V., 1994. A benthic storm in the northeastern tropical Pacific over the fields of manganese nodules. *Deep-Sea Research I*, **41**, 1069–89.
Kukert, H. & Smith, C.R., 1992. Disturbance, colonization and succession in a

deep-sea sediment community: Artificial-mound experiments. *Deep-Sea Research*, **39**, 1349-71.

Lambshead, P.J.D., Elce, B.J., Thistle, D., Eckman, J.E. & Barnett, P.R.O., 1995. A comparison of the biodiversity of deep-sea marine nematodes from three stations in the Rockall Trough, Northeast Atlantic, and one station in the San Diego Trough, Northeast Pacific. *Biodiversity Letters*, **2**, 21-33.

Lampitt, R.S., 1985. Evidence for the seasonal deposition of detritus to the deep-sea floor and its subsequent resuspension. *Deep-Sea Research*, **32**, 885-97.

Levin, L.A., 1984. Life history and dispersal patterns in a dense infaunal polychaete assemblage: Community structure and response to disturbance. *Ecology*, **65**, 1185-200.

Levin, L.A. & Smith, C.R., 1984. Responses of background fauna to disturbance and enrichment in the deep sea: A sediment tray experiment. Deep-Sea Research, **31A**, 1277-85.

Lochte, K. & Turley, C.M., 1988. Bacteria and cyanobacteria associated with phytodetritus in the deep sea. *Nature*, **333**, 67-9.

Lonsdale, P. & Hollister, C.D., 1979. A near-bottom traverse of Rockall Trough: Hydrographic and geologic inferences. *Oceanologica Acta*, **2**, 91-105.

Magurran, A.E., 1988. *Ecological Diversity and its Measurement*. London: Chapman & Hall.

May, R.M., 1994. Biological diversity: Differences between land and sea. *Philosophical Transactions of the Royal Society of London, Series B*, **343**, 105-11.

McCave, I.N., 1986. Local and global aspects of the bottom nepheloid layers in the world ocean. *Netherlands Journal of Sea Research*, **20**, 167-81.

MacDougall, N. & Edelsten, D.J., 1987. *Current Measurements in the Rockall Channel 1979-1981*. Marine Physics Group Report no. 18. Dunstaffnage Marine Research Laboratory, Oban: Scottish Marine Biological Association.

Mercier, H. & de Verdière, A.C., 1985. Space and time scales of mesoscale motions in the eastern North Atlantic. *Journal of Physical Oceanography*, **15**, 171-83.

Miller, M.C., McCave, I.N. & Komar, P.D., 1977. Threshold of sediment motion under unidirectional currents. *Sedimentology*, **24**, 507-27.

Mougenot, D., 1988. Geologie de la Marge Portugaise. Thèse de Doctorat d'Etat. Université Pierre et Marie Curie, Paris.

Mulchern, P.J., Filloux, J.H., Lilley, F.E., Bindorf, N.L. & Ferguson, I.J., 1986. Abyssal currents during the formation and passage of the warm-core ring in the East Australian Current. *Deep-Sea Research*, **33**, 1563-76.

Nowell, A.R.M., Jumars, P.A. & Eckman, J.E., 1981. Effects of biological activity on the entrainment of marine sediments. *Marine Geology*, **42**, 133-53.

Nyffeler, F. & Godet, C.-H., 1986. The structural parameters of the benthic nepheloid layer in the northeast Atlantic. *Deep-Sea Research*, **33**, 195-207.

Osman, R.W. & Whitlach, R.B., 1978. Patterns of species diversity: Fact or artifact? *Paleobiology*, **4**, 41-54.

Paterson, G.L.J. & Lambshead, P.J.D., 1995. Bathymetric patterns of polychaete diversity in the Rockall Trough, northeast Atlantic. *Deep-Sea Research*, **42**, 1199-214.

Petraitis, P.S., Latham, R.E. & Niesenbaum, R.A., 1989. The maintenance of species diversity by disturbance. *Quarterly Review of Biology*, **64**, 393-417.

Pingree, R.D. & Le Cann, B., 1992. Three anticyclonic Slope Water Oceanic eDDIES (SWODDIES) in the southern Bay of Biscay in 1990. *Deep-Sea Research*, **39**, 1147-75.

Pingree, R.D. & New, A.L., 1989. Downward propagation of internal tidal energy into the Bay of Biscay. *Deep-Sea Research*, **36**, 735–58.

Prior, D.B., Coleman, J.M. & Doyle, E.H., 1984. Antiquity of the continental slope along the Middle-Atlantic margin of the United States. *Science*, **223**, 926–8.

Rice, A.L. & Lambshead, P.J.D., 1994. Patch dynamics in the deep-sea benthos: The role of a heterogeneous supply of organic matter. In *Aquatic Ecology: Scale, Pattern and Process*, ed. P.S. Giller, A.G. Hildrew & D.G. Raffaelli, pp. 469–97. Oxford: Blackwell Scientific Publications.

Richards, K.J., 1990. Physical processes in the benthic boundary layer. *Philosophical Transactions of the Royal Society of London, Series A*, **331**, 3–13.

Richardson, M.J., Wimbush, M. & Mayer, L., 1981. Exceptionally strong near-bottom flows on the continental rise of Nova Scotia. *Science*, **213**, 887–8.

Richardson, M.J., Weatherly, G.L. & Gardner, W.D., 1993. Benthic storms in the Argentine Basin. *Deep-Sea Research II*, **40**, 975–87.

Riemann, F., 1989. Gelatinous phytoplankton detritus on the Atlantic deep-sea bed, structure and mode of formation. *Marine Biology*, **100**, 533–9.

Robb, J.M., Hampson, J.C. & Twichell, D.C., 1981. Geomorphology and sediment stability of a segment of the USA continental slope off New Jersey. *Science*, **211**, 935–7.

Sanders, H.L., 1968. Marine benthic diversity: A comparative study. *American Naturalist*, **102**, 243–82.

Saunders, P.M., 1988. Bottom currents near a small hill on the Madiera Abyssal Plain. *Journal of Physical Oceanography*, **18**, 868–79.

Schmitz, W.J., 1984. Abyssal eddy kinetic energy in the North Atlantic. *Journal of Marine Research*, **42**, 509–36.

Schmitz, W.J., 1988. Exploration of the eddy field in the midlatitude North Pacific. *Journal of Physical Oceanography*, **18**, 459–68.

Shepard, F.P., Marshall, N.F., McLoughlin, P.A. & Sullivan, G.G., 1979. *Currents in Submarine Canyons and Other Seavalleys*. AAPG Studies in Geology, vol. 8. Tulsa, OK: The American Association of Petroleum Geologists.

Smith, C.R., 1985. Colonization studies in the deep sea: Are results biased by experimental design? In *Proceedings of the Nineteenth European Marine Biology Symposium*, ed. P.G. Gibbs, pp. 183–90. Cambridge: Cambridge University Press.

Smith, C.R., 1986. Nekton falls, low intensity disturbance and community structure of infaunal benthos in the deep sea. *Deep-Sea Research*, **44**, 567–600.

Smith, C.R., 1994. Tempo and mode in deep-sea benthic ecology: Punctuated equilibrium revisited. *Palaios*, **9**, 3–13.

Smith, C.R. & Hessler, R.R., 1987. Colonization and succession in deep-sea ecosystems. *Trends in Ecology and Evolution*, **2**, 358–63.

Smith, C.R., Jumars, P.A. & DeMaster, D.J., 1986. *In situ* studies of megafaunal mounds indicate rapid sediment turnover and community response at the deep-sea floor. *Nature*, **323**, 251–3.

Snelgrove, P.V.R. & Butman, C.A., 1994. Animal–sediment relationships revisited: Cause versus effect. *Oceanography and Marine Biology: An Annual Review*, **32**, 111–77.

Southard, J., Young, R. & Hollister, C.D., 1971. Experimental erosion of calcareous ooze. *Journal of Geophysical Research*, **76**, 5903–9.

Stork, N.E., 1988. Insect diversity: Facts, fiction and speculation. *Biological Journal of the Linnean Society*, **35**, 321–37.

Taft, B.A., Ramp, S.R., Dworski, J.G. & Holloway, G., 1981. Measurements of deep currents in the central North Pacific. *Journal of Geophysical Research*, **86**, 1955–68.
Thiel, H., 1983. Meiobenthos and nanobenthos of the deep sea. In *The Sea*, vol. **8**, ed. G.T. Rowe, pp. 167–230. New York: Wiley-Interscience.
Thiel, H., Pfannkuche, O., Schriever, G., *et al.*, 1989. Phytodetritus on the deep-sea floor in a central oceanic region of the Northeast Atlantic. *Biological Oceanography*, **6**, 203–39.
Thistle, D., 1979. Harpacticoid copepods and biogenic structures: Implications for deep-sea diversity maintenance. In *Ecological Processes in Coastal and Marine Systems*, ed. R.J. Livingstone, pp. 217–31. New York: Plenum Press.
Thistle, D., 1981. Natural physical disturbances and communities of marine soft bottoms. *Marine Ecology Progress Series*, **6**, 223–8.
Thistle, D., 1988. A temporal difference in harpacticoid–copepod abundance at a deep-sea site: Caused by benthic storms? *Deep-Sea Research*, **35**, 1015–20.
Thistle, D. & Eckman, J., 1988. Response of harpacticoid copepods to habitat structure at a deep-sea site. *Hydrobiologia*, **167/168**, 143–9.
Thistle, D. & Wilson, G.D.F., 1987. A hydrodynamically modified, abyssal isopod fauna. *Deep-Sea Research*, **34**, 73–87.
Thistle, D., Yingst, J.Y. & Fauchald, K., 1985. A deep-sea benthic community exposed to strong near-bottom currents on the Scotian Rise (Western Atlantic). *Marine Geology*, **66**, 91–112.
Thistle, D., Ertman, S.C. & Fauchald, K., 1991. The fauna of the HEBBLE site: Patterns in standing stock and sediment-dynamic effects. *Marine Geology*, **99**, 413–22.
Thorpe, S.A., Hall, P. & White, M., 1990. The variability of mixing at the continental slope. *Philosophical Transactions of the Royal Society of London, Series A*, **331**, 183–94.
Thorson, G., 1971. *Life in the Sea*. New York: McGraw-Hill.
Tyler, P.A., Grant, A., Pain, S.L. & Gage, J.D., 1982. Is annual reproduction in deep-sea echinoderms a response to variability in their environment? *Nature*, **300**, 747–9.
Vangriesheim, A., 1985. Hydrologie et circulation profonde. In *Peuplements Profonds du Golfe de Gascogne*, ed. L. Laubier & C. Monniot, pp. 43–70. Brest: Institut Français de Recherche pour l'Exploitation de la Mer.
Weatherly, G.L., 1984. An estimate of bottom frictional dissipation by Gulf Stream fluctuations. *Journal of Marine Research*, **42**, 289–301.
Weatherly, G.L. & Kelley, E.A., 1985. Storms and flow reversals at the HEBBLE site. *Marine Geology*, **66**, 205–18.
Wheatcroft, R.A., Smith, C.R. & Jumars, P.A., 1989. Dynamics of surficial trace assemblages in the deep sea. *Deep-Sea Research*, **36**, 71–91.
Wyrtki, K.E., Magaard, L. & Hager, J., 1976. Eddy energy in the oceans. *Journal of Geophysical Research*, **81**, 2641–6.
Yingst, J.Y. & Aller, R.C., 1982. Biological activity and associated sedimentary structures in HEBBLE-area deposits, western North Atlantic. *Marine Geology*, **48**, 7–15.
Zajac, R.N. & Whitlach, R.B., 1991. Demographic aspects of marine, soft sediment patch dynamics. *American Naturalist*, **31**, 808–20.

Chapter 8
Diversity and structure of tropical Indo-Pacific benthic communities: Relation to regimes of nutrient input

JOHN D. TAYLOR
Department of Zoology, The Natural History Museum, Cromwell Road, London, SW7 5BD, UK

Abstract

Major regional and local differences in the composition and structure of shallow-water communities in the Indo-Pacific marine province may be related to differences in nutrient availability. Animals in eutrophic environments tend to be fast-growing, with rapid population turnover and generalist habits. Bivalves, for instance, are much more diverse and abundant in eutrophic 'continental' environments, than on oceanic, oligotrophic atolls and reefs. Neogastropods with generalist diets include *Babylonia*, which are restricted to continental shores, and Nassariidae species, which are much more diverse in eutrophic environments. A comparison of food webs, involving predatory gastropods from oligotrophic and eutrophic environments, shows that the oligotrophic webs are based upon benthic algae and detritus, whilst the eutrophic webs are based upon phytoplankton. There is some evidence that rates of evolutionary diversification may be related to nutrient regimes.

8.1 Introduction

The shallow-water biota of the Indo-Pacific marine province is characterised both by the very high species diversity, but also by the very broad geographical ranges of many species, some encompassing the entire area of the province (Kohn, 1983; Kohn & Perron, 1994). Considerable attention has been given to the analysis of the biogeographical patterns and origins of this diversity in relation to plate and local tectonic history, dispersal barriers, habitat diversity, temperature, and so on (Kohn, 1983; Kay, 1984; and general review by Rosen, 1988). Probably because of the broad distributional ranges of many species, there has been a tendency, particularly in textbooks, to regard the tropical Indo-Pacific as a rather uniform environment. However, there are major environmental differences between, for example, high and low islands, or between oceanic islands and continental margins. One of the most important factors may be the variations in nutrient supply; some recent studies have presented strong evidence that local and regional differences in nutrient availability may exert important controls both on the distribution of tropical organisms, and also upon the organisation of benthic communities (Birkeland, 1977, 1987, 1988a,b; Hallock & Schlager, 1986; Hallock, 1988; Vermeij, 1990; Taylor, 1993a).

Results from the satellite imagery of the Coastal Zone Colour Scanner (CZCS) have vividly confirmed the large differences in primary productivity across the oceans (Lewis, 1989). As is well known, the open ocean waters of the tropical Indo-Pacific usually have very low levels of primary production (Ryther, 1969). High levels of phytoplankton production are found only in relatively restricted areas. These may be upwelling areas, such as around the southern Arabian peninsula (Savidge et al., 1990), or on continental margins, where high monsoonal rainfall brings pulses of nutrients from terrestrial run-off, as around most of South-East Asia or around high oceanic islands, such as the granitic Seychelles (Littler et al., 1991). Rates of nutrient supply vary both on a local scale, as for example, between an enclosed lagoon and the seaward reefs of an atoll; or on a regional scale as between Micronesia and South-East Asia. Nutrient supply will also vary seasonally with wind direction controlling upwelling, or seasonal rainfall controlling terrestrial run-off.

Differences in the composition and diversity of marine faunas between oceanic islands and continental margins have long been known (some examples for the Indian Ocean are given in Taylor, 1971), but the causes of these differences were not understood. Birkeland's (1987) analysis showed that many regional differences in community composition and organisation might be related to the rates of nutrient supply. His studies provided a conceptual framework for the regional comparison of reefs and other shallow-water communities of the tropics.

The nutrient control model is briefly reviewed below, followed by some examples of how differences in the nutrient supply may influence the diversity and distribution of some major groups of animals both on the regional and local scale. Following this, an analysis of some coral reef food webs involving predatory gastropods shows how the organisation of the webs varies between oligotrophic and eutrophic conditions.

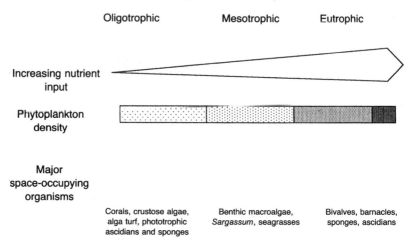

Fig. 8.1. Diagram illustrating the effect of increasing nutrient input on tropical benthic communities. Only major groups of organisms are mentioned. All transitional states may be seen in natural communities. See the text for further details.

8.2 The nutrient control model

A model of how tropical shallow-water communities may be controlled by different regimes of nutrient input has been developed by Birkeland (1977, 1987, 1988a,b). A summary of the model is illustrated in Fig. 8.1 and briefly outlined below.

In nutrient-poor, phytoplankton-poor, oligotrophic waters, such as surround oceanic atolls, the major space-occupying organisms are largely autotrophic and phototrophic, scleractinian corals, alcyonaria, sponges with symbiotic cyanobacteria, ascidians with *Prochloron* and the *Tridacna*. It is believed that the prevalent symbioses are an adaptation for efficient nutrient recycling (Muscatine & Porter, 1977; DeAngelis et al., 1989). Crustose coralline algae are also important. Phytoplankton-feeding organisms such as barnacles and bivalves are usually uncommon in these environments. Despite being situated in nutrient-poor oceanic waters, the reefs themselves have generally high gross productivities sustaining high biomass levels (Grigg et al., 1984; Birkeland, 1988a), but net productivities are relatively low (Hallock & Schlager, 1986). The high productivities are believed to be maintained because of various types of efficient nutrient recycling mechanisms within the communities (Muscatine & Porter, 1977; Birkeland, 1988a). For example, in a study of a Pacific reef Grigg et al. (1984) estimated that 85% of net production was taken up amongst the heterotrophic benthos and reef fish.

At increasing nutrient levels, benthic algae become more important and will overgrow and outcompete corals for space (Littler et al., 1991). Experiments have shown that rapidly-growing, filamentous algae inhibit the settlement of coral planulae (Birkeland, 1977). Coralline algae may also be overgrown by other algae. Phytoplankton productivity is higher and the water may be less transparent. Where sufficient sediment is available, sea-grass beds develop and are characteristic of mesotrophic environments (Wiebe, 1987). Although juvenile corals are often outcompeted for space by the algae, mixed coral/fleshy-algal communities often occur, particularly on reefs at higher latitudes such as the Houtman Abrolhos Islands, Western Australia (Crossland et al., 1984).

At the extreme end of the scale, with the highest levels of nutrient input, as for example in the in-shore waters of monsoonal continental margins, most primary production occurs in the water column via the phytoplankton. High densities of phytoplankton reduce the light levels reaching the benthos and also reduce the amount of nutrients in the water available for the growth of benthic algae. Algal growth is reduced and the bottom community becomes dominated by suspension-feeding animals such as epifaunal bivalves, barnacles, massive sponges and ascidians without symbionts and bryozoa. Only very tolerant corals such as *Porites* and *Goniopora* can survive at the low light levels in these turbid waters.

Birkeland (1987) presents much empirical and experimental evidence to support this outline model and only a few examples are cited here. An instructive example is the well-known case of Kaneohe Bay in Hawaii (Kinsey, 1988) where eutrophication of a coral community was caused by sewage discharge. With increasing nutrient levels in the southern bay, the corals were swamped by benthic algae and populations of suspension-feeding sponges, ascidians, barnacles, bivalves and polychaetes. Also the physical substrate was destroyed by boring, filter-feeding infauna. Since diversion of the sewage in 1979, the algae and filter feeding populations have declined, new corals have appeared and the biota is reverting to a more typical, fringing reef community.

Transects from oligotrophic to eutrophic environments have also demonstrated such changes in community structure. Wilkinson (1986) and Wilkinson & Cheshire (1989) described the distribution of sponges in a transect across the Great Barrier Reef. In the oligotrophic conditions of the Outer Barrier, the sponges were dominantly phototrophic, deriving most of their nutrition from symbiotic cyanobacteria. In the mid- and inner parts of the reef where phytoplankton concentrations are greater, the sponges were larger, more abundant and dominantly suspension-feeding. In similar transects across the Great Barrier Reef, Williams et al. (1986) found the greatest biomass of planktivorous fish at mid-reef sites. Similarly, in the Sudanese Red Sea, the coastal fringing reefs have more algae and larger populations of herbivorous gastropods than the off-shore patch reefs (Taylor & Reid, 1984).

Typically, suspension-feeding bivalves in eutrophic regimes have high fecundity, early maturity and fast growth rates, features that enable the animals to respond to pulses of nutrient and phytoplankton growth. These are life-history characters possessed by many bivalves in Hong Kong waters, for example (Morton, 1991).

Various evidence suggests that the bioerosion of coral reefs is generally proportional to nutrient input and productivity (Highsmith, 1980; Hallock, 1988; Kinsey, 1988) in discussing the eastern Pacific reef fauna, Glynn (1988) concluded that disturbance events on coral reefs had the greatest long-term consequences in waters of high productivity. The high rates of bioerosion in eutrophic water may explain the meagre development of reefs in areas of high productivity (Hallock & Schlager, 1986). In Tolo Channel, Hong Kong, Scott and Cope (1990) correlated the increased erosion of the corals lining the Channel with the progressive eutrophication of the seaway.

Increased nutrient input tends to shorten the number of trophic levels in food webs (Ryther, 1969; Birkeland, 1988a). Small organisms with high fecundity, rapid growth rates and short population turnover rates can respond rapidly to pulses of nutrient input. These types of organism tend to prevail in eutrophic environments and the populations, although abundant, tend to fluctuate and be short-lived. Mortalities from periodic low salinities, eutrophication and anoxia may be frequent (Wu, 1982). Thus the grazers and predators that exploit these populations would be expected to have generalist feeding habits and diets. In Tolo Channel, Hong Kong, the predatory gastropods with more specialised diets are being eliminated by progressive eutrophication and increased disturbance (Taylor & Shin, 1990; Taylor, 1992). By contrast, in oligotrophic regimes, organisms tend to be slow-growing, with slower rates of population turnover, resources are predictable and animals with highly specialised diets are common (Taylor, 1984, 1989). In a general review of sublittoral, soft-substrate benthos, Pearson & Rosenberg (1978) have shown how increasing eutrophication/disturbance results in the dominance of short-lived, generalist, r-selected, species.

8.3 Distribution of organisms in relation to nutrient input

Although many animals have very broad distributional ranges within the Indo-Pacific Province, there are large differences between the faunas of oceanic coral islands and the faunas of high islands and continental margins. In many groups of molluscs there are examples of suites of species restricted to high islands or the margins of continents, as in the Strombidae (Abbott, 1960), Littorinidae (Reid, 1986), *Bullia* (Taylor, 1971) or *Conus* (Taylor, 1971; Kohn & Perron, 1994). One of the most striking examples concerns bivalve molluscs.

8.3.1 Bivalve diversity and abundance

Most bivalve molluscs are suspension feeders on phytoplankton and they are usually neither very diverse nor abundant on coral reefs in oligotrophic environments (Morton, 1983; Vermeij, 1990). However, they are both more diverse and abundant around high islands and at localities along continental margins where primary productivity is higher (Table 8.1). Although the areas for which estimates of species numbers are available vary considerably in areal extent, the differences are very marked. For instance, 635 species of bivalve are recorded from Taiwan, and 426 from Okinawa. These figures may be compared with oceanic atolls such as Enewetak Atoll where there are only 115 species of bivalve out of 1116 species of molluscs, or the Hawaiian islands where there are 141 species of bivalve out of 1007 molluscan species; 109 bivalve species have been recorded from Cocos-Keeling Atoll and only 105 species from the entire French Polynesia. Overall diversity may be related to habitat variety and space. But if individual habitats are considered, for example rocky shores, then on continental shores, suspension-feeding bivalves such as *Saccostrea*, *Brachidontes*, *Septifer*, *Isognomon*, *Perna*, *Chama* and *Barbatia* are the dominant space-occupying organisms (Lee & Morton, 1985; Tong, 1986; Tsuchiya & Lirdwitayapsit, 1986). By contrast, on oceanic atolls and reefs, bivalves are much less abundant and, for example, are barely mentioned in accounts of the intertidal reef-flat fauna of Enewetak Atoll (Kohn, 1987), while on the Fanning Island seaward reef-flat, 97 gastropod species were reported, but only three bivalves (Kay, 1971). Similarly, bivalves, except for *Tridacna* with photosymbionts, are relatively uncommon on the seaward shores of atolls and reefs in French Polynesia (Salvat, 1967; Richard, 1982). Bivalves are often much more abundant in lagoons (Kay

184   Tropical Indo-Pacific benthic communities

Table 8.1. Comparison of the numbers of species of bivalve molluscs recorded from oligotrophic oceanic sites compared with eutrophic continental margins

Location	No. of species	Reference
Oceanic sites		
Cocos-Keeling	109	Wells (1994)
French Polynesia	105	Richard (1982)
Enewetak	115	Kay & Johnson (1987)
Hawaii	141	Kay (1979)
Continental margins		
Taiwan	635	Wu (1980)
Okinawa	426	Kuroda (1960)
Hainan	287	Qi Zhongyan et al. (1984)
Hong Kong	273	Tsi & Ma (1982)
Red Sea	411	Oliver (1992)

& Switzer, 1974) and strikingly so in the closed lagoons of French Polynesia, such as Takapoto, where enormously high densities and biomasses have been recorded (Salvat, 1967; Richard et al., 1979). The main species involved are *Tridacna maxima*, *Pinctada margaritifera*, *Chama iostoma* and *Arca ventricosa*. Nutrient levels both in the lagoon and surrounding ocean waters are low, but it is believed that the nearly enclosed nature of the atoll increases the residence time of the water in the lagoon, allowing the recycling of nutrients into the biomass of the bivalves (Sournia & Ricard, 1975; Birkeland, 1987).

Species of the bivalve family Mytilidae are strongly associated with eutrophic environments; they are four to seven times more diverse on continental shores than on oceanic atolls (Table 8.2). There are, for instance, twice as many species recorded from Hong Kong as for the entire Hawaiian Islands. Mytilid species are frequently major space-occupying organisms on hard substrates on shores in eutrophic areas (Lee & Morton, 1985; Tong, 1986; Tsuchiya & Lirdwitayapsit, 1986; Morton, 1995), but are generally subordinate in oceanic coral-reef environments. An interesting example of an individual distribution is the mussel *Perna* (Siddall, 1980), which often inhabits highly eutrophic waters as for example in Hong Kong harbour (Lee, 1986; Cheung, 1993). The distribution of the two tropical Indo-Pacific species (*P. viridis* and *P. perna*), shows a close association with continental margins and with areas of high productivity, such as the southern Arabian Peninsula upwelling, and the monsoonal coasts of India and South-East Asia

Table 8.2. *Numbers of species of the bivalve family Mytilidae recorded from some islands situated in oligotrophic environments compared with some localities situated in eutrophic environments of continental margins*

Location	Non-lithophagine species	Lithophagine species	Reference
Oceanic islands:			
Fanning I.	2	1	Kay & Switzer (1974)
Cocos-Keeling	2	2	Wells (1994)
Enewetak	6	1	Kay & Johnson (1987)
Hawaii	9	2	Kay (1979)
French Polynesia	3	4	Richard (1982)
Continental margins:			
Hong Kong	22	7	Lee & Morton (1985)
Taiwan	26	7	Wu (1980)
S. Japan	21	5	Azuma (1960)
Hainan	14	5	Qi Zhongyan et al. (1984)
Red Sea	20	8	Oliver (1992)

Mytilidae have been divided into lithophagine and non-lithophagine taxa.

(Taylor, 1993a). Coral and rock-boring lithophagid bivalves are also more abundant and diverse on continental margins and upwelling situations (Highsmith, 1980).

It is now becoming increasingly recognised that many coral-reef bivalves are dependant on nutritional sources other than phytoplankton and detritus. Although the association of photosymbionts with the reef-associated bivalves *Tridacna* and *Corculum* is well-known (Morton, 1983), other members of the Fraginae such as *Fragum, Hemicardia* and *Lunulicardia*, which are often very abundant on shallow water reef flats, also have the reduced and modified gut that is typical of known chemosymbionts (Schneider, 1993), although photosymbionts have as yet only been recorded in *Fragum* species (Kawaguti, 1983). Lucinid bivalves such as *Codakia, Ctena, Fimbria* and *Anodontia* are also often abundant in sea-grass beds (Taylor, 1968; Jackson, 1973) and species of these genera are known to possess sulphide-oxidising bacteria in their highly modified gills, and also to have reduced guts and labial palps (Reid, 1990).

Barnacles show a similar diversity and abundance pattern to suspension-feeding bivalves, in that they are usually rare and inconspicuous animals on oceanic atolls compared to their abundance and diversity in the eutrophic waters of continental margins (Newman, 1960; Jones, 1994).

8.3.2 Gastropod distribution

It has been argued in the preceding section that animals tend to have more generalist diets in areas with pulses of high nutrient input and/or high disturbance. The feeding habits of predatory gastropods have been extensively studied in parts of the Indo-Pacific and the following are two examples of benthic gastropods that have generalist diets, and are either restricted to, or exhibit greater diversity on, continental margins.

The buccinid gastropod genus *Babylonia* comprises some 12 living species (Altena & Gittenberger, 1981), which inhabit sublittoral silt and muddy habitats around the continental margins of the Indo-Western Pacific, from the Red Sea to Japan, but are absent from oceanic islands. The gastropods can be very abundant and are widely fished in South-East Asia and Japan. The feeding habits of two species (*B. lutosa* and *B. areolata*) have been investigated in Hong Kong and they have been found to be generalist feeders, eating molluscs and polychaetes, but also including a high proportion of carrion in their diet (Morton, 1990; Taylor & Shin, 1990).

Species of Nassariidae are renowned for their generalist feeding habits, although there is little known in detail about most tropical species (Morton, 1990; Britton & Morton, 1992). Most species will take carrion when available, and also many probably feed upon organic detritus and diatoms, as does the Atlantic species *Ilyanassa obsoleta* (Curtis & Hurd, 1979). The few examples in Table 8.3 show that nassariids are at least twice as diverse on continental margins as on oceanic islands. Qualitative observations suggest that nassariids are also more abundant on continental margins, for example high abundances have been recorded on mud flats at Phuket, Thailand (Frith *et al.*, 1976) and mean densities of 66 m^{-2} have been recorded from beaches in Hong Kong (Britton & Morton, 1992). A trawl survey of sublittoral gastropods around southern Hong Kong showed that although more than 70 species of gastropod were recovered, 6 species of Nassariidae totalled 45% of the gastropods found (Taylor, 1994). Moreover, reference to Cernohorsky (1984) shows that most species of Indo-Pacific Nassariidae are found either around continental margins or high islands, with those species found on oceanic coral islands comprising a small suite of very widespread species (e.g. Wells, 1994).

Table 8.3. *Numbers of species of the neogastropod family Nassariidae recorded from some oceanic islands in oligotrophic situations, compared with localities with monsoonal eutrophic regimes*

Location	Nassariid species	Reference
Oceanic islands:		
Enewetak	7	Kay & Johnson (1987)
Cocos-Keeling	5	Wells (1994)
Fanning I.	3	Kay & Switzer (1974)
Hawaii	7	Kay (1979)
Continental margins:		
Hong Kong	14	Cernohorsky (1984)
Phuket	14	Tantanasiriwong (1978)
Sagami Bay, Japan	18	Kuroda *et al.* (1971)
Philippines	30	Springsteen & Leobrera (1986)

8.4 Structure of food webs in different nutrient regimes

Although there are major differences in the composition of the communities between oligotrophic and eutrophic environments, do these differences affect the ways that communities are organised? Evidence from studies of predatory gastropods shows large differences in the structure of food webs from communities inhabiting oligotrophic as compared to eutrophic environments.

For example, Kohn (1987) illustrated the food subweb involving gastropods on the intertidal rock platform at Enewetak Atoll, in the Marshall Islands (Fig. 8.2). This mid-Pacific atoll lies in highly oligotrophic waters of very low productivity. The food web was derived from evidence from the analysis of gut contents, faeces and field observations. Twenty-one species of predatory gastropod (families Conidae, Muricidae, Mitridae, Buccinidae and Vasidae) are represented in the web, of which 17 species ate polychaetes, sipunculans and other gastropods. These prey are known to feed in turn upon benthic algae or upon deposited detritus. Only four species of predatory gastropods fed entirely upon polychaete or gastropod prey that are known to eat phytoplankton or suspended detritus. No barnacles or bivalves were recorded as prey in this web, which is a reflection of their relative scarcity in the environment.

Similarly, Taylor (1984) described a food subweb for predatory gastropods from another Pacific reef at Guam, Marianas Islands. Here 42 species of predatory gastropods were investigated and only 2 species,

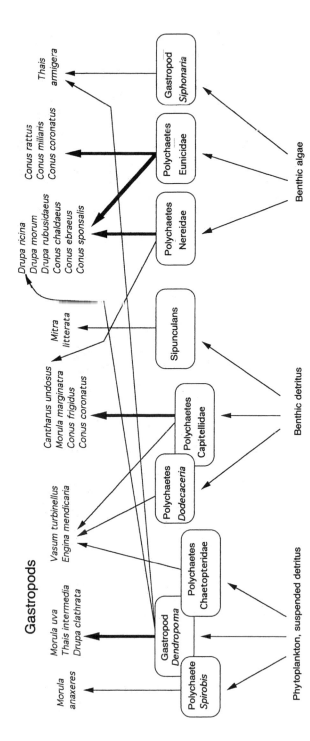

Fig. 8.2. Food subweb involving predatory gastropods on a windward fringing reef at Enewetak Atoll. Redrawn from Kohn (1987).

Morula granulata and *Gutturnium muricinum*, regularly consumed bivalve prey, while barnacles were a small part of the diet of 2 *Cantharus* species. Most of the other species of predators ate herbivorous gastropods, polychaetes and sipunculans. A similar pattern was seen in a food web involving predatory gastropods obtained from a sandy reef-flat also on Guam (Taylor, 1986). Here, 17 species of predatory gastropod from the families Terebridae, Costellariidae and Mitridae ate mainly deposit-feeding polychaetes and sipunculans.

These coral-reef food webs have a number of common features. There are diverse assemblages of predators, with many species having specialised diets. The lowest levels in the food web consist mainly of species that graze on benthic algae or feed on deposited detritus. Suspension-feeding animals such as barnacles and bivalves are either very minor items in the diet of the gastropods, or are completely absent from the webs.

A food web involving predatory gastropods from Rottnest Island, off Perth, Western Australia is of interest. At the western end of the island, broad, low, intertidal limestone platforms are similar in morphology to coral reef-flats. The platforms are inhabited by a molluscan fauna comprising mostly warm-temperate species, but with a significant proportion of tropical species normally associated with coral reefs. These survive at this latitude by virtue of the warm southward-flowing Leeuwin Current that washes the western end of the island. The continental shelf waters are nutrient-poor (Johannes *et al.*, 1994) and the situation thus resembles that of oceanic islands. The gastropod fauna inhabiting the platforms consists dominantly of species of the families Muricidae, Mitridae and Conidae, which are usually the most abundant and diverse families on seaward reef-flats of Indo-Pacific oceanic reefs (Taylor, 1978, 1984). The feeding habits of the most abundant predatory gastropods of the platforms have recently been investigated by Kohn & Almasi (1993), Morton & Britton (1993) and Taylor (1993b) and a food web constructed from the results of these studies is shown in Fig. 8.3. The three species of Mitridae are all sipunculan specialists, the *Conus* species feed largely on polychaetes, particularly the families Eunicidae and Polynoidae. By comparison, the two common muricid species, *Dicathais orbita* and *Cronia avellana*, are both food generalists. Gastropods are the dominant component (57%) in the diet of *Dicathais*, with the bivalve *Septifer* (16%) and carrion (27%) as lesser items. Another generalist, *Bursa granularis*, feeds mainly on polynoid polychaetes, with ophiuroids as a minor item. Suspension-feeding organisms, notably bivalves and barnacles, are uncommon at the site and are minor items in the diets of

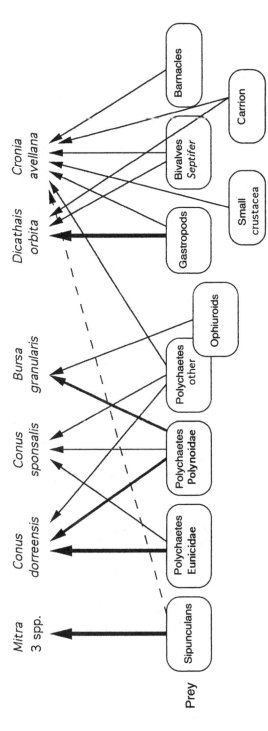

Fig. 8.3. Food subweb involving predatory gastropods on an intertidal platform at Cape Vlamingh, Rottnest Island, Western Australia. Data from Kohn & Almasi (1993), Morton & Britton (1993) and Taylor (1993b).

these gastropods. The food web from this oligotrophic, continental margin site is thus similar to that from oceanic atolls in being based largely on benthos-feeding, rather than suspension-feeding, prey.

By contrast, food webs involving gastropods in eutrophic waters are different in structure. The first example concerns the food web involving predatory gastropods on an intertidal rocky shore at Wu Kwai Sha, in Tolo Channel. This site lies in a fjord-type inlet and the water is highly eutrophic with high nutrient input from terrestrial run-off, farming and sewage (Morton, 1989). This web has been constructed from the studies of Tong (1986) and Taylor (1990). The shore at this site supports a rich assemblage of barnacles, epifaunal bivalves and herbivorous gastropods. The dominant space-occupying organism is the oyster *Saccostrea cucullata*. There are two common predatory gastropods *Thais clavigera* and *Morula musiva* and their diets were established using a combination of gut content analysis and field observations. The results (Fig. 8.4) show that most of the food of the two predators consists of the suspension-feeding bivalves *Saccostrea*, *Brachidontes*, *Barbatia* and the barnacle *Balanus amphrite*. Algal-feeding gastropods, such as *Siphonaria* and *Monodonta*, form only a minor part of the diet of *M. musiva*.

Interestingly, Tong (1986) and Taylor & Morton (1997) have also studied the diet of *T. clavigera* and *M. musiva* at Cape d'Aguilar, on the extreme southern tip of Hong Kong Island where the water is less eutrophic than in Tolo Channel. On the exposed seaward side of the Cape, barnacles and the mussel *Septifer virgatus* are the dominant space-occupying organisms and these are the dominant prey of *T. clavigera* and *T. luteostoma*. In a small bay in the shelter of the Cape, barnacles and bivalves are less abundant and the *T. clavigera* and *M. musiva* include in their diet a high proportion of the grazing gastropods *Patelloida* and *Siphonaria*.

From slightly deeper water in Tolo Channel, Hong Kong, Taylor (1980, 1993a) presented dietary data on an assemblage of predatory gastropods living on the bouldered and coral sublittoral shores lining the Channel. This fauna consisted of a diverse assemblage of muricid gastropods, particularly *Chicoreus microphyllus* (Lamarck), *Ch. brunneus*, *Mancinella echinata* and *Cronia margariticola*. In the food web derived for gut content and field observations it was found that suspension-feeding bivalves and barnacles were the main prey items. A few herbivorous gastropods were eaten by *Cronia margariticola*. The latter species along with *Drupella rugosa* and *Coralliophila costularis* also ate coral tissue.

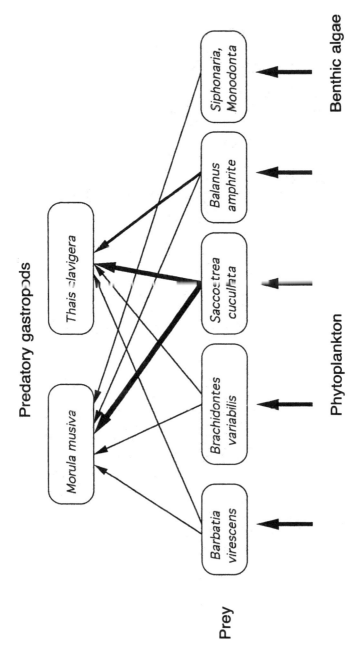

Fig. 8.4. Food subweb involving predatory gastropods from an intertidal rocky shore at Hoi Si Wan, Tolo Channel, Hong Kong. Data from Tong (1986) and Taylor (1990).

Lastly, the diets of the sublittoral, soft-bottom gastropods from Hong Kong have been investigated in Mirs Bay (Taylor & Shin, 1990) and off southern Hong Kong Island (Taylor, 1994). Many of the gastropods have generalist diets feeding upon a variety of prey with *Babylonia* and *Nassarius* species commonly feeding upon carrion, whilst the conoidean *Turricula nelliae* ate a wide variety of polychaetes. There were a few food specialists in the fauna, notably *Bursa rana*, which preyed upon ophiuroids, and *Distorsio reticulata*, which ate chaetopterid polychaetes. Most of the prey species in the web were deposit-feeding polychaetes.

The two food webs for Hong Kong rocky shores show that the primary consumers at the base of the webs are mainly phytoplankton-feeding bivalves and barnacles and prey that graze on the benthic flora are not so commonly eaten. A similar pattern was described by Abe (1989) for a guild of muricid predators on a Japanese shore. These results confirm a prediction of Birkeland (1988a,b) that levels of nutrient input cause a switch from benthic algal-based food webs to phytoplankton-based webs. The sublittoral, soft-substrate web for Hong Kong, with many generalist and carrion-feeding species, may be contrasted with the sandy bottom web for Guam (Taylor, 1986), where there is just a single generalist-feeding species.

8.5 Evolutionary implications

In this review some differences in the composition and structure of benthic communities subjected to different regimes of nutrient input have been highlighted. In the food webs described, the primary consumers that feature as the main prey of the predatory gastropods are quite different in oligotrophic and eutrophic environments. Thus, bivalves and barnacles are the main prey in eutrophic environments and herbivorous gastropods, polychaetes and detritus feeders are the main prey at oligotrophic sites. Thus it would appear that any co-evolutionary interactions between predators and prey, leading for example to selection for predator-avoidance adaptations of armament, behaviour or life-history traits (Vermeij, 1978, 1987), will involve completely different groups of animals in the different nutrient regimes.

Vrba (1987) has discussed the evolutionary patterns of 'specialist' and 'generalist' clades. Various lines of argument predict that generalist clades will tend to be species-poor, having low speciation rates coupled with low extinction rates. On the other hand, specialist clades will be species-rich, with both high speciation and extinction rates.

Diverse assemblages of gastropods with specialist diets are strongly associated with oligotrophic, coral-reef environments (Kohn, 1983; Taylor, 1989), whilst regimes characterised by pulses of nutrient input support larger numbers of generalist-feeding species. Unfortunately, there are no phylogenetic analyses of Indo-Pacific predatory gastropods that have been made in sufficient detail to make an adequate test of any hypothesis about evolutionary rates in different nutrient regimes. However, there are some small pieces of evidence that suggest that such differences may exist. Altena & Gittenberger (1981) have documented Recent and fossil species of *Babylonia*. This is a genus of generalist-feeding gastropods that inhabit nutrient-rich, continental margins of the northern Indo-Western Pacific. The genus, with 12 living species, has never been very diverse, with a more or less steady-state of species from the Miocene to the Recent. By contrast, the speciose, specialist-feeding, gastropod genus *Conus*, which is strongly associated with coral-reef habitats, shows a high species turnover with a massive diversification in the Miocene and during the Pleistocene to the Recent, with around 500 living species (Kohn, 1990). A similar pattern is seen in the highly specialist, sipunculan-feeding, gastropod subfamily Mitrinae, which is also diverse in coral-reef habitats (Cernohorsky, 1976; Taylor, 1989, 1993b).

Vrba (1987) has suggested that the factors controlling the speciation and extinction rates of clades are environmentally controlled. If this is generally true, then the different nutrient regimes found in the Indo-Pacific may be generating important differential rates of speciation and evolutionary change. Different groups of animals are more diverse under different nutrient regimes. Bivalves are both more diverse and abundant along eutrophic continental margins, where elevated levels of nutrient input result in high phytoplankton production. On oligotrophic reefs, primary, suspension-feeding animals tend to be both less abundant and less diverse. By contrast, there are more food-specialised predatory gastropods in oligotrophic regimes than in eutrophic areas, where generalist-feeding species are more abundant.

References

Abbott, R.T., 1960. The genus *Strombus* in the Indo-Pacific. *Indo-Pacific Mollusca*, 1, 831–999.
Abe, N., 1989. Interactions between carnivorous gastropods and their sessile animal prey at a rocky intertidal shore. *Physiological Ecology, Japan*, 26, 1–38.
Altena, C.O. van R. & Gittenberger, E., 1981. The genus *Babylonia* (Prosobranchia, Buccinidae). *Zoologische Verhandelingen*, 188, 1–57.

Azuma, M., 1960. *A Catalogue of the Shell-bearing Mollusca of Okinoshima, Kasiwajima and the Adjacent Area (Tosa Province), Shikoku, Japan*. Osaka: Nishinomiya.
Birkeland, C., 1977. The importance of rate of biomass accumulation in early successional stages of benthic communities to the survival of coral recruits. In *Proceedings of the Third International Coral Reef Symposium, Miami*, vol. **1**, pp. 15–21. Miami, FL: University of Miami.
Birkeland, C., 1987. Nutrient availability as a major determinant of differences among coastal hard-substratum communities in different regions of the tropics. *UNESCO Reports in Marine Science*, **46**, 45–97.
Birkeland, C., 1988a. Second-order ecological effects of nutrient input into coral communities. *Galaxea*, **7**, 91–100.
Birkeland, C., 1988b. Geographic comparisons of coral-reef community processes. In *Proceedings of the Sixth International Coral Reef Symposium, Townsville*, vol. **1**, pp. 211–20. Townsville, Queensland: Sixth International Coral Reef Symposium Executive Committee.
Britton, J.C. & Morton, B., 1992. The ecology and feeding behaviour of *Nassarius festivus* (Prosobranchia: Nassariidae) from two Hong Kong bays. In *Proceedings of the Fourth International Marine Biological Workshop: The Marine Flora and Fauna of Hong Kong and Southern China (III), Hong Kong 1989*, ed. B. Morton, pp. 395–416. Hong Kong: Hong Kong University Press.
Cernohorsky, W.O., 1976. The Mitridae of the world. *Indo-Pacific Mollusca*, **3**, 273–528.
Cernohorsky, W.O., 1984. Systematics of the family Nassariidae. *Bulletin of the Auckland Institute and Museum*, **14**, 1–356.
Cheung, S.G., 1993. Population dynamics and energy budgets of green-lipped mussel *Perna viridis* (Linnaeus) in a polluted harbour. *Journal of Experimental Marine Biology and Ecology*, **168**, 1–24.
Crossland, C.J., Hatcher, B.G., Atkinson, M.J. & Smith, S.V., 1984. Dissolved nutrients of a high-latitude coral reef, Houtman Abrolhos Islands, Western Australia. *Marine Ecology Progress Series*, **14**, 159–63.
Curtis, L.A. & Hurd, L.E., 1979. On the broad nutritional requirements of the mud snail, *Ilyanassa (Nassarius) obsoleta* (Say), and its polytrophic role in the food web. *Journal of Experimental Marine Biology and Ecology*, **41**, 289–97.
DeAngelis, D.L., Mulholland, P.J., Palumbo, A.V., Steinman, A.D., Huston, M.A. & Elwood, J.W., 1989. Nutrient dynamics and food-web stability. *Annual Review of Ecology and Systematics*, **20**, 71–95.
Frith, D., Tantansiriwong, R. & Bhatia, O., 1976. Zonation of macrofauna on a mangrove shore, Phuket island. *Research Bulletin Phuket Marine Biological Center*, **10**, 1–37.
Glynn, P.W., 1988. El Nino warming, coral mortality and reef framework distribution by echinoid bioerosion in the eastern Pacific. *Galaxea*, **7**, 129–60.
Grigg, R.W., Polovina, J.J. & Atkinson, M.J., 1984. Model of a coral reef ecosystem. III. Resource limitation, community regulation, fisheries yield and resource management. *Coral Reefs*, **3**, 23–7.
Hallock, P., 1988. The role of nutrient availability in bioerosion: Consequences to carbonate buildups. *Palaeogeography, Palaeoclimatology, Palaeoecology*, **63**, 275–91.
Hallock, P. & Schlager, W., 1986. Nutrient excess and the demise of coral reefs and carbonate platforms. *Palaios*, **1**, 389–98.
Highsmith, R.C., 1980. Geographic patterns of coral bioerosion: A productivity

hypothesis. *Journal of Experimental Marine Biology and Ecology,* **46**, 177–96.
Jackson, J.B.C., 1973. The ecology of molluscs of *Thalassia* communities, Jamaica, West Indies. I. Distribution, environmental physiology, and ecology of common shallow-water species. *Bulletin of Marine Science,* **23**, 313–50.
Johannes, R.E., Pearce, A.F., Wiebe, W.J., Crossland, C.J., Rimmer, D.W., Smith, D.F. & Manning, C., 1994. Nutrient characteristics of well-mixed coastal waters off Perth, Western Australia. *Estuarine, Coastal and Shelf Science,* **39**, 273–85.
Jones, D.S., 1994. Barnacles (Cirripedia, Thoracica) of the Cocos (Keeling) islands. *Atoll Research Bulletin,* **413**, 1–5.
Kawaguti, S., 1983. The third record of association between bivalve molluscs and zooxanthellae. *Proceedings of the Japan Academy, Series B,* **59**, 17–20.
Kay, E.A., 1971. The littoral marine mollusks of Fanning Island. *Pacific Science,* **25**, 260–81.
Kay, E.A., 1979. *Hawaiian Marine Shells.* Honololu: Bernice P. Bishop Museum Press.
Kay, E.A., 1984. Patterns of speciation in the Indo-West Pacific. In *Biogeography of the Tropical Pacific,* ed. F.J. Radovsky, P.H. Raven & S.H. Sohmer, pp. 15–31. Bernice P. Bishop Museum Special Publication, no. 72. Honolulu: Bernice P. Bishop Museum Press.
Kay, E.A. & Johnson, S., 1987. Mollusca of Enewetak Atoll. In *The Natural History of Enewetak Atoll,* ed. D. Devaney, E.S. Reese, B.L. Burch & P. Helfrich, vol. 2, pp. 105–46. Oak Ridge, TN: U.S. Department of Energy.
Kay, E.A. & Switzer, M.F., 1974. Molluscan distribution patterns in Fanning Island lagoon and a comparison of the mollusks of the lagoon and seaward reefs. *Pacific Science,* **28**, 275–95.
Kinsey, D.W., 1988. Coral reef system response to some natural and anthropogenic stresses. *Galaxea,* **7**, 113–28.
Kohn, A.J., 1983. Marine biogeography and evolution in the tropical Pacific: Zoological perspectives. *Bulletin of Marine Science,* **33**, 528–35.
Kohn, A.J., 1987. Intertidal ecology at Enewetak Atoll. In *The Natural History of Enewetak Atoll,* ed. D. Devaney, E.S. Reese, B.L. Burch & P. Helfrich, vol. **1**, pp. 139–57. Oak Ridge, TN: U.S. Department of Energy.
Kohn, A.J., 1990. Tempo and mode of evolution in Conidae. *Malacologia,* **32**, 55–66.
Kohn, A.J. & Almasi, K.N., 1993. Comparative ecology of a biogeographically heterogeneous *Conus* assemblage. In *The Marine Flora and Fauna of Rottnest Island, Western Australia. Proceedings of the Fifth International Marine Biological Workshop,* ed. F.E. Wells, D.I. Walker, H. Kirkman & R. Lethbridge, pp. 523–38. Perth: Western Australian Museum.
Kohn, A.J. & Perron, F.E., 1994. *Life History and Biogeography: Patterns in Conus.* Oxford: Oxford University Press.
Kuroda, T., 1960. *A Catalogue of Molluscan Fauna of Okinawa Islands (exclusive of Cephalopoda).* Okinawa: University of Okinawa.
Kuroda, T., Habe, T. & Oyama, K., 1971. *The Sea Shells of Sagami Bay.* Tokyo: Maruzen Co. Ltd.
Lee, S.Y., 1986. Growth and reproduction of the green mussel *Perna viridis* (L.) (Bivalvia: Mytilacea) in contrasting environments in Hong Kong. *Asian Marine Biology,* **3**, 111–28.
Lee, S.Y. & Morton, B., 1985. The Hong Kong Mytilidae. In *Proceedings of the Second International Workshop on the Malacofauna of Hong Kong and*

Southern China, Hong Kong, 1983, ed. B. Morton & D. Dudgeon, pp. 49–76. Hong Kong: Hong Kong University Press.

Lewis, M.R., 1989. The variegated ocean: A view from space. *New Scientist*, 124 (1685), 37–40.

Littler, M.M., Littler, D.S. & Titlyanov, E.A., 1991. Comparisons of N- and P-limited productivity between high granitic islands versus low carbonate atolls in the Seychelles Archipelago: A test of the relative-dominance paradigm. *Coral Reefs*, 10, 199–209.

Morton, B., 1983. Coral-associated bivalves of the Indo-Pacific. In *The Mollusca*, vol. 6. *(Ecology)*, ed. W.D. Russell-Hunter, pp. 139–224. New York and London: Academic Press.

Morton, B., 1989. Pollution of the coastal waters of Hong Kong. *Marine Pollution Bulletin*, 20, 310–18.

Morton, B., 1990. The physiology and feeding behaviour of two marine scavenging gastropods in Hong Kong: The subtidal *Babylonia lutosa* (Lamarck) and the intertidal *Nassarius festivus* (Powys). *Journal of Molluscan Studies*, 56, 275–88.

Morton, B., 1991. Do the Bivalvia demonstrate environment specific sexual strategies? A Hong Kong model. *Journal of Zoology*, 223, 131–42.

Morton, B., 1995. The population dynamics and reproductive cycle of *Septifer virgatus* (Bivalvia: Mytilidae) on an exposed rocky shore in Hong Kong. *Journal of Zoology*, 235, 485–500.

Morton, B. & Britton, J.C., 1993. The ecology, diet and foraging strategy of *Thais orbita* (Gastropoda: Muricidae) on a rocky shore of Rottnest island, Western Australia. In *The Marine Flora and Fauna of Rottnest Island, Western Australia. Proceedings of the Fifth International Marine Biological Workshop*, ed. F.E. Wells, D.I. Walker, H. Kirkman & R. Lethbridge, pp. 539–63. Perth: Western Australian Museum.

Muscatine, L. & Porter, J.W., 1977. Reef corals: Mutualistic symbiosis adapted to nutrient-poor environments. *Bioscience*, 27, 454–60.

Newman, W.A., 1960. The paucity of intertidal barnacles in the tropical Western Pacific. *Veliger*, 2, 89–94.

Oliver, P.G., 1992. *Bivalved Seashells of the Red Sea*. Wiesbaden and Cardiff: Verlag Christa Hemmen and the National Museum of Wales.

Pearson, T.H. & Rosenberg, R., 1978. Macrobenthic succession in relation to organic enrichment and pollution of the marine environment. *Oceanography and Marine Biology Annual Review*, 16, 229–311.

Qi Zhongyan, Ma Xiutong, Xie Yukan & Lin Biping, 1984. Preliminary list of marine molluscs from Hainan Island. In *Research on Tropical Ocean-Marine Environment and Experimental Marine Biology of Tropical Ocean*, pp. 1–22. Xiamen, South China Sea Institute of Oceanology: China Ocean Press.

Reid, D.G., 1986. *The Littorinid Molluscs of Mangrove Forests in the Indo-Pacific Region*. London: British Museum (Natural History).

Reid, R.G.B., 1990. Evolutionary implications of sulphide-oxidising symbioses in bivalves. In *The Bivalvia. Proceedings of a Memorial Symposium in Honour of Sir Charles Maurice Yonge, Edinburgh 1986*, ed. B. Morton, pp. 127–40. Hong Kong: Hong Kong University Press.

Richard, G., 1982. Mollusques Lagunaires et Récifaux de Polynesie Française. D.Sc. Thesis, Université Pierre et Marie Curie, Paris.

Richard, G., Salvat, B. & Millous, O., 1979. Mollusques et faune benthique du lagon de Takapoto. *Journal de la Societé des Océanistes*, 62, 59–68.

Rosen, B.R., 1988. Progress, problems and patterns in the biogeography of reef

corals and other tropical reef organisms. *Helgoländer Meeresuntersuchungen*, **42**, 269–301.
Ryther, J.H., 1969. Photosynthesis and fish production in the sea. *Science*, **166**, 72–6.
Salvat, B., 1967. Importance de la faune malacologique dans les atolls polynesiens. *Cahiers du Pacifique*, **11**, 1–49.
Savidge, G., Lennon, J. & Matthews, A.J., 1990. A shore based survey of upwelling along the coast of Dhofar region, southern Oman. *Continental Shelf Research*, **10**, 259–75.
Schneider, J.A., 1993. Phylogenetic relationships of advanced cardiid bivalves. In *Origin and Evolutionary Radiation of the Mollusca*. Centenary Symposium of the Malacological Society of London. London, September, 1993 (Abstract).
Scott, P.J.B. & Cope, M., 1990. Tolo revisited: A resurvey of the corals in Tolo Harbour and Channel six years and half a million people later. In *Proceedings of the Second International Marine Biological Workshop: The Marine Fauna and Flora of Hong Kong and Southern China, Hong Kong, 1986*, ed. B. Morton, pp. 1203–20. Hong Kong: Hong Kong University Press.
Sidall, S.E., 1980. A clarification of the genus *Perna* (Mytilidae). *Bulletin of Marine Science*, **30**, 858–70.
Sournia, A. & Ricard, M., 1975. Phytoplankton production and primary production in Takapoto Atoll, Tuamotu Islands. *Micronesia*, **11**, 159–66.
Springsteen, F.J. & Leobrera, F.M., 1986. *Shells of the Philippines*. Manila, Philippines: Carfel Shell Museum.
Tantanasiriwong, R., 1978. An illustrated checklist of marine shelled gastropods from Phuket Island, adjacent mainland and offshore islands, western peninsular Thailand. *Research Bulletin Phuket Marine Biological Center*, **21**, 1–22.
Taylor, J.D., 1968. Coral reef and associated invertebrate communities (mainly molluscan) around Mahé, Seychelles. *Philosophical Transactions of the Royal Society of London, Series B*, **254**, 129–206.
Taylor, J.D., 1971. Reef associated molluscan assemblages in the western Indian Ocean. *Symposium of the Zoological Society of London*, **28**, 501–34.
Taylor, J.D., 1978. Habits and diet of predatory gastropods at Addu Atoll, Maldives. *Journal of Experimental Marine Biology and Ecology*, **31**, 83–103.
Taylor, J.D., 1980. Diets and habitats of shallow water predatory gastropods around Tolo Channel, Hong Kong. In *Proceedings of the First International Workshop on the Malacofauna of Hong Kong and Southern China, Hong Kong, 1977*, ed. B. Morton, pp. 163–80. Hong Kong: Hong Kong University Press.
Taylor, J.D., 1984. A partial food web involving predatory gastropods on a Pacific fringing reef. *Journal of Experimental Marine Biology and Ecology*, **74**, 273–90.
Taylor, J.D., 1986. Diets of sand-living predatory gastropods at Piti Bay, Guam. *Asian Marine Biology*, **3**, 47–58.
Taylor, J.D., 1989. The diet of coral-reef Mitridae (Gastropoda) from Guam; with a review of other species of the family. *Journal of Natural History*, **23**, 261–78.
Taylor, J.D., 1990. Field observations of prey selection by the muricid gastropods *Thais clavigera* and *Morula musiva* feeding upon the intertidal oyster *Saccostrea cucullata*. In *Proceedings of the Second International Marine Biological Workshop: The Marine Flora and Fauna of Hong Kong and Southern China (II), Hong Kong, 1986*, ed. B. Morton, pp. 837–55. Hong Kong: Hong Kong University Press.
Taylor, J.D., 1992. Long-term changes in the gastropod fauna of Tolo Channel and Mirs Bay, Hong Kong: The 1989 survey. In *Proceedings of the Fourth International Marine Biological Workshop: The Marine Flora and Fauna of*

Hong Kong and Southern China (III), Hong Kong, 1989, ed. B. Morton, pp. 557–73. Hong Kong: Hong Kong University Press.

Taylor, J.D., 1993a. Regional variation in the structure of tropical benthic communities: Relation to regimes of nutrient input. In *The Marine Biology of the South China Sea*, ed. B. Morton, pp. 337–56. Hong Kong: Hong Kong University Press.

Taylor, J.D., 1993b. Dietary and anatomical specialization of mitrid gastropods (Mitridae) at Rottnest Island, Western Australia. In *The Marine Flora and Fauna of Rottnest Island, Western Australia. Proceedings of the Fifth International Marine Biological Workshop*, ed. F.E. Wells, D.I. Walker, H. Kirkman & R. Lethbridge, pp. 583–99. Perth: Western Australian Museum.

Taylor, J.D., 1994. Sublittoral benthic gastropods around southern Hong Kong. In *The Malacofauna of Hong Kong and Southern China, III. Proceedings of the Third International Workshop on the Malacofauna of Hong Kong and Southern China, Hong Kong, 1992*, ed. B. Morton, pp. 479–95. Hong Kong: Hong Kong University Press.

Taylor, J.D. & Reid, D.G., 1984. The abundance and trophic classification of molluscs upon coral reefs in the Sudanese Red Sea. *Journal of Natural History*, **18**, 175–209.

Taylor, J.D. & Morton, B. 1997. The diets of predatory gastropods in the Cape d'Aguilar Marine Reserve, Hong Kong. *Asian Marine Biology* (in press).

Taylor, J.D. & Shin, P.K.S., 1990. Trawl surveys of sublittoral gastropods in Tolo Channel and Mirs Bay; a record of change from 1976–1986. In *Proceedings of the Second International Marine Biological Workshop: The Marine Flora and Fauna of Hong Kong and Southern China (II), Hong Kong, 1986*, ed. B. Morton, pp. 857–81. Hong Kong: Hong Kong University Press.

Tong, L.K.Y., 1986. The feeding ecology of *Thais clavigera* and *Morula musiva* (Gastropoda: Muricidae) in Hong Kong. *Asian Marine Biology*, **3**, 163–78.

Tsi, C.Y. & Ma, S.T., 1982. A preliminary checklist of the marine Gastropoda and Bivalvia (Mollusca) of Hong Kong and southern China. In *The Marine Flora and Fauna of Hong Kong and Southern China. Proceedings of the First International Workshop on the Marine Flora and Fauna of Hong Kong and Southern China, 1980*, ed. B.S. Morton & C.K. Tseng, pp. 431–58. Hong Kong: Hong Kong University Press.

Tsuchiya, M. & Lirdwitayapasit, T., 1986. Distribution of intertidal animals on rocky shores of the Sichang Islands, the Gulf of Thailand. *Galaxea*, **5**, 15–25.

Vermeij, G.J., 1978. *Biogeography and Adaptation*. Cambridge, MA: Harvard University Press.

Vermeij, G.J., 1987. *Evolution and Escalation. An ecological History of Life*. Princeton, NJ: Princeton University Press.

Vermeij, G.J., 1990. Tropical Pacific pelecypods and productivity: A hypothesis. *Bulletin of Marine Science*, **47**, 62–7.

Vrba, E.S., 1987. Ecology in relation to speciation rates: Some case histories of Miocene–Recent mammal clades. *Evolutionary Ecology*, **1**, 293–300.

Wells, F.E., 1994. Marine mollusks of the Cocos (Keeling) islands. *Atoll Research Bulletin*, **410**, 1–22.

Wiebe, W.J., 1987. Nutrient pools and dynamics in tropical, marine, coastal environments, with special reference to the Caribbean and Indo-West Pacific regions. *UNESCO Reports in Marine Science*, **46**, 19–42.

Wilkinson, C.R., 1986. The nutritional spectrum of coral reef benthos. *Oceanus*, **29**, 68–75.

Wilkinson, C.R. & Cheshire, A.C., 1989. Patterns in the distribution of sponge populations across the central Great Barrier Reef. *Coral Reefs*, **8**, 127–34.
Williams, D.McB., Russ, G. & Doherty, P.J., 1986. Reef fish. *Oceanus*, **29**, 76–82.
Wu, R.S.S., 1982. Periodic defaunation and recovery in a sub-tropical epibenthic community in relation to organic pollution. *Journal of Experimental Marine Biology and Ecology*, **64**, 253–69.
Wu, Wen-Lung, 1980. The list of Taiwan bivalve fauna. *Quarterly Journal of the Taiwan Museum*, **33**, 1–208.

Chapter 9
Why are coral reef communities so diverse?

ALAN J. KOHN
Department of Zoology, University of Washington, Seattle, Washington 98195, USA

Abstract

Extrinsic and intrinsic factors, physical and biological factors, and the interactions of all these are hypothesised to affect the diversity of coral reef-associated faunas. Extrinsic factors include environmental change at large spatial (regional) and temporal (geological) scales, and small (local or landscape) scale variation over ecologically relevant time. The latter also impinge on aspects intrinsic to the organisms, e.g. ecological specialisation, key innovations and life-history attributes, especially larval biology of sedentary benthic invertebrates. Certain species-rich genera and families of invertebrates and fishes contribute importantly to high species diversity of Indo-Pacific coral reefs. Naturally occurring gradients in diversity, ecology and life-history within these taxa can serve as dependent variables to test hypotheses of diversification. Gradient analyses of the most diverse genus of Indo-Pacific reef invertebrates, the gastropod *Conus*, support the ecological determinism and life-history hypotheses of diversity. The data do not refute other hypothesised mechanisms, which probably also operate. Broader application of gradient analysis to diverse taxa in the future may sharpen understanding of processes leading to modern patterns of diversity.

9.1 Introduction

Although geographically restricted to tropical seas and occupying only 0.1% of the earth's surface, coral reefs have globally important implications for marine biodiversity. Reefs support unusually diverse animal communities with distinctive taxonomic structure and geographical distribution patterns. Coral reefs are oases of high primary productivity in barren tropical seas, and reef-building organisms have changed the face of the earth by creating entire archipelagoes of islands and, over geological time, landforms that have become incorporated in continents.

Here I will examine hypotheses of why the present diversity of reef-associated animal assemblages is high, and what factors and processes have fostered diversification. To exemplify patterns of tropical marine biodiversity, I will focus on *Conus*, a species-rich, conspicuous and ecologically important genus of gastropod molluscs represented by more than 100 species on Indo-Pacific (IP) coral reefs. I will suggest that analysis of taxonomic, ecological and geographical gradients in diverse taxa can provide data that may help to evaluate alternative hypotheses of how diversity is generated and maintained.

Two striking and enigmatic patterns of IP reef-associated invertebrates and fishes that contribute importantly to biodiversity are (a) large numbers of species in certain genera and families, probably due to their rapid evolutionary radiation and (b) very broad geographical ranges of many of these species. An important long-term goal of research in marine biodiversity is to understand these patterns, which contrast markedly to those of the most diverse tropical terrestrial environments (Kohn, 1983a). In the latter, each species tends to occupy a much smaller fraction of the total range of its genus or family, and species turnover as a function of distance is thus much more rapid (e.g. Cody, 1993). The species richness of several reef taxa is impressive (Table 9.1), but many groups remain poorly documented. Recent efforts to improve systematic knowledge of reef-associated taxa indicate sharp increases in numbers of known species of several groups (Fig. 9.1) and suggest that we may severely underestimate the diversity of many that have not been subjected to recent scrutiny. Another reason for the underestimation of coral reef community diversity is that sibling species are proving to be much more common than previously realised, especially among sessile, clonal reef invertebrates such as corals and bryozoans (Knowlton, 1993; Knowlton & Jackson, 1994), suggesting that many more species remain to be discovered.

The different hypotheses advanced to explain diversification emphasise factors and processes that are either extrinsic (**H1** and **H2**) or intrinsic (**H2–H4**) to the organisms and that may differentially affect survival, distribution and radiation of different taxa and hence determine diversity patterns. These hypotheses are not mutually exclusive, and they are difficult to refute. I have adapted the following grouping from Kohn *et al.* (1991) and Ricklefs & Schluter (1993):

Table 9.1. *Species richness of coral reef-associated faunas*

Taxon	Number of species	Reference
General Indo-Pacific reef faunas		
Fishes	>2200	Sale (1980)
Corals	390	Sebens (1994)
Nudibranchs	>500	Gosliner (1993)
Estimates of faunas of single reefs		
Fishes	860	Sale *et al.* (1994)
Molluscs	150–500	Sorokin (1993)
Crustacea	100–250	Sorokin (1993)
Polychaetes	100–200	Sorokin (1993)
Echinoderms	50–100	Sorokin (1993)
Sponges	>50	Sorokin (1993)

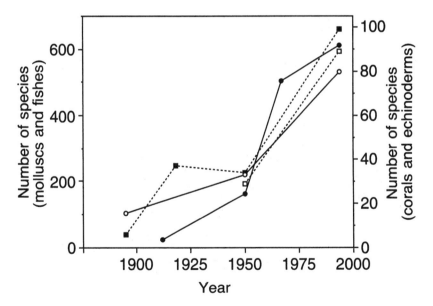

Fig. 9.1. An example of increasing knowledge of the diversity of coral reef-associated animals over time. Cumulative numbers of species of molluscs, fishes, corals and echinoderms recorded from Cocos (Keeling) Islands, based on data from reports in Woodroffe (1994). The initial values for fishes and corals indicate the number of species reported before 1900. -●-, molluscs; -○-, fishes; -■-, corals, -□-, echinoderms.

H1. *Historical contingency.* Large-scale spatial (i.e. regional) and temporal (i.e. geological) environmental stability versus change, for example changing patterns of the geographical configurations of continents, sea-level fluctuations and major extinction events.

H2. *Ecological determinism.* Local (i.e. community- or landscape-level) environmental complexity or spatial heterogeneity, ecological opportunism and specialisation, rate and stability of primary production, and amplitude of physical environmental variables, including episodic disturbance, in ecologically relevant time. The ability to invade new or empty local habitats, and to accomplish 'niche diversification' depend both on habitat structure and heterogeneity and on the ability of organisms to take advantage of opportunities to enhance resource specialisation.

H3. *Key innovations.* Novel attributes associated with diversification early in clade evolution, including the roles of genetic flexibility and developmental and structural constraints.

H4. *Breeding systems, life-history strategies and population structure.* These attributes all affect dispersal ability and thus geographical distribution and the likelihood of range fragmentation, speciation and extinction.

9.2 Diversity and community trophic structure

Coral reefs are biogenic environments; their habitat results entirely from photosynthetically aided calcium carbonate skeletogenesis, primarily of corals and algae. The diversity of hermatypic corals (trophic composites since they are both primary producers and predators) is better understood than that of other reef-associated organisms (Huston, 1985; Sebens, 1994). Coral species diversity is typically low on the shallowest parts of the reef, increases with depth from the crest to about 15–30 m, and then decreases with increasing depth. This pattern results mainly from the interaction of local extrinsic and intrinsic factors. The former include gradients of light and disturbance, both of which decrease with depth. The main intrinsic factors are colony growth following disturbance and concomitant competition for space, both with each other, and with other, mainly modular primary space-occupying organisms. Depth and the frequency and severity of disturbance also affect these attributes of the reef-building organisms. The evidence (Huston, 1985) thus favours primarily **H2** and, secondarily because of colonial growth via asexual reproduction, **H4**.

Although herbivores, most importantly sea urchins and fishes, may indirectly affect coral diversity by reducing the cover of algae that compete with corals for primary space on the reef (Huston, 1985), the factors that determine herbivore and carnivore diversity on reefs are much more poorly understood than for corals, and they largely remain a challenge for the future.

9.3 *Conus* as a model taxon

With more than 500 extant species, the benthic prosobranch gastropod genus *Conus* is probably the largest among marine invertebrates. More than 300 species occur in the tropical IP region, about 45% of which occupy coral reef-associated habitats (Röckel *et al.*, 1995). The genus is thus an important contributor to the high biotic diversity of reefs. Species of *Conus* are ecologically important, large, often abundant and ecologically better known than most reef-associated animals.

A coral reef is a heterogeneous mosaic or landscape of substrate types including calcareous sands, smooth reef rock, living and dead coral, and calcareous and fleshy algae, as well as intermediates and combinations of these that are variably suitable as habitat for gastropods. Various species of *Conus* use all of these except living coral and fleshy algae as substrates (Kohn, 1983b). Trophically, most *Conus* species occupy a characteristic level in their communities as primary carnivores, preying on worms, other gastropods or fishes that are generally either herbivorous or feed on particulate organic matter. The heavy, durable shells of *Conus* are well preserved in the fossil record, and broad patterns of its evolutionary history have been worked out (Kohn, 1990).

Biogeographically, many *Conus* species appear to be very widely distributed, some occupying all or most of the IP region (e.g. Kohn & Perron, 1994, fig. 5.2). This attribute enhances the similarity of species assemblages from place to place in this vast area. For subtidal coral reef assemblages across the entire IP region, the mean difference (defined as percentage species turnover of Cody, 1993) is only 55%, the curve of difference versus log distance asymptotes at 60%, and the most different communities are at moderate distances from each other, not the farthest apart (Fig. 9.2(a)). The mean percentage turnover of intertidal bench assemblages is only 39%, and the curve asymptotes at 45%, despite the inclusion of sites more than 15 000 km apart (Fig. 9.2(b)). In contrast, animal and plant species turnover rates in terrestrial communities often attain 100% in a few hundred km (see Cody, 1993 for

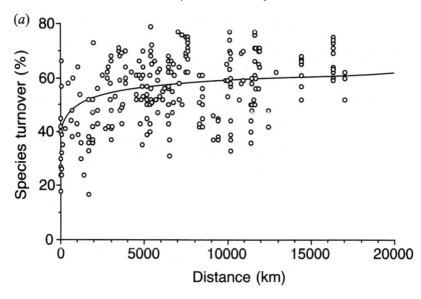

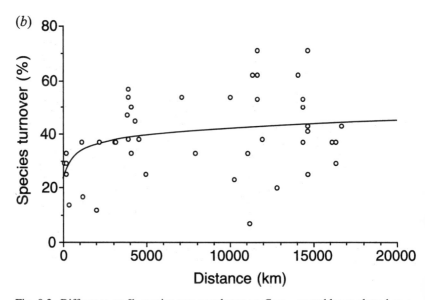

Fig. 9.2. Difference or % species turnover between *Conus* assemblages plotted as a function of distance apart of the sites. Percent difference between two sites is determined as $100(1 - [2c/(a + b)])$, where c = the number of species occurring in both samples, a = the total number in one sample and b = the total number in the other sample. The expression in square brackets is Sørensen's (1948) coefficient of community (Whittaker, 1967; Cody, 1993). (*a*) subtidal reef assemblages; (*b*) intertidal bench assemblages.

Table 9.2. *Salient gradients in* Conus

SPECIES DIVERSITY: From 1 species along geographical periphery of genus range to 25 or more on complex coral reefs
HABITAT: On and in fine to coarse sediments, algal turfs, rubble to reef limestone
GENERALIST–SPECIALIST: Markedly differential food specialisation by co-occurring species; some differential habitat use
REPRODUCTION AND DEVELOPMENT: <100 to >1.5 × 10^6 eggs/clutch; eggs 125 μm to 1 mm diameter; pre-hatching period 8–25 days; pre-competent planktonic period 0–30 days
GEOGRAPHICAL DISTRIBUTION: Single archipelagoes to entire tropical Indo-Pacific region

---

Based on data in Kohn (1967, 1983b), Kohn & Nybakken (1975) and Kohn & Perron (1994).

examples) and the proportion of narrowly endemic species is very high in the most diverse tropical communities (Gentry, 1986).

Indo-Pacific coral reef-associated *Conus* assemblages are further characterised by the presence of striking gradients in species diversity, resource utilisation patterns, reproduction and development, and geographical ranges (Table 9.2). In the next section I will show how the data from analyses of these gradients may be used as dependent variables to test the hypotheses of diversity causes listed above.

9.4 Gradient analysis in *Conus*

9.4.1 Diversity gradients

Species diversity in IP *Conus* has long been known to differ strikingly among the major habitat types associated with coral reefs (Kohn, 1967, 1969). Extensive patches of sand substrate support the least diverse assemblages, with a mean of 3 species, although species richness is quite variable. A larger (mean = 8) and much less variable number of species co-occur on topographically simple, smooth intertidal reef rock, with variable cover of low algal turf, throughout the IP region. The most diverse assemblages (mean = 14), again with a broad range, occupy the topographically most complex and physically more benign subtidal reef platforms (Table 9.3). The more recent information incorporated in Table 9.3 confirms the patterns shown in the earlier studies (Kohn, 1967, 1969).

Table 9.3. *Species diversity of* Conus *assemblages in Indo-Pacific coral reef-associated habitats*

Habitat type	Number of assemblages studied	Mean sample size	Number of co-occurring species Mean	Range	Species diversity (H') Mean	Range
I. Extensive sand areas	6	96	3.3	1–8	0.5	0.0–1.4
II. Intertidal benches	10	250	7.9	6–9	1.3	0.4–1.5
III. Subtidal reef platforms	24	112	14.4	9–24	2.1	1.6–2.7

Data from Kohn (1967, 1969, and unpublished results).
Geographical ranges of data: I. 21°28'N–4°38'S; 55°30'E–157°49'W; II. 22°16'N–5°26'S; 55°31'E–156°23'W; III. 21°25'N–23°27'S; 47°34'E–157°49'W.

At the finer within-habitat level, substrate factors also appear to be important determinants of diversity. Kohn (1983b, figs. 1–4) showed that both species diversity and population density increase significantly with the area of algal-bound sand on reef platforms, in both Micronesia and the Australian Great Barrier Reefs. Conversely, both diversity and density decrease precipitously as the least favourable microhabitat, living coral, increases in cover. The former microhabitat provides both food and shelter, while living coral stings the feet of gastropods that tread on it and causes an avoidance response (my unpublished observations), and it is unsuitable habitat for the predominantly polychaete prey organisms. The results of diversity gradient analysis thus mainly support **H2**, the dependence of diversity on combinations of local extrinsic and intrinsic ecological factors.

9.4.2 Ecological gradients

Co-occurring species of *Conus* on coral reefs typically vary considerably along a generalist–specialist spectrum of space and food resource use. As implied above, microhabitats range widely from hard to soft substrates. Some species are almost always found on or in the same substrate type, while others utilise a broad range (e.g. Table 9.4). Use of food resources follows a similar pattern. In a rapid-strike predation method, *Conus* paralyses its prey by injecting potent neuroactive peptides (Olivera *et al.*, 1990; Terlau *et al.*, 1996). With few exceptions, each

Table 9.4. *Use of different substrate types during daytime by* Conus *species on Great Barrier Reef platforms (Heron, One Tree, Green and Low Isles reefs)*

Species	Substrate type					
	Large sand patches	Small sand-filled depressions	Sand under coral rock	Rubble, rubble + sand	Reef limestone	Algae–sand on reef limestone
C. eburneus (12)	1.00					
C. arenatus (13)	0.84	0.08	0.08			
C. frigidus (27)		0.48		0.30		0.22
C. chaldaeus (15)		0.40	0.33		0.20	0.07
C. miles (17)		0.35		0.29	0.35	
C. flavidus (82)		0.33		0.30	0.26	0.11
C. miliaris (74)		0.41	0.03	0.35	0.11	0.10
C. sanguinolentus (43)		0.44		0.28	0.23	0.05
C. sponsalis (24)		0.46		0.21	0.04	0.29
C. lividus (14)				0.50	0.50	
C. textile (15)	0.07	0.07	0.80	0.07		
C. marmoreus (16)	0.19	0.19	0.19	0.31	0.12	
C. ebraeus (81)		0.23	0.01	0.31	0.31	0.14
C. musicus (10)		0.30		0.50	0.10	0.10
C. coronatus (72)	0.03	0.25	0.08	0.42	0.14	0.07

Species are arranged to grade from occupation of softest to hardest substrates. Numbers in body of Table are proportions of species at left using substrate type at top. Sample sizes are in parentheses.

species preys only on polychaetes, gastropods, or fishes, but within these categories the number of species taken varies considerably, from one or two to ten or more. The clear pattern that emerges from these studies is that co-occurring species differentially utilise food or microhabitat resources or both (Kohn & Nybakken, 1975; Reichelt & Kohn, 1985). Differential specialisation on prey species is typically much more pronounced than on microhabitat type, that is between-species overlap or similarity is considerably lower, as illustrated in Kohn & Nybakken (1975, fig. 12) and in new analyses of data from the study of Reichelt & Kohn (1985; shown in Fig. 9.3). The results of ecological gradient analyses thus support the ecological opportunism and specialisation component of **H2**.

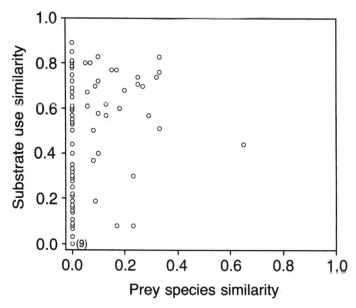

Fig. 9.3. Relationship between similarity of substrate-type occupation and prey species utilisation by *Conus* on Australian Great Barrier Reef Platforms (listed in Table 9.4). Symbols indicate values of proportional similarity (PS_I) (Whittaker, 1952; Kohn & Riggs, 1982) for all pairwise comparisons of species. For 9 of the 99 species pairs, $PS_I = 0$ on both axes. Data from Table 9.4 and Reichelt & Kohn (1985).

9.4.3 Disturbance

The effects of disturbances on reef organisms other than corals are poorly known, and *Conus* is no exception. However, two opportunistic studies at Enewetak in the Marshall Islands suggest that it is important. In 1979, a typhoon reduced *Conus* diversity and population density in less protected sites but not those with secure refuges, and an unusually severe rainstorm coinciding with low tide reduced diversity on the inshore portion of the seaward, windward reef platform of the atoll in 1972 (Kohn, 1980; Leviten & Kohn, 1980).

9.4.4 Life history and biogeographic gradients

Finally, how do gradients in reproductive biology and life-history relate to the geographical distribution of biodiversity? Kohn & Perron (1994) recently documented a broad range of attributes, and strong correlations among them, in 62 primarily reef-associated IP *Conus* species. Egg diameter in this set of species ranges from 125–825 μm, equivalent to a nearly 300-fold volume range. Egg size is a critical variable that enables prediction of several subsequent life-history attributes, including time of development to hatching, size at hatching, larval growth rate and pre-competent planktonic larval period. Dispersal ability must depend on the time larvae spend in the plankton, and analysis of this gradient shows that the pre-competent duration explains a highly significant 43% of the variance in absolute geographical range area of IP *Conus* species (Kohn & Perron, 1994, fig. 5.4). After a short benthic, intracapsular embryonic period, species with small eggs hatch into small veliger larvae. These must feed and grow for several weeks in the plankton before attaining competence to metamorphose to benthic juveniles. Such larvae are subject to extensive passive dispersal by ocean currents, and their species ranges tend to be very broad, some embracing the entire IP region (Kohn & Perron, 1994, fig. 5.2). In contrast, *Conus* species with eggs 0.5 mm in diameter or larger hatch as larger veliconcha larvae ready to metamorphose immediately with a very brief or no free-swimming stage. The ranges of these species are typically narrowly restricted to archipelagoes or to reefs along continental shores. The results of life-history gradient analysis thus indicates that geographical patterns of diversity depend on life-history strategies, thus supporting a component of **H4**.

9.5 Discussion

The genus *Conus* exemplifies those taxa of reef-associated invertebrates and fishes that are characterised by high rates of both speciation and survivorship of species and by broad geographical ranges, and thus contribute importantly to the high diversity of coral reef faunas in the IP tropics. Naturally occurring gradients of species diversity and of ecological and life-history attributes among such congeners provide dependent variables to test hypotheses of the causes of high diversity. This is especially important because experimentally manipulating these factors is difficult or impossible. The data from gradient analyses summarised here (Table 9.5) most strongly support the hypotheses that ecological factors and life-history strategies are important determinants of diversity on reefs.

Studies of the better-known reef-building corals and other sessile invertebrates also support the predominance of local ecological determination of diversity. The most important factors are both extrinsic, mainly the light gradient with depth and the severity and frequency of disturbance, and intrinsic attributes such as growth rate and competitive ability (Huston, 1985; Jackson & Hughes, 1985; Sebens, 1994).

Table 9.5. *Gradient analyses of coral reef-associated Conus*

Gradient (dependent variable)	Independent variable	+ or − association	Significance of relationship	Reefs in study
Beta species diversity	Habitat complexity	+	Types I–II: *[a] I–III: ***[a] II–III: ***[a]	Trans-Indo-Pacific ($n = 40$ habitats)
Alpha species diversity	Prevalence of living coral	− −	**[b] **[b]	Enewetak, Guam Great Barrier Reefs
Alpha species diversity	Prevalence of favourable substrate	+ +	**[b] **[b]	Enewetak, Guam Great Barrier Reefs
Differential substrate use	Substrate type distribution	−	$\pm$[c]	Hawaii, Indian Ocean, Great Barrier Reefs
Differential food use	Food type distribution	−	+[c]	Hawaii, Indian Ocean, Great Barrier Reefs
Egg size, developmental mode	Extent of geographical range	−	***[d]	Trans-Indo-Pacific ($n = 61$ species)

*$p < 0.05$; ** $p < 0.01$; *** $p < 0.001$.
[a]Mann–Whitney U-tests; data from Kohn (1967, 1969 and unpublished results).
[b]Tests of correlation analyses (Kohn, 1983).
[c]General relationships; data not amenable to statistical test; data from Kohn (1959, 1968), Kohn & Nybakken (1975), Reichelt & Kohn (1985) and this chapter.
[d]Test of regression analysis (Kohn & Perron, 1994).

Analyses of diversity gradients in IP reef fishes also support the ecological determinism hypothesis. In a decade-long study of the fish assemblages of patch reefs at One Tree Reef, Australian Great Barrier Reefs (Sale *et al.*, 1994), species richness related directly to reef diameter, area, volume and topographic complexity, and inversely to living coral cover. A step-up multiple regression with those factors accounted for 83% of the variance. Galzin *et al.* (1994) showed significant correlations of numbers of species and genera with atoll lagoon size, which they interpreted as a proxy for habitat complexity. Reef fishes in some families favour living coral, and their species richness was directly correlated with percentage living coral cover (Bell & Galzin, 1984).

The results presented here by no means refute the other hypotheses, however, such as environmental change over time, and key innovations that may open new adaptive zones. Indeed evidence could be marshalled from other studies of the same genus in support of both of these (Kohn, 1982, 1990). In addition, historical contingency has profoundly affected reef diversity over the broad sweep of Phanerozoic time (Kauffman & Fagerstrom, 1993). Many factors and processes are thus important, all of the hypotheses discussed above explain some components of observed patterns, and it is not now possible to ascribe primacy to any single explanation. I conclude that gradient analysis is a promising but under-appreciated and hence under-utilised approach to the types of integrated studies at different spatial and temporal scales and different levels of biological organisation that are necessary to increase our understanding of marine biodiversity.

Acknowledgements

I thank Gustav Paulay for helpful comments. National Science Foundation grants including BSR-8700523 supported the research on *Conus* described here.

References

Bell, J.D. & Galzin, R., 1984. Influence of live coral on coral reef fish communities. *Marine Ecology Progress Series*, 15, 265–74.

Cody, M.L., 1993. Bird diversity components within and between habitats in Australia. In *Species Diversity in Ecological Communities*, ed. R.E. Ricklefs & D. Schluter, pp. 147–58. Chicago: University of Chicago Press.

Galzin, R., Planes, S., Dufour, V. & Salvat, B., 1994. Variation in diversity of coral reef fish between French Polynesian atolls. *Coral Reefs*, 13, 175–80.

Gentry, A.H., 1986. Endemism in tropical versus temperate plant communities. In *Conservation Biology*, ed. M.E. Soulé, pp. 153–81. Sunderland, MA: Sinauer Associates.

Gosliner, T.M., 1993. Biotic diversity, phylogeny and biogeography of opisthobranch gastropods of the north coast of Papua New Guinea. In *Proceedings of the 7th International Coral Reef Congress*, ed. R.H. Richmond, vol. 2, pp. 702–9. Mangilao, Guam: University of Guam Press.

Huston, M.A., 1985. Patterns of species diversity on coral reefs. *Annual Review of Ecology and Systematics*, 16, 149–77.

Jackson, J.B.C. & Hughes, T.P., 1985. Adaptive strategies of coral-reef invertebrates. *American Scientist*, 73, 265–74.

Kauffman, E.G. & Fagerstrom, J.A., 1993. The Phanerozoic evolution of reef diversity. In *Species Diversity in Ecological Communities*, ed. R.E. Ricklefs & D. Schluter, pp. 315–29. Chicago: University of Chicago Press.

Knowlton, N., 1993. Sibling species in the sea. *Annual Review of Ecology and Systematics*, 24, 189–216.

Knowlton, N. & Jackson, J.B.C., 1994. New taxonomy and niche partitioning on coral reefs: Jack of all trades or master of some? *Trends in Ecology and Evolution*, **9**, 7–9.

Kohn, A.J., 1959. The ecology of *Conus* in Hawaii. *Ecological Monographs*, **29**, 42–90.

Kohn, A.J., 1967. Environmental complexity and species diversity in the gastropod genus *Conus* on Indo-West Pacific reef platforms. *American Naturalist*, **101**, 251–9.

Kohn, A.J., 1968. Microhabitats, abundance and food of *Conus* on atoll reefs in the Maldive and Chagos Islands. *Ecology*, **49**, 1046–62.

Kohn, A.J., 1969. Diversity, utilization of resources, and adaptive radiation in shallow-water marine invertebrates of tropical oceanic islands. *Limnology and Oceanography*, **16**, 332–48.

Kohn, A.J., 1980. Populations of intertidal gastropods before and after a typhoon. *Micronesia*, **16**, 215–28.

Kohn, A.J., 1982. Gastropod paleoecology and the evolution of taxonomic diversity. In *Third North American Paleontological Convention Proceedings*, ed. B. Mamet & M.J. Copeland, vol. 2, pp. 313–17. Toronto, Ontario: Third North American Paleontological Convention.

Kohn, A.J., 1983a. Marine biogeography and evolution in the tropical Pacific: Zoological perspectives. *Bulletin of Marine Science*, **33**, 528–35.

Kohn, A.J., 1983b. Microhabitat factors affecting abundance and diversity of *Conus* on coral reefs. *Oecologia*, **60**, 293–301.

Kohn, A.J., 1990. Tempo and mode of evolution in Conidae. *Malacologia*, **32**, 55–67.

Kohn, A.J. & Nybakken, J.W., 1975. Ecology of *Conus* on eastern Indian Ocean fringing reefs: Diversity of species and resource utilization. *Marine Biology*, **29**, 211–34.

Kohn, A.J. & Perron, F.E., 1994. *Life History and Biogeography: Patterns in Conus.* Oxford: Oxford University Press.

Kohn, A.J. & Riggs, A.C., 1982. Sample size dependence in measures of proportional similarity. *Marine Ecology Progress Series*, **9**, 147–51.

Kohn, A.J., Mitter, C. & Farrell, B.D., 1991. Diversification: Patterns, rates, causes and consequences. Introduction to the symposium. In *The Unity of Evolutionary Biology*, ed. E. Dudley, vol. **1**, pp. 207–9. Portland, OR: Dioscorides Press.

Leviten, P.J. & Kohn, A.J., 1980. Microhabitat resource use, activity patterns and periodic catastrophe: *Conus* on tropical intertidal reef rock benches. *Ecological Monographs*, **50**, 56–75.

Olivera, B.M., Rivier, J., Clark, C., Ramilo, C.A., Corpuz, G.P., Abogadie, F.C., Mena, E.E., Woodward, S.R., Hillyard, D. & Cruz, L.J., 1990. Diversity of *Conus* neuropeptides. *Science*, **249**, 257–63.

Reichelt, R.E. & Kohn, A.J., 1985. Feeding and distribution of predatory gastropods on some Great Barrier reef platforms. In *Proceedings of the Fifth International Coral Reef Congress*, ed. M. Harmelin Vivien & B. Salvat, vol. **5**, pp. 191–6. Moorea, French Polynesia: Antenne Museum-EPHE.

Ricklefs, R.E. & Schluter, D., 1993. Species diversity: Regional and historical influences. In *Species Diversity in Ecological Communities*, ed. R.E. Ricklefs & D. Schluter, pp. 350–63. Chicago: University of Chicago Press.

Röckel, D., Korn, W. & Kohn, A.J., 1995. *Manual of the Living Conidae*, vol. **1**. Wiesbaden: Christa Hemmen Verlag.

Sale, P.F., 1980. The ecology of fishes on coral reefs. *Oceanography and Marine Biology Annual Review*, **18**, 367–421.

Sale, P.F., Guy, J.A. & Steel, W.J., 1994. Ecological structure of assemblages of coral reef fishes on isolated patch reefs. *Oecologia*, **98**, 83–99.

Sebens, K.P., 1994. Biodiversity of coral reefs: What are we losing and why. *American Zoologist*, **34**, 115–33.

Sørensen, T.A., 1948. A method of establishing groups of equal amplitude in plant sociology based on similarity of species content, and its application to analyses of the vegetation on Danish commons. *Kongelige Danske Videnskabernes Selskab, Biol. Skr.*, **5**, 1–34.

Sorokin, Yu.I., 1993. *Coral Reef Ecology*. Berlin: Springer-Verlag.

Terlau, H., Shon, K., Grilley, M., Stocker, M., Stühmer, W. & Olivera, B.M., 1996. Strategy for rapid immobilization of prey by a fish-hunting marine snail. *Nature*, **381**, 148–51.

Whittaker, R.H., 1952. A study of summer foliage insect communities in the Great Smoky Mountains. *Ecological Monographs*, **22**, 1–44.

Whittaker, R.H., 1967. Gradient analysis of vegetation. *Biological Reviews*, **42**, 207–64.

Woodroffe, C.D. (ed.), 1994. *Ecology and Geomorphology of the Cocos (Keeling) Islands*. Atoll Research Bulletin, nos 399–414. Washington: Smithsonian Institution.

Chapter 10
The biodiversity of coral reef fishes

RUPERT F.G. ORMOND† and CALLUM M.
ROBERTS‡
†*Tropical Marine Research Unit, Department of Biology, University of York,
York, YO1 5DD, UK*
‡*Department of Environmental Economics and Environmental Management,
University of York, York, YO1 5DD, UK*

Abstract

Reef fish show patterns of distribution and diversity at all scales, from global to regional to local. At the largest scale diversity appears to be principally controlled by the interaction of tectonic and geomorphological events with evolutionary and dispersion processes. At a regional level, water column characteristics such as turbidity and current strength, perhaps requiring differential adaptation at the larval stage, may control species distribution patterns. At a local level, diversity is influenced by habitat characteristics such as depth, heterogeneity and complexity. Because of the vagaries of larval distribution and recruitment, the assemblages to be found at a site are quite variable, although the extent of this variation probably differs among species and areas. It is not clear that stochasticity *per se* promotes local or regional diversity, but the fact that currents must frequently carry larvae of a species from areas where it is a superior competitor to ones where it is an inferior competitor seems very likely to do so. But regarding the tantalising issue of the high biodiversity of reef fishes – 'why is it that the species richness of coral reef fish assemblages is so high?' – we have no clear answer. There are various plausible mechanisms, but as yet no conclusive evidence to indicate their relative importance. However, efforts to conserve coral reefs and coral reef fishes should not be delayed by such lack of full understanding.

10.1 Introduction

Anyone who has dived, or snorkelled, on a coral reef cannot help but be impressed by the almost overwhelming variety of fish species that are present, many of them in considerable numbers. For the non-specialist it can seem many weeks or months before one stops seeing new species on almost every dive in a particular area or region. This very evident richness is generally assumed to represent the highest species diversity of any vertebrate assemblage, at least on a local scale (i.e. within hundreds of metres); coral reefs certainly support the highest number of fish species found anywhere. This impressive diversity has long fascinated marine ecologists, and immediately prompts the question 'why are there so many species of fish on coral reefs?' This is the ultimate problem that we seek to address, but on route we need to consider various other questions about how and why fish species richness varies between different reefs and different regions.

In fact coral fish communities show marked variation in species diversity at all scales. In the following sections we describe this variation and discuss its causes, examining all three components of diversity, alpha, beta and gamma. Gamma diversity, or the total number of species present in a region, is covered first in the section on global patterns of fish diversity. We then turn our attention to more local variation in diversity between and among reefs (beta diversity), and finally consider aspects of the community structure of fishes that have been proposed as important in the maintenance of diversity within individual reef habitats (alpha diversity). Finally we consider the implications of our conclusions for the conservation of coral reef fishes, particularly of rarer species.

10.2 Global patterns of diversity

Global patterns of coral reef fish diversity have recently been examined in detail by McAllister *et al.* (1994). Their analyses were based on a geographical information system data base of distribution maps for 799 species of reef fishes from seven families (Gobiidae, Serranidae, Labridae, Chaetodontidae, Zanclidae, Acanthuridae and Tetraodontidae). The first part of this chapter will draw extensively on the results of their study and extend their analyses in several ways.

10.2.1 Latitudinal patterns

At the global scale there is a strong latitudinal gradient in the species diversity of marine fishes, especially in the northern hemisphere (Rohde, 1992). This pattern reflects the much greater number of fishes supported by tropical coral reef habitats compared to temperate and polar regions. There are at least 4000, perhaps even 4500, species found on coral reefs,

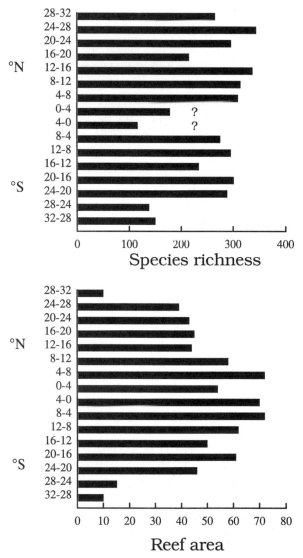

Fig. 10.1. Latitudinal patterns in the distribution of fish species richness (*a*) and coral reef area (*b*). For this analysis the globe was divided into equal area grid squares whose edge corresponds to 2° of latitude at the equator, each square covering approximately 50 000 km^2. Reef occurrence within each grid square was mapped from UNEP/IUCN (1988a,b,c) and from data obtained from a questionnaire survey of coral reef scientists throughout the world. Reef area was then calculated as the number of grid squares occupied by reefs in each latitudinal 'slice'. Data on fish species richness are from McAllister *et al.* (1994). Latitudes adjacent to the equator, marked with question marks, were undersampled compared to others and their fish species richness is almost certainly higher than depicted.

corresponding to more than 25% of all marine fish species (McAllister, 1991). When attention is restricted to tropical latitudes, the latitudinal gradient in fish diversity is much less clear (Fig. 10.1), despite a pronounced gradient in coral reef area. At a global scale reefs appear to support high diversities of fish almost regardless of latitude.

In some parts of the world, pronounced latitudinal clines in fish diversity do occur. For example, there are roughly 2000 species of fish present in the northern Great Barrier Reef (Paxton *et al.*, 1989) compared to closer to 100 species 1000 km further south in the Capricorn Group (Russell, 1983). Richness continues to fall off toward the southernmost reefs of this region with 314 reported from the Elizabeth and Middleton Reefs (Hutchings, 1992) and 447 at Lord Howe Island (Allen *et al.*, 1976). However, latitudinal patterns may be confounded by effects of the distribution of reefs in other regions. The Red Sea covers a similar latitudinal range to the Great Barrier Reef (16° versus 12°) and its northernmost reefs rank among the highest latitude reefs in the world. In a detailed analysis of butterflyfish distributions, Roberts *et al.* (1988, 1992) found that species richness initially increased with latitude, falling off only in the northernmost part of the area (Fig. 10.2). This pattern was very closely linked to the distribution of hard coral cover on the reefs. The central part of the Red Sea contains very well-developed reefs extending into deep water and supporting high coral cover. In the shallow, turbid southern and the high latitude northern regions reef distribution is more sparse with lower coral cover.

10.2.2 Longitudinal patterns

Longitudinal gradients in fish diversity are much steeper than latitudinal ones (Fig. 10.3) and their causes have long interested biogeographers. Longitudinal diversity gradients are matched by many other taxa found on reefs, for example corals (Veron, 1993) and gastropod molluscs (Kohn, 1985). Coral diversity correlates very closely with reef fish diversity (Fig. 10.4) suggesting strongly that these patterns share similar origins (Jokiel & Martinelli, 1992). However, there has been much controversy over causal mechanisms. The central part of the Indo-Pacific region, lying between Australia and Asia, clearly supports the highest diversity of many marine taxa (Briggs, 1974, 1992). This area has often been referred to as the 'centre of origin', and patterns of declining diversity with increasing distance have been explained as a consequence of dispersal away from this centre.

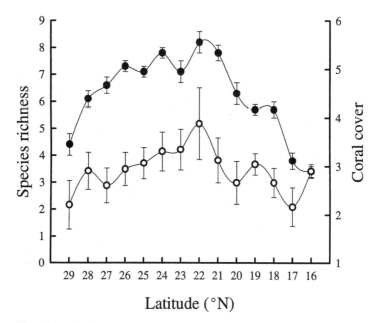

Fig. 10.2. Distribution of butterflyfish (Chaetodontidae) species richness from north to south in the Red Sea (filled circles). The figures indicate mean number of species observed per 10 minute count per degree of latitude (±S.E.) at 341 sites distributed from north to south. Species richness corresponds closely with the pattern of latitudinal variation in coral cover (open circles, 95% C.I.). Coral cover was estimated on a six point scale from 0 (none) to 5 (81–100%). Data are used with permission from Roberts et al. (1988, 1992).

There is a rapid rate of drop-off in species number from west to east across the Pacific, and a slower drop-off from east to west across the Indian Ocean. This difference has been conveniently explained as being due to differences in dispersal ability across each region. The Indian Ocean has currents that reverse seasonally under monsoonal influence, and these would be expected to support widespread dispersal. By contrast, currents in the Pacific predominantly flow from east to west, and the East Pacific Barrier, a huge and geologically ancient area of deep water with no 'stepping stone' reefs, represents a final formidable obstacle to dispersal to the eastern shores of Central America, explaining low species richness there.

High rates of speciation within the centre of origin are argued to have been due to frequent division of the region into isolated basins by sea-level fluctuations in the Plio–Pleistocene (e.g. McManus, 1985). This process is assumed to have allowed allopatric speciation to take

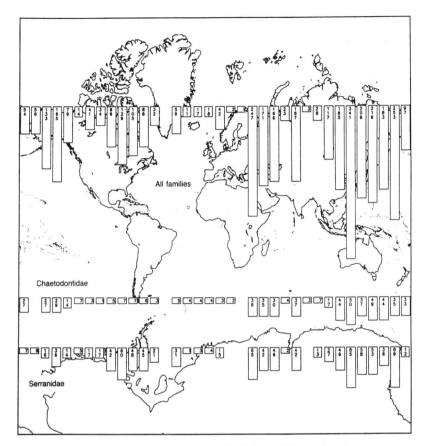

Fig. 10.3. Longitudinal gradients in fish species richness. The top bar chart represents the total numbers of fish species (from a sample of 799 species) that occur in each 10° wide longitudinal 'slice' through the globe. The bottom two bar charts show numbers of butterflyfishes (Chaetodontidae) and seabasses (Serranidae) separately. Numbers in each column show the numbers of species present. The figure is reproduced with permission from McAllister *et al.* (1994).

place, an interpretation supported for some groups of fish by high levels of endemism in the South-East Asian region (e.g. anemonefishes, Pomacentridae; Fautin & Allen, 1992). These groups suggest considerable speciation in the centre of origin.

However the centre of origin model is inadequate to explain patterns completely. There has been very substantial fish speciation in peripheral areas. For example, endemism in the Red Sea reaches 10–15% (Ormond & Edwards, 1987), in the Arabian Gulf 10% (Smith *et al.*, 1987), and

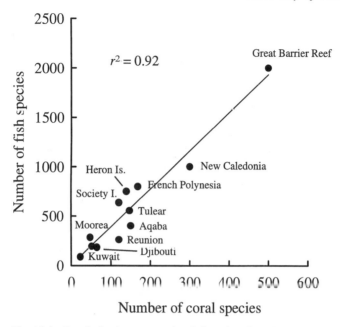

Fig. 10.4. Correlation between regional diversity of corals and coral reef fishes. Figure redrawn from Harmelin-Vivien (1989).

in Hawaii 30% (Hourigan & Reese, 1987). Clearly, the central Indo-Pacific has not been the only place characterised by high rates of speciation. Contrasting explanations have been provided to explain exactly the same pattern of diversity distribution in corals. Potts (1985) has suggested that sea-level fluctuations have done precisely the opposite for speciation rates in reef corals that they have supposedly done for fishes, that is inhibit species formation. He attributed this to the order of magnitude longer generation times that characterise corals compared to fishes. According to Potts' model, sea-level fluctuations were so rapid during the Plio–Pleistocene that there was not enough time for coral speciation. He argued that the broad and diverse shallow-water habitats of the western Pacific acted as a refuge from extinction for coral species, whereas oceanic islands suffered repeated extinctions when sea levels fell. Island populations have thus been maintained by repeated long-distance dispersal from the western Pacific.

Rosen's (1984) argument also invoked the central Indo-Pacific as a refuge from extinction but suggested that species formation took place largely on islands of the western Pacific where populations could become sufficiently isolated. The species that dispersed from these islands to the central Indo-Pacific were the ones most likely to survive periods of falling sea level, and so this area accumulated species over geological time. However, it does not represent the centre of origin for these species.

A similar argument has been advanced by Woodland (1983) for fish. Woodland noticed that there were two areas with high levels of endemism for rabbitfishes (Siganidae), one in the western Indian Ocean and the other in the central Pacific. The central Indo-Pacific had higher levels of species richness but low levels of endemism, suggesting that it represented an area of confluence of species dispersing from these peripheral areas. Donaldson (1986) found support for this hypothesis in the similar distributions of hawkfishes (Cirrhitidae).

The plethora of different explanations for diversity patterns, each of which is plausible in itself, lacks the elegance of a general theory. It would be more satisfying to find a general principle applicable across the broad range of species groups sharing common patterns of diversity distribution. Jokiel & Martinelli (1992) have provided the first component of such a general theory. They have developed a null model of diversity distribution based solely on dispersal patterns, which are a critical component of all of the different explanations that have been advanced. In their model, which they called the 'vortex model' because it is based on prevailing surface circulation patterns within oceans, they created a grid of equally spaced islands and assigned probabilities of dispersal among them based on strength and direction of oceanic circulation. Speciation and extinction rates did not differ among islands and took place at random throughout their archipelago. After a large number of iterations, the distribution of diversity became heavily skewed towards the western side of the ocean at central latitudes. The model shows that by constraining patterns of dispersal, oceanic circulation patterns alone can produce a fair approximation of present global marine diversity patterns.

If dispersal is so critical to distribution then dispersal ability should reflect geographical range size; long-distance dispersers should have larger ranges than short-distance ones. However, in the limited number of fish families studied so far there has been little evidence of a correlation between range size and the duration of the planktonic dispersal phase (Victor, 1991). It is only at the extremes of dispersal ability that a relationship is apparent at all. Fishes with a very short pelagic larval duration tend to have smaller ranges than those with a very long one. At broader levels, there is a trend towards a reduced proportion of demersal spawning fishes compared to pelagic spawners in fish assemblages moving west to east across the Pacific (Thresher, 1991). Pelagic spawned eggs tend to have a longer pelagic larval duration than demersal and, reflecting this, demersal spawners appear to have found it harder to disperse across the Pacific.

Victor (1991) argued that correlations of dispersal ability with range are so weak because studies to date have ignored all but pelagic larval duration. Reef distribution clearly plays a critical role in facilitating or inhibiting colonisation across a region. Jokiel & Martinelli (1992) took no account in their model of differences in reef distribution, and it is this that must form the second component of a general model to explain diversity distribution. Fig. 10.5 shows the relationship between reef area and longitudinal variation in fish species richness. There is a close relationship between the amount of reef present in any area and the number of fish species that occur there. This species/area relationship has long been recognised by ecologists and it is surprising that it has been passed over by biogeographers so often, although McCoy & Heck (1976) drew attention to it to explain the high diversity of mangroves, sea-grasses and corals in the central Indo-Pacific.

In combination, oceanic circulation patterns and patterns in the distribution of reef area explain the great majority of global variation in the distribution of species richness. This explanation applies not only to fish and corals but also across the broad range of marine taxa that share similar patterns of diversity distribution. Speciation takes place throughout the globe and does not have to be confined to any particular region (although rates may vary among them, with the possible existence of minor centres of origin). New species subsequently disperse and accumulate in areas supporting extensive reef systems. Such a model explains the common observation that adjacent regions are most similar biogeographically (e.g. Sheppard, 1987; Pandolfi, 1992) rather than similarity decreasing with distance from a putative centre of origin. It can also accommodate the observed high rates of species formation throughout many peripheral regions of the oceans.

It reflects a strange foible of the human mind that the subtleties and nuances of species distributions have led to the erection of countless evolutionary and biogeographical hypotheses to explain them. At the same time, the fact that dispersal and reef area alone can account for more than 90% of longitudinal variation in species richness has been neglected. It is the latter that ought to hold our attention!

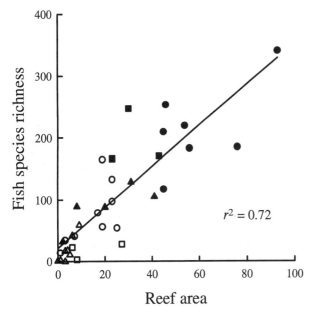

Fig. 10.5. Relationship between longitudinal variation in fish species richness and area of coral reef. Reef area alone explains 72% of inter-regional variation in fish species richness. Numbers of fish species are taken from McAllister *et al.* (1994) (see Fig. 10.3), and area of reef calculated using the same approach as described in the legend to Fig. 10.1. Open symbols represent areas from the central and eastern regions of the three oceans, filled symbols western areas. According to Jokiel & Martinelli's (1992) 'vortex model', eastern regions will be comparatively impoverished compared to western areas as a consequence of dispersal alone. The graph supports this prediction with most of the open (eastern) symbols falling below the calculated regression and the majority of the filled (western) symbols above. △, East Atlantic; ▲, West Atlantic; □, East and Central Indian Ocean; ■, West Indian Ocean/Red Sea; ○, East and Central Pacific; ●, West Pacific/Indonesia.

10.2.3 Reasons for Atlantic impoverishment compared to Indo-Pacific

A number of authors have remarked on the lower diversity of the Atlantic compared to the Indo-Pacific. Such a pattern holds for the majority of taxa, including fishes. If you regress the data in Fig. 10.5 separately for Atlantic and Indo-Pacific regions then the relationship between coral species richness and fish species richness for the Atlantic has a very similar slope but lies slightly below that for the Indo-Pacific (Atlantic: fish species richness = 15.152.98 × coral species richness; Indo-Pacific: fish species richness = 32.523.14 × coral species richness). A frequent explanation offered for the low diversity of Atlantic species is that the region provided fewer and smaller refuges from extinction due to fluctuating sea levels and lowered water temperatures during Pleistocene glaciations than did the Indo-Pacific. Several authors have suggested that life history has moderated the influence of these harsh conditions for some species groups. For example, Jackson et al. (1985) argued that reef corals suffered greater extinction rates in the Atlantic than Cheilostome Bryozoa because they were specialist reef inhabitants. The Bryozoa, by contrast, were able to inhabit a much wider range of habitats so reducing their probability of extinction. Bellwood (1994) suggested a similar explanation for the observation that most Atlantic parrotfish (Scaridae) species inhabit sea-grass beds as well as coral reefs, whereas most Indo-Pacific species are restricted to reefs. Loss of Atlantic reef habitat during the Plio–Pleistocene forced parrotfishes to adapt to different habitats to avoid extinction and only those that did survive to the present day.

Kikkawa & Green (1977) have offered a similar explanation for the observation that Atlantic wrasse species (Labridae) tend to have more r-selected life histories than wrasses of the Great Barrier Reef in the western Pacific. They contend that this indicates that the Atlantic has suffered greater disturbance on evolutionary time-scales than the western Pacific, and that this has inhibited speciation in the former region while greater environmental stability has promoted it in the latter. Their observations were restricted to wrasses, and at the level of entire reef assemblages their argument has been undermined by a recent study by Kulbicki (1992). Kulbicki found that assemblages of fishes in the western Pacific contained a significantly higher proportion of r-selected species than those of the eastern Pacific. If Kikkawa & Green's logic is correct then this would indicate a greater level of environmental disturbance in the western Pacific! This example highlights the problems inherent in generalising from small samples and of deriving evolutionary hypotheses from present-day patterns.

10.3 Speciation and endemicity

Allopatric speciation, that is speciation through subdivision of an ancestral species' range, is widely accepted as the most important form of speciation. The diversity of marine communities is held to be lower than that of comparable terrestrial communities because the greater dispersal capabilities of most marine taxa make isolation less likely in the sea.

Throughout evolutionary time there have been many tectonic and eustatic events that have subdivided and often re-united regions of the circumtropical Tethys Sea, which formed during the Mesozoic. Among them, we have already mentioned the subdivision of the central Indo-Pacific into more-or-less isolated basins, which may have promoted speciation there. As we have argued above, speciation has by no means been restricted to this region. In Recent times of very high sea levels many of the land barriers isolating regions have gone and speciation rates accordingly are probably depressed. Nevertheless, the presence of high levels of endemism in many faunas suggest that some isolating mechanisms are still active.

First among these is simple isolation from other faunas by distance. We mentioned above the high levels of endemicity in fish of the Hawaiian archipelago (30%). Hawaii appears to be sufficiently isolated from other island groups of the region such that dispersal from ancestral populations is insufficient to inhibit speciation (Hourigan & Reese, 1987). The extremely isolated Isla de Pacques (Easter Island) is also characterised by high endemicity (25%: Randall & Cea Egaña, 1989), presumably for the same reason.

Many endemic species are also found on oceanic islands from which non-returning current flows appear to have prevented their onward dispersal to other regions with reefs. For example, Lord Howe Island lies at the southernmost extremity of the area that is reached by the warm waters of the East Australian Current. Only 4% of its species are endemic (Allen *et al.*, 1976), probably due to relatively high rates of genetic mixing from larval inputs from sources further north. However, those species that have evolved have remained limited to Lord Howe, since currents sweep their larvae south to oblivion in the cold waters of the South Pacific. If the rates of gene flow to outlying islands on the paths of non-returning currents are high, then species formation may never take place. This may be the case with Bermuda, which has a very low level of endemism despite being the last outpost of reef in the North Atlantic. Larval transport to Bermuda is virtually one-way, carried by warm core rings from the Gulf Stream.

Peripheral seas have also acted as areas with intensive speciation. For example, the Red Sea has been isolated from the Indian Ocean several times throughout the Plio–Pleistocene due to falling sea levels. However, it is unlikely that this separation led to high rates of speciation within the Red Sea, since it is thought that it resulted in either complete drying or, at the very least, an increase in salinity to levels that would have killed the great majority of the fauna (Ormond & Edwards, 1987; Sheppard *et al.*, 1992). Nor is the narrow entrance at the southern end of the Red Sea thought to constitute a sufficient barrier to dispersion. The Gulf of Aden and Arabian Sea however are characterised by intensive seasonal upwelling so that they are largely devoid of reefs and could represent a barrier to dispersal. This has been the most frequent suggestion to explain the major biogeographical discontinuity between the Red Sea and Indian Ocean, equivalent to an 'underwater Wallace Line' (Sheppard *et al.*, 1992).

Roberts *et al.* (1992) noticed that many Red Sea endemics are confined to the central and northern part of the Red Sea. For these species the barrier to colonisation of the Indian Ocean seems to lie in the southern Red Sea rather than in southern Arabia. Roberts (1991) argued that this was a consequence of the very different conditions prevailing in the southern Red Sea water column presenting a barrier to larval survival. Thus the isolating mechanism could operate on larvae rather than adult fishes. If this hypothesis proves correct, then complete isolation of populations by physical barriers alone may not be a prerequisite of speciation. If water conditions in areas of exchange between regions are not conducive to survival of larvae (for at least some species) then these areas can be just as effectively isolated as if there were a land-mass between them.

Speciation of coral reef fishes is related to body size. Fig. 10.6 reveals what appears to be an upper boundary to the level of diversification within fish families, which is an inversely declining function of body size. The common observation of more species in smaller body size classes found throughout the Animal Kingdom holds also for reef fishes. There are two principal reasons for this: firstly, body size in itself is related to generation time (Peters, 1983; Thresher, 1991), the smallest species tending to be the shortest-lived, most r-selected ones (Kulbicki, 1992); short generation times provide for more rapid evolution, allowing speciation even where intervals of lowered sea level that split species' ranges are relatively brief. Secondly small body size offers greater opportunities for ecological specialisation than for larger species (Morse *et al.*, 1985); high levels of diversification have been possible only for fishes <50 cm long and the majority of reef fish species are smaller than 20 cm long.

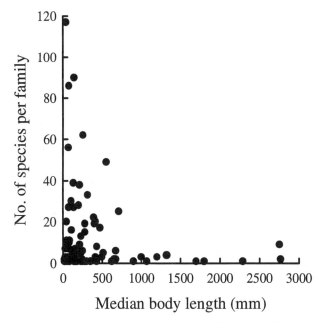

Fig. 10.6. Relationship between median body size and the number of species in each of 94 families of fishes occurring on reefs of Micronesia. Data are from Myers (1991).

10.4 Local patterns of diversity

Local species diversity, while depending on regional species diversity, is also influenced by ecological factors such as habitat type, and by processes influencing community structure. To examine these links, we will next describe the types of subregional and local distribution patterns shown by reef fish and identify at least some of the factors that appear to affect local diversity.

10.4.1 Subregional patterns

Reef fish show subregional patterns in distribution and diversity. Anderson *et al.* (1981) first described systematic changes in the abundance of different butterflyfish (Chaetodontidae) species occurring across the width (continental shelf) of the Great Barrier Reef, and subsequently Williams (1982, 1983), Williams & Hatcher (1983) and Russ (1984a,b) undertook similar broad-scale surveys of the distribution of other fish families at a series of different latitudes. In most families marked cross-shelf patterns in species composition were found, which were comparable at different latitudes, and large compared to differences between reefs in similar locations at different latitudes. These cross-shelf patterns in species composition have been stable for more than 8 years (Williams, 1991). Cross-shelf variation was also apparent in trophic structure and species diversity: species richness was greatest on the middle shelf, intermediate on the outer shelf, and lowest near the shore.

A few similar studies have been undertaken elsewhere. Little latitudinal variation in assemblage composition is apparent along the 800 km Vanuatu archipelago, although some consistent variation in species composition was detected between different reef types (Williams, 1989). Greater latitudinal variation occurs along the 1500 km Hawaiian archipelago (Hobson, 1984), and, as mentioned above, consistent major latitudinal changes in species composition and diversity occur along the 2000 km eastern seaboard of the Red Sea (Roberts et al., 1992). Taken together these studies all show that both the occurrence and species diversity of reef fishes can be influenced at a subregional level by environmental factors and reef type. The southern in-shore Red Sea resembles the inner shelf of the Great Barrier Reef in being shallower and characterised by a heavier sediment load, suggesting as a general finding that some reef fish species may be characteristic of more turbid reefs, others of typical clear water reefs, and yet others of more oceanic reefs where more variable current regimes or greater wave action may be important.

Such distributions could reflect adaptation of the larval stage to feeding in turbid water or distribution by ocean currents, as well as or instead of adaptation of the adult. On the Great Barrier Reef cross-shelf patterns in species composition appear linked to cross-shelf patterns of recruitment that appear remarkably consistent from year to year. Williams and co-workers (see Williams, 1991) have concluded from this that, since hydrographical studies suggest that enough cross-shelf mixing of currents occurs for larval distribution patterns not to be solely explicable in terms of the distribution of reproducing adults, therefore the patterns in adult distribution must be determined by patterns of larval distribution and survival. Following Roberts (1991) they suggested that differential adaptation of larvae to different pelagic habitats may be the major factor. By contrast Anderson et al. (1981) suggested that the systematic cross-shelf replacement of one butterflyfish species by another on the Great Barrier Reef fitted the classical notion of the balance of competition between one species and another being tipped by environmental factors, although Sale & Williams (1982) have discounted this interpretation, pointing to several methodological problems in the original study. In fact there is no reason why adaptation to different pelagic environments by otherwise similar species should not also be an outcome of interspecific competition; and indeed one might expect that the behaviour and ecology of both larval and adult stages should become adapted to minimise interference with competitor species.

We can speculate as to why species diversity should be greatest on off-coast clear-water reefs, less on oceanic reefs, and least on in-shore sedimented reefs – this may reflect the difficulties of adapting adult or larval behaviour to these environments, or to the total extent and spatial relationships of these reef types.

10.4.2 Between zone patterns

There have been many studies describing how within most (if not all) families different species, and hence different fish assemblages, tend to be characteristic of different reef (physiographical) zones (Talbot & Goldman, 1972; Allen, 1975; Clarke, 1977). Fisheries data indicate that this is also true down to depths well beyond those studied by SCUBA diving (400 m or more: Brouard & Grandperrin, 1985). There is some evidence that more abundant species tend to be distributed across more zones than do less abundant species (Bouchon-Navaro, 1980; Fowler, 1988), but both abundance and the range of zones occupied may vary between sites. Again substrate selection by settling fish has been considered important, at least among damselfish, in determining the zone that adult fish occupy (Williams, 1979). However, adults and even juveniles of larger species are more mobile than are adult damselfish, so that habitat selection by adults may also be involved, when interactions such as interference competition between species may influence behaviour (Robertson & Gaines, 1986).

Fish species diversity is also strongly linked to this zonation, diversity changing with depth. In general diversity increases with depth down to a certain point, and then decreases again. The same pattern has been described for many other marine groups, including corals. The zone or depth of maximum species diversity varies between families (e.g. Russ, 1984a,b). However, the exact depth of maximum species diversity depends on the scale at which the phenomenon is being examined. Since the abundance of reef fishes decreases with depth below about 10 m, the depth of maximum diversity appears greater the larger the sampling unit. Interestingly species diversity within reef zones has been found to vary less between geographical areas, for example the Caribbean and Indo-Pacific (Bohnsack & Talbot, 1980), or Tulear in Madagascar and Moorea in French Polynesia (Harmelin-Vivien, 1989), than does the overall species diversity of these reefs. This can only be so if between-zone differences in species composition are less, and the number of zones occupied by individual species more, in those regions (e.g. the Caribbean) where the total species pool is smaller. Again reasons why species diversity should vary with depth are easy to imagine, but there seems to be little hard evidence. In general, resource abundance will be greatest in shallow water, although reduced in very shallow water and in the intertidal zone, and will decline with depth as photosynthetic rates are reduced. Reef fish abundance seems approximately to mirror resource abundance, and intensity of interspecific competition may be expected to follow the same pattern. Feeding in very shallow water becomes more difficult if not impossible as wave action increases, and only some species such as the surf zone surgeonfishes (*Acanthurus lineatus* and *A. sohal*) seem well adapted to cope with this problem (R.F.G.O., personal observation). At greater depth, as resources become more dispersed foraging efficiency will become increasingly critical, but the larger areas of reef slope present at intermediate depth may also serve to promote species diversity.

10.4.3 Within-zone patterns

Within a reef zone the distribution of resident fish species is often patchy or variable. Various factors have been implicated in accounting for some of this variation including food availability (Williams, 1979), the amount of live coral cover (Bell & Galzin, 1984), and topographical complexity and the availability of refuges (Luckhurst & Luckhurst, 1978; Roberts & Ormond, 1987). Site-attached, typically smaller fish species such as damselfishes (Pomacentridae) often display preferred microhabitats, e.g. for coral growth forms or substrate character (Gladfelter & Gladfelter, 1978; Ormond *et al.*, 1996); but these microhabitats are not evenly distributed, and fish are typically not evenly distributed among similar patches. Again, as considered in more detail below, patterns of larval recruitment have been considered responsible for this variability.

Previous reviews (see Sale, 1977, 1980a; Talbot *et al.*, 1978; Bohnsack, 1983; Abrams, 1984), have agreed in considering that high fish species diversity is in part associated with the spatial heterogeneity of reefs and the spatial and temporal heterogeneity of their resources.

There is often, for example, a striking relationship between fish species richness and habitat structural complexity or heterogeneity (Risk, 1972; Luckhurst & Luckhurst, 1978; Roberts & Ormond, 1987; Ormond *et al.*, 1996). Such a relationship is well known from terrestrial (McArthur, 1972; Pearson, 1977) as well as other marine studies. However, perhaps surprisingly, reef fish diversity is not strongly linked to coral abundance (i.e. coral cover), except for very low values of coral cover when fish diversity also tends to be low. A decrease in live coral cover has been observed to have effects on butterflyfish assemblages but not on other fish families (Williams, 1986).

Sale (1980a) has pointed out that the relationship between the number of species present and the number of fish present on small patch reefs is remarkably consistent across geographical regions. Since over a larger area local species number reflects regional species number, this observation implies either that putative differences cannot be detected at the smallest scales, or that at small scales another process limits the number of species present. This observation resembles the finding that for some families at least within-site species richness may differ relatively little between regions that have very different regional or subregional species pools. Thus we have found (R.F.G.O., unpublished data; Righton *et al.*, 1996) that mean number of butterflyfish species per 200 m length of reef face differs relatively little between the northern Red Sea, East Africa, central Indian Ocean and Great Barrier Reef, even though the total number of species seen during comparable surveys varied by a factor of more than three.

10.5 Community structure and species diversity

The mechanisms influencing reef fish biodiversity cannot be approached without at least some consideration of the long-standing debate over processes determining reef fish community structure. We will briefly review and comment on the critical features of this debate, which is primarily about whether reef fish assemblages are essentially predictable, the species present within a particular habitat within a particular area being largely pre-determined as a result of competition and resource limitation and other ecological processes (predation, mutualism etc); or whether partly or strongly stochastic oceanographical processes result in a significant element of chance affecting the composition of the assemblage. Based on these two views contrasting hypotheses have been proposed to account for the high diversity of fishes on coral reefs. The older (classic) model is that tropical reefs have experienced a stable environment over long periods of geological time during which competition and resource limitation have resulted in finer and finer resource partitioning between an increasing number of species (Smith & Tyler, 1972; Smith, 1978; Gladfelter et al., 1980). The more recent and more fashionable model has been that reef fish assemblages are essentially non-equilibrium systems in which competition is averted, thus allowing co-existence of guilds of essentially similar species that are in effect thus able to occupy a single ecological niche (Russell et al., 1974; Sale, 1974, 1977, 1978; Sale & Dybdahl, 1975). This latter view has been referred to as the 'lottery' (Sale, 1974), 'equal chance' (Connell, 1978) or 'multi-species equilibrium' (Doherty, 1983) hypothesis. Interestingly and significantly, this divergence of views about the determinants of community structure mirrors comparable debate over community structure in other terrestrial and aquatic communities (May, 1984; Wiens, 1984).

In fact different adaptations of the two basic views of reef fish community structure have resulted in four or more alternative paradigms being distinguished. These are conveniently summarised by the diagram (Fig. 10.7) devised by Jones (1991). This demonstrates how where fish assemblages are proposed as being determined principally by post-recruitment processes, these may be predominantly either competition or predation. Similarly, where variable recruitment is proposed *not* to be modified by post-recruitment processes, but instead to be the primary factor determining the fish assemblage, then either larval supply may be below the carrying capacity of the reef, thus avoiding competition (the recruitment limitation model (Victor, 1983)), or larval supply may exceed the carrying capacity of the reef, in which case competition will ensue, perhaps with larvae lucky enough to recruit to vacant locations being able to prevent subsequent settlement of other larvae to the same location (the lottery model (Sale, 1978)).

10.5.1 Variation in assemblage structure and recruitment

In looking at the evidence, a first question to consider is obviously whether fish assemblage composition is, as the older competition and predation models would suppose, moderately constant (i.e. predictable) across comparable sites, or whether it is relatively variable, as recruitment driven models would predict. There is no clear-cut answer, for while some workers (for example Gladfelter et al., 1980) have considered that species assemblages on comparable sites are fairly similar, others have

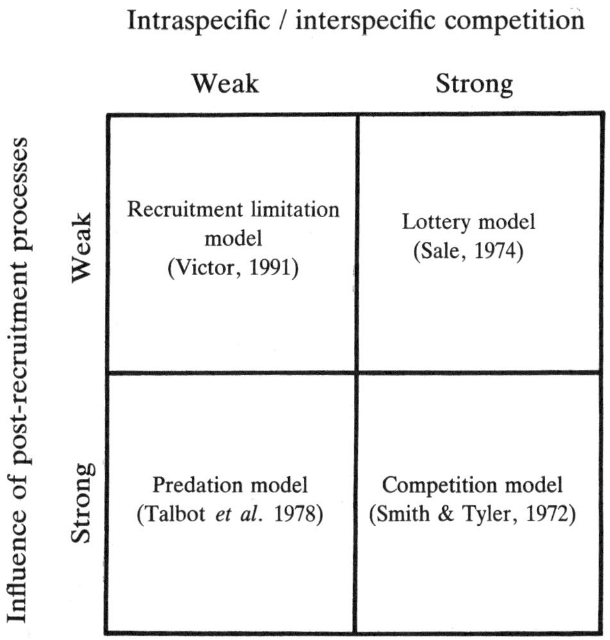

Intraspecific / interspecific competition

	Weak	Strong
Weak (Influence of post-recruitment processes)	Recruitment limitation model (Victor, 1991)	Lottery model (Sale, 1974)
Strong	Predation model (Talbot et al. 1978)	Competition model (Smith & Tyler, 1972)

Fig. 10.7. Diagram illustrating the relationships between the four principal hypotheses of reef fish community structure. The two upper models assume that community structure is largely determined by processes (competition or predation) operating after successful larval settlement. The two lower models assume that community structure is largely determined by larval dispersion and recruitment processes, and differ only over the question of whether fish populations are determined entirely by larval supply (recruitment limitation) or are also limited by available habitat space (lottery model). After Jones (1991).

concluded, based on very similar data, that such differences in species composition are in fact considerable (Sale & Dybdahl, 1975, 1978; Sale & Douglas, 1984). These differences in interpretation are due in part to different approaches to the calculation and interpretation of similarity indices. A mean similarity index of about 0.7 has been interpreted by some workers as confirming a relatively high degree of similarity, and by others as indicating a surprising degree of variability. Moreover the different indices (Jaccard Index and Czekanowski Index) that have been used can give quite different values (e.g. 0.9 compared with 0.5) when applied to the same data sets (see Sale, 1991). What is clear is that detailed quantitative studies show that species composition and abundance does vary more between similar sites than either casual observation or expectations based on the classical view might lead one to expect.

Fish assemblage structure also varies over time at any one site. Thus Sale (1980b) denuded three similar patch reefs and compared the recovery of their fish assemblage with that of three control reefs. Not only were the experimental reefs recolonised by a partially different set of species to give eventual similarity indices (Czekanowski) of no more than about 0.6, but the fauna of the control reefs also changed over time until they also had a comparable degree of similarity with their initial species set. Subsequently during a study of 20 patch reefs whose fish populations were determined about twice a year over 10 years the assemblages recorded at different times on the same reef showed a mean Czekanowski similarity index to one another of 0.51 (Sale, 1991).

The interpretation of recruitment driven models is that this variation in assemblage is linked to variation in recruitment. Recruitment to coral areas has indeed been found to be highly variable and unpredictable, in both time and space, and on a number of scales (Doherty & Williams, 1988; Doherty, 1991). Such variability is of course not peculiar to reef fishes but has been regarded as characteristic of marine organisms with long pelagic larval stages, including both invertebrates (Dayton, 1984) and non-reef fishes (Steele, 1984). Reef fishes show, superimposed on seasonal patterns of reproduction, much variation in size of recruitment pulses to a single site, many if not most species showing inter-annual variation in recruitment of at least one order of magnitude (Doherty & Williams, 1988), and sometimes of two (Kami & Ikehara, 1976). Recruitment to sites within one area has often, though not always, been found to be synchronised and correlated (Allen, 1975), even when sites or groups of sites were as far apart as 50 km (Victor, 1984). In general these patterns of recruitment have been found to match patterns of larval supply (Milicich *et al.*, 1992), suggesting that larval supply is being influenced by some broad-scale process, probably involving oceanographical conditions. However, numbers recruiting to different sites also differ, to some extent systematically, probably linked to the oceanographical or habitat characteristics of each site and the larval behavioural preferences of at least some species (Milicich & Doherty, 1994).

Broad-scale studies both of damselfishes (Pomacentridae) on the Great Barrier Reef (Williams, 1979; Doherty, 1980; Doherty & Fowler, 1994) and of wrasses (Labridae) in the Caribbean (Victor, 1983, 1986) have indicated that marked changes in abundance of common species can follow marked variation in recruitment. Doherty & Fowler (1994) for example, after monitoring recruitment to patches on seven different reefs in the southern Great Barrier Reef over 9 years, concluded that for the common damselfish, *Pomacentrus mollucensis*, variation in recruitment explained over 90% of the spatial variation in abundance of the adult fish. There have, however, also been recent studies in which adult populations appear not to have been directly determined by recruitment. Thus Robertson (1988a,b,c) found that densities of adult surgeonfish, damselfish and triggerfish were not correlated with settlement patterns over a 6 year period, and Jones (1990), who compared densities of adult *Pomacentrus amboinensis* on the Great Barrier Reef with between-year average recruitment success, found that breeding population size plateaued above a certain level of recruitment, suggesting that in those years population level was not recruitment limited.

Possible additional evidence for the view that reef fish communities are essentially random assemblages follows from the assessment that both the taxonomic and trophic composition of assemblages vary much more between areas and regions than might be expected if comparable structuring forces were at work (Sale, 1991). Harmelin-Vivien (1989) on the other hand, on comparing the fish faunas on reefs at Tulear, Madagascar and at Moorea, French Polynesia, found notable similarities in a whole series of functional characteristics, such as size distribution, activity rhythms, distribution of species among ecological and trophic categories etc, despite the widely separated geographical locations, and major differences in diversity (552 species compared with 280 species) and taxonomic composition.

10.5.2 Resource partitioning and limitation

A second key issue is whether or not there is evidence for specialisation and partitioning of resources between species. In fact there is plenty of evidence for resource partitioning between species (e.g. Roughgarden, 1974; Gladfelter & Gladfelter, 1978; Luckhurst & Luckhurst, 1978; Robertson & Lassig, 1980; Gladfelter & Johnson, 1983; Harmelin-Vivien & Bouchon-Navaro, 1983; Bouchon-Navaro, 1986; Robertson & Gains, 1986), although its significance has been called into doubt. Ebeling & Hixon (1991) in reviewing the available research concluded that in 82% out of 38 studies of coral reef fishes results were obtained that were consistent with resource partitioning. Most frequently investigators have examined partitioning of food type or of habitat or microhabitat. These however are only two of the dozen or more dimensions along which species may be diverged. These include form, size, colour, feeding behaviour, foraging strategy, digestive physiology, temperature

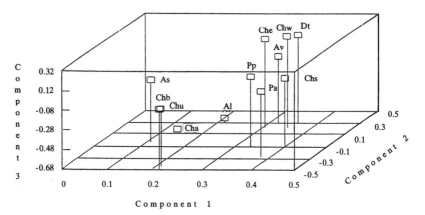

Fig. 10.8. A plot of the first three components obtained from a principal component analysis of the values for each pair of species of the combined overlap index for the reef zones and substrates used by those species for spawning and nesting among 12 non-territorial species of damselfish (Pomacentridae) in the Maldives. The species codes next to each symbol are abbreviations of the scientific names as follows: Al, *Amblyglyphidon leucogaster*; As, *Abudefduf sordidus*; Av, *Abudefduf vaigiensis*; Cha, *Chromis atripectoralis*; Chb, *Chrysiptera biocellata*; Chc, *Chromis caeruleus*; Chs, *Chrysiptera* sp.; Chu, *Chrisiptera unimaculata*; Chw, *Chromis weberi*; Dt, *Dascyllus trimaculatus*; Pa, *Pomacentrus albicaudatus*; Pp, *Pomacentrus phillippinus*. The figure illustrates how, with the exception of *Chrysiptera biocellata* and *Ch. unimaculata*, most species are well distributed in resource space (from Ormond et al., 1996).

preference and tolerance, anti-predator behaviour, social behaviour, reproductive behaviour, reproductive strategy (egg mass, type and number), larval period, larval behaviour, settlement behaviour, and various behavioural attributes of the juvenile, which in some species are markedly different from those of the adult. Ormond *et al.* (1996) for example describe how in the Maldives different non-territorial species of damselfish use different substrates and zones for nesting and spawning (Fig. 10.8). The scope for resource partitioning is in itself however a difficulty since Sale (1979) and Connell (1980) have expressed the view that given the many aspects of a species' biology it would be surprising if any pair of fishes did not differ in some aspect of their ecology. One approach to this point has been to test statistically whether mean overlap in resource use between species within a guild was less than a null value calculated on the assumption that species overlap varied randomly. Only Gladfelter & Johnson's (1983) study on resource partitioning in squirrelfishes has both used this approach and rejected the null hypothesis of random overlap.

Another argument advanced by Sale (1984) against the idea that resource partitioning is important in structuring reef fish communities is that there is no indication that reef fishes divide their resources any more finely than do less diverse temperate fish assemblages. In fact there is evidence that food resource utilisation by fish is more extensive in the tropics, fish not only feeding on some food resources unique to the tropics, but also consuming some food types present but unused in temperate waters (Harmelin-Vivien, 1983). Strict specialists feeding on only one prey species are scarce on reefs as they are elsewhere, but diffuse specialists, feeding on only one class of prey are more abundant on reefs than in the western Mediterranean, although so too are very generalist feeders (Harmelin-Vivien, 1989). However, such generalists are often specialised, and sometimes highly specialised, in the way in which they obtain their food (see e.g. Ormond, 1980); we would argue that often on reefs resource use is effectively finely partitioned by the way in which different species use different specialist behaviours to capture food, rather than by different species feeding on different types of food. For example from observation in the Red Sea and Indian Ocean we have noted that some groupers (Serranidae), as well as hawkfishes (Cirrhitidae) and lizardfishes (Synodontidae), constitute a guild of sit-and-wait ambush predators (Fig. 10.9); there is considerable overlap in the food species caught, but different predators are differently patterned and coloured to be cryptic against different corals and other substrates. Since the success of each predator species will be limited by the extent of the substrate(s) against which it is effectively camouflaged, the different predator species are effectively partitioning the prey resource between them.

Particularly relevant to the issue of whether resources or recruitment limit abundance and control community structure are a number of studies that have sought to assess whether a species' abundance is altered in response to altered resource level (resource limitation). In fact, even though such an effect has often been denied (Sale, 1980a, 1984), a clear majority of the few pertinent studies have found evidence for resource limitation. For example Jones (1986) found an increased rate in growth of juvenile *Pomacentrus amboinensis* when their food supply was supplemented by additional plankton. A change in growth rate will be significant for the population dynamics of a species, since this may be expected to have a more than proportional effect on the reproductive output of individuals.

(a)

(b)

Fig. 10.9. Photographs of two species from the guild of sit-and-wait ambush predators found on Red Sea reefs. (a) A scorpionfish, *Scorpaenopsis oxycephala*, which is pink and orange camouflaged against a cluster of pink soft corals, *Xenia* sp. (b) The sandperch, *Parapercis hexopthalma*, which is mottled black and tan on pale grey, camouflaged against a substrate of coarse sand and fine rubble. Although such predators may take a wide range of prey species, they are specialist species to the extent that they can only ambush prey most effectively when resting against suitable substrate.

10.5.3 Competition and predation

There are also studies suggesting that at least some effects of competition and predation are evident on reefs. For example competition between territorial damselfishes (*Pomacentrus* and *Stegastes* spp.) or surgeonfishes (*Acanthurus* spp.) and other territorial or non-territorial herbivores is common (Myrberg & Thresher, 1974; Robertson et al., 1976, 1979). As yet no large-scale experimental work has been undertaken on reefs to demonstrate conclusively that such competition is influencing the abundance or distribution of species, but such studies, involving reciprocal removal of each species among closely related pairs, have been undertaken for both surfperches (Hixon, 1980; Holbrook & Scmitt, 1989) and rockfishes (Larson, 1980) on temperate rocky reefs in California. When the more aggressive species of a pair was removed, the less aggressive species expanded into its range. The most similar observations to have been made on coral reefs probably arise from unintended experiments. For example Watson & Ormond (1994) compared fish assemblages in Kenya between similar fished and unfished areas and found that although the abundances of most commercial species were much higher in the latter, the small grouper *Cephalopholis* spp. was significantly more abundant at fished sites than at unfished ones, which they attributed to release from competition with, or perhaps predation by, larger grouper species.

Predation has also been observed to influence survivorship and assemblage composition. Thresher (1983a,b) found that across 26 patch reefs the survivorship of both the damselfish *Acanthochromis polyacanthus* and four nocturnal cardinalfishes (Apogonidae), but not of some other species, was inversely related to the density of the grouper *Plectropomus leopardus*. Shulman et al. (1983) found that recruitment and survival of grunts and drums (*Haemulon* and *Equetus* spp.) on artificial reefs was significantly lower on artificial reefs that had also been colonised by piscivorous snappers (Lutjanidae). Also Hixon & Beets (1989), by using artificial reefs with different numbers and sizes of shelter-holes that influenced the abundance of larger piscivores, demonstrated that the presence of piscivores had a predictable effect on the resulting species composition of the fish community establishing on these reefs. Even Doherty & Sale (1985) observed that for certain species survivorship of new recruits during the first few weeks was considerably greater on plots from which piscivores had been excluded by cages, than on plots from which they had not; surprisingly however, as noted by Hixon

(1991), they interpreted this pattern as not so much supporting a role for predation, but as indicating the relative importance of the recruitment limitation hypothesis.

10.5.4 The lottery model and species diversity

While the lottery hypothesis and competing models have been principally concerned with understanding the composition of fish assemblages at any location, Sale (1977, 1980b, 1984) in particular has argued that stochastic recruitment processes could account for the high species diversity of fish on coral reefs. It was proposed that if populations of superior competitors were frequently limited by recruitment, competition due to them would be correspondingly weak, thus enabling competitively inferior species to co-exist, so boosting species diversity. At a regional level stochastic variation in the distribution and abundance of competing species, by allowing competitively inferior species to avoid competition in at least some locations, would allow species to survive in the regional species pool that otherwise would have been eliminated.

Although at first sight plausible, a little further consideration makes it difficult to see how stochastic processes *per se* can promote either local or regional species diversity. The supply of recruits, however stochastic, must depend on the presence somewhere of a population of surviving adults; and however stochastic, the net population of a species must either tend to track upwards towards some limiting ceiling, or to track downwards towards extinction. In fact the model will only work in the long term if any species is able to restore its population size once it has become rare (Abrams, 1984). This could in theory happen if co-occurring species have suitably hump-shaped stock-recruitment relationships (Sale, 1982) and have overlapping generations so that strong cohorts can safeguard the population in the face of stochastically low recruitment (Chesson & Warner, 1981). But the inverse components of the stock-recruitment relationships, depending on larval survivorship, would need to operate independently for the competing species, which seems unlikely if they are closely related. Even were such a situation shown to exist, we would argue that co-existence of these species would then depend on this differential aspect of the species' larval ecology, not on the mere stochasticity of recruitment. What can be conceded is that at a regional level the effect of stochastic recruitment in weakening the effect of competition between superior and inferior competitor, might delay the full replacement of the inferior by the superior, although it could not do so indefinitely. Assuming an independent rate of speciation within a region this would increase the equilibrium size of the regional species pool.

This analysis does not deny that stochastic larval distribution and recruitment processes can greatly influence species composition and abundance at any one locality. Stochastic distribution and recruitment processes will greatly obscure any direct link between the conditions under which a particular species may be superior, in the sense of being able to survive against competition indefinitely, and its observed distribution pattern. Also, at a subregional scale species richness may be higher than would otherwise be possible because of regular occurrence of species that are not able in the long term to maintain populations within an area, but which are present because of the repeated influx into the region of larvae produced by viable populations in neighbouring areas (Dale, 1978; Abrams, 1984). For many species a reef region may effectively constitute a mosaic of sources within which populations are viable and from which larvae are dispersed, and of sinks, where populations only survive because of restocking by propagules from source areas. In such a system species may co-exist because of their differential distribution of source and sink areas (Pulliam, 1988). Stevens (1989) has even suggested that this mechanism alone may be sufficient to account for the higher species diversity of tropical as opposed to temperate fish communities. In fact there is only circumstantial evidence for possible sink and source areas among coral reef fish, for cardinal fishes (Apogonidae) in the Caribbean (Dale, 1978) and for one species of damselfish (Pomacentridae), also in the Caribbean (Waldner & Robertson, 1980). But there is better evidence for marginal or non-breeding fish populations being supported by current-mediated transport of larvae from more northerly or southerly populations among temperate reefs and kelp beds in both California (Cowen, 1985; Ebeling & Hixon, 1991) and New Zealand (Choat *et al.*, 1988; Jones, 1988).

10.5.5 Synthesis

In the end it may be that the most plausible model for the high diversity of reef fish communities must find room for several of the apparently conflicting explanations of their community structure. Part of the resolution between contrasting hypotheses of community structure is likely to be that different mechanisms may predominate in different scales, of space and of time, in different reef areas, and for different sets of species.

Firstly, as emphasised by many authors including Wiens (1984), May (1984) and Harmelin-Vivien (1989), variability in species composition will be most apparent at a small scale, the scale of the artificial reef assembled from blocks, and also of the coral patches studied by Sale & Douglas (1984). Although comparable variation in recruitment may be detectable across reef systems, at subregional and regional levels community composition, including the characteristic subcommunities of different physiographical zones, will appear much more consistent. Secondly, as demonstrated by Bohnsack (1983), species composition may appear more stable if averaged over time, or sampled at yearly intervals, rather than at monthly or seasonal intervals as has been undertaken in some studies of recruitment. Thirdly, as Sale (1991) comments 'it is not unreasonable to suppose that reef fish communities are organised differently in different biogeographic regions'. Thus the importance of stochastic recruitment is likely to differ markedly between, for example, the mid-ocean reefs of the central Pacific, and the reefs of say, the Gulf of Aqaba, which is an enclosed appendix of an enclosed sea, within which (we suspect) recruitment may be much more predictable. Fourthly, there will be differences between species; some, particularly those that are more abundant and widespread, are more likely to be limited by resources, rather than by recruitment. By contrast other species may be limited by and essentially have their populations determined by low and unpredictable levels of recruitment across much or most of their range. Preliminary analysis of data from Egypt is compatible with such a trend (R.F.G.O., unpublished results).

According to Sale (1991, p. 586) the evidence is that both regional patterns and local physiographical zonation are not dependent on present-day competition since the juveniles settle out into the reef areas and physiological zones occupied by the adults (Doherty & Williams, 1988), so there remains little scope for competition; he states that while such differences in recruitment pattern could be 'the ghost of competition past' (*sensu* Connell, 1980), there is no direct evidence to support such an explanation. There is however, as described above, strong circumstantial evidence that competition may have shaped guild structure *via* resource partitioning, and increasing evidence of cases where competition is occurring or where predation has been demonstrated to influence post-recruitment survivorship.

Sale (1991) sometimes seems too ready to dismiss the view that these other processes might influence community composition along with oceanographical variation in recruitment, as if these different mechanisms were mutually incompatible. Elsewhere he does however accept that patterns of spatial and temporal variation in populations of adults of some species could be due at least as much if not more to post-recruitment processes as to variation in recruitment (Sale, 1991, p. 593), and agrees with Hixon (1991) and Jones (1991) on the need 'to move beyond attempts to identify a single organising process to assessment of relative roles of processes, being aware always that relative importance may itself vary from place to place or through time'.

10.6 Conserving reef fish biodiversity

One principal reason for seeking to understand reef fish diversity and community structure is to facilitate the protection and conservation of that diversity. Naturally, particular attention is focused on rarer species.

10.6.1 Rarity

There are many different forms of rarity (Gaston, 1994). For example, a species may be represented by very few individuals within any given assemblage but be geographically widespread. Alternatively, a species may have a very restricted geographical range throughout which it is abundant. The species considered most vulnerable in conservation terms are both geographically restricted and locally rare throughout their range. By the various criteria reef fish assemblages contain numerous rare species. Many species are locally rare, being represented by very few individuals within assemblages. Harmelin-Vivien (1989) examined rarity in reef fish assemblages in Madagascar and French Polynesia. She found that in both areas over 85% of species were each represented by less than 1% of the total sample, and 25% of species were represented by only a single individual per reef habitat (<0.1% of the sample sizes). In French Polynesia 13.6%, and in Madagascar 17% of species were represented by only a single individual in the entire reef tract.

Many of the species in Harmelin-Vivien's study were only locally rare but are widespread geographically. Rarity within assemblages is often a consequence of a species being at the edge of its distributional range. Such species may be very common in areas closer to the centre of their ranges. For example, southern Red Sea sites have very few individuals of the endemic butterflyfish *Chaetodon paucifasciatus*, while the species is very common further north (Fig. 10.10).

While it is clear that geographical differentiation of marine species

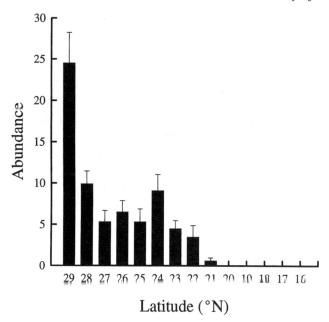

Fig. 10.10. Abundance distribution of the Red Sea endemic butterflyfish, *Chaetodon paucifasciatus*, throughout its entire geographical range. Abundance is expressed as numbers observed per 10 minute count (S.E.) along reef edge transects at 341 sites distributed from the far northern to the far southern Red Sea. Redrawn from Roberts et al. (1992).

has been far less than for terrestrial organisms (e.g. ICBP, 1992), McAllister et al. (1994) have shown that there is a higher frequency of restricted range species than previously suspected. They found that 17% of species were known only from a single grid square (50 000 km^2) on their world map (Fig. 10.11). These species consist mainly of island endemics. A total of 21.9% had a range span of less than 1100 km representing the island endemics and species endemic to peripheral seas or smaller archipelagoes. Long-distance dispersal is the norm for marine fishes but evidently there are limits to this process. The factors restricting the ranges of these species are the barriers to dispersal described earlier: distance, currents flowing from the tropics to higher latitudes, and isolation in peripheral seas, such as is observed for *Chaetodon paucifasciatus* in the Red Sea (Fig. 10.10).

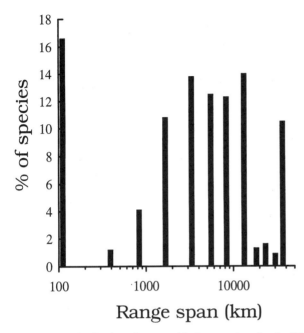

Fig. 10.11. Distribution of geographical range sizes for the 799 coral reef fish species analysed by McAllister et al. (1994). Range sizes were calculated as the angular distance between the extremes of the known range and these values converted to kilometres. While many reef fishes are widespread, 17% of species were restricted to an area <50 000 km^2, a range span of only 225 km or less.

10.6.2 Implications for conservation

Coral reefs are under threat worldwide as a consequence of increased coastal development and exploitation by expanding human populations (Wilkinson, 1992; Roberts, 1993). It is frequently argued that the broad geographical ranges of marine species can prevent their global extinction. However, as we showed earlier not all marine species are widespread, and even for those that are, it has proven all too easy to cause widespread losses. For many species these could end in global extinction. Overfishing has been a potent source of local and regional species loss, especially of large-bodied, long-lived, late maturing species such as the Caribbean jewfish (*Epinephelus itajara*: Bullock et al., 1992). Our growing appreciation of the consequences of metapopulation dynamics underscores these observations and shows that human activities can drive even seemingly robust populations to extinction. Populations inhabiting coral reefs correspond closely to 'ideal' metapopulations. The metapopulation consists of subpopulations inhabiting patches of habitat (reefs) among which there is dispersal. Metapopulations persist through a dynamic balance between colonisation of patches and extinction from them. Recent models provide alarming predictions for conservation (Lawton, 1993). For example, a reduction in population density throughout a species range or a reduction in the proportion of patches occupied can cause range contraction or even collapse of the entire metapopulation. From this it is easy to see how species very vulnerable to overfishing could become threatened. Metapopulation models also imply that loss of populations of a species in some areas, for example through reef degradation, can result in a loss of diversity in sites remote from them (Nee & May, 1992). Such sites might otherwise give the impression of being completely unaffected by human impacts.

Degradation of reef habitat and overharvesting are now leading to heavy losses of biodiversity both locally and regionally. Our understanding of the roles species play within ecosystems is still rudimentary making it very difficult to predict the impacts of biodiversity loss at these scales (Done et al., 1995). However, it is clear that fishes are important to many processes on reefs such as control of algal populations, bioerosion and sediment production, nutrient cycling between adjacent habitats and control of populations of other important species such as echinoderms (Ormond et al., 1991; Roberts, 1995). It is probable that functional complementarity of species in diverse reef systems can ameliorate the impacts of species loss. But where reef systems are already impoverished, such as in the eastern Pacific or Arabian Gulf, other species may not be available to take over important roles from those eliminated by human activities. In such areas species loss may have serious consequences at the ecosystem level. Given our level of uncertainty the most prudent course to adopt is to make strenuous efforts to conserve biodiversity at all spatial scales.

Acknowledgements

We would like to thank the Darwin Initiative of the UK Department of the Environment, the Sir Peter Scott IUCN Species Survival Commission Action Plan Fund, the Sir Peter Scott Trust for Education in Research and Conservation, the British Ecological Society/Coalbourn Trust, the Curtis and Edith Munson Foundation, the Norcross Wildlife Foundation, Ocean Voice International and the World Conservation Monitoring Centre for supporting our work on reef fish biodiversity.

Julie Hawkins, Don McAllister, Fred Schueler and Noel Alfonso have worked extensively with one of us (C.R.) on reef fish biodiversity and we would like to thank them for their contributions to the ideas developed in this paper. The other of us (R.F.G.O.) has been greatly assisted by collaboration with Jeremy Kemp, David Righton, Maggie Watson, Joanna Macmillan and the staff of National Parks in Egypt and in Kenya; we wish to thank them also for their support. We also extend our appreciation to the many people who have given freely of their time and expertise to the biodiversity mapping projects of the IUCN SSC Coral Reef Fish Specialist Group.

References

Abrams, P.A., 1984 Recruitment, lotteries and coexistence in coral reef fish. *American Naturalist*, **123**, 44-55.

Allen, G.R., 1975. *Damselfishes of the South Seas*. Neptune City, NJ: TFH Publications.

Allen, G.R., Hoese, D.F., Paxton, J.R., Randall, J.E., Russell, B.C., Starck, W.C., Talbot, F.H. & Whitley, G.P., 1976. Annotated checklist of the fishes of Lord Howe Island. *Records of the Australian Museum*, **30**, 366-454.

Anderson, G.R.V., Ehrlich, A.H., Ehrlich, P.R., Roughgarden, J.D., Russell, B.C. & Talbot, F.H., 1981. The community structure of coral reef fishes. *American Naturalist*, **117**, 476-95.

Bell, J.D. & Galzin, R., 1984. Influence of live coral cover on coral reef fish communities. *Marine Ecology Progress Series*, **15**, 265-74.

Bellwood, D.R., 1994. A phylogenetic study of the parrotfishes, family Scaridae (Pisces: Labroidei), with a revision of genera. *Records of the Australian Museum Supplement*, **20**, 1-86.

Bohnsack, J.A., 1983. Species turnover and the order versus chaos controversy concerning reef fish community structure. *Coral Reefs*, **1**, 223-8.

Bohnsack, J.A. & Talbot, F.H., 1980. Species packing by reef fishes on Australian and Caribbean reefs: An experimental approach. *Bulletin of Marine Science*, **30**, 710-23.

Bouchon-Navaro, Y., 1980. Quantitative distribution of the Chaetodontidae on a fringing reef of the Jordanian coast (Gulf of Aqaba, Red Sea). *Tethys*, **9**, 247-51.

Bouchon-Navaro, Y., 1986. Partitioning of food and space resources by chaetodontid fishes on coral reefs. *Journal of Experimental Marine Biology and Ecology*, **103**, 21-40.

Briggs, J.C., 1974. *Marine Zoogeography*. New York: McGraw Hill.

Briggs, J.C., 1992. The marine East Indies: Centre of origin? *Global Ecological and Biogeographical Letters*, **2**, 149-56.

Brouard, F. and Grandperrin, R., 1985. Deep-bottom fishes of the outer reef slope of Vanuatu. Working Paper South Pacific Commission (SPC), 17th Regional Technical Meeting on Fisheries. *SPC/Fisheries*, vol. **17**, *Working Paper no. 12*, pp. 1-127.

Bullock, L.H., Murphy, M.D., Godcharles, M.F. & Mitchell, M.E., 1992. Age, growth, and reproduction of jewfish *Epinephelus itajara* in the eastern Gulf of Mexico. *Fisheries Bulletin, USA*, **90**, 243-9.

Chesson, P.L. & Warner, R.R., 1981. Environmental variability promotes coexistence in competitive lottery systems. *American Naturalist*, **17**, 923–43.

Choat, A.H., Ayling, A.M. & Scheil, D.R., 1988. Temporal and spatial variation in an island fish fauna. *Journal of Experimental Marine Biology and Ecology*, **121**, 91–111.

Clarke, R.D., 1977. Habitat distribution and species diversity of chaetodontid and pomacentrid fishes near Bimini, Bahamas. *Marine Biology*, **40**, 277–89.

Connell, J.H., 1978. Diversity in tropical rain forests and coral reefs. *Science*, **199**, 1302–10.

Connell, J.H., 1980. Diversity and coevolution of competitors, or the ghost of competition past. *Oikos*, **35**, 131–8.

Cowen, R.K., 1985. Large-scale pattern of recruitment by the labrid, *Semicossyphus pulcher*: Causes and implications. *Journal of Marine Research*, **43**, 719–42.

Dale, G., 1978. Money-in-the-bank: A model for coral reef fish coexistence. *Environmental Biology of Fishes*, **3**, 103–8.

Dayton, P.K., 1984. Processes structuring some marine communities: Are they general? In *Ecological Communities: Conceptual Issues and the Evidence*, ed. D.R. Strong Jr, D. Simberloff, L.G. Abele & A.B. Thistle, pp. 181–97. Princeton, NJ: Princeton University Press.

Doherty, P.J., 1980. Biological and Physical Constraints on the Populations of Two Sympatric Territorial Damselfishes on the Southern Great Barrier Reef. Ph.D. thesis, University of Sydney, Sydney, Australia.

Doherty, P.J., 1983. Tropical territorial damselfishes: Is density limited by aggression or recruitment? *Ecology*, **64**, 176–90.

Doherty, P.J., 1991. Spatial and temporal patterns in recruitment. In *The Ecology of Coral Reef Fishes*, ed. P.F. Sale, pp. 261–93. San Diego, CA: Academic Press.

Doherty, P.J. & Fowler, A.J., 1994. An empirical test of recruitment limitation in a coral reef fish. *Science*, **263**, 935–9.

Doherty, P.J. & Sale, P.F., 1985. Predation on juvenile coral reef fishes: An exclusion experiment. *Coral Reefs*, **4**, 225–34.

Doherty, P.J. & Williams, D.McB., 1988. The replenishment of coral reef fish populations. *Oceanography and Marine Biology*, **26**, 487–551.

Donaldson, T.J., 1986. Distribution and species richness patterns of Indo-West Pacific Cirrhitidae: Support for Woodland's hypothesis. In *Indo-Pacific Fish Biology: Proceedings of the Second International Conference on Indo-Pacific Fishes*, ed. T. Uyeno, R. Arai, T. Taniuchi & K. Matsuura, pp. 623–8. Tokyo: Ichthyological Society of Japan.

Done, T.J., Ogden, J.C. & Wiebe, W.J. (eds), 1995. Biodiversity and ecosystem function of coral reefs. In *Global Biodiversity Assessment*. Nairobi: UNEP; Cambridge University Press.

Ebeling, A.W. & Hixon, M.A., 1991. Tropical and temperate reef fishes: Comparison of community structures. In *The Ecology of Coral Reef Fishes*, ed. P.F. Sale, pp. 475–563. San Diego, CA: Academic Press.

Fautin, D.G. & Allen, G.R., 1992. *Field Guide to Anemonefishes and Their Host Sea Anemones*. Perth: Western Australian Museum.

Fowler, A.J., 1988. Aspects of the Population Biology of Three Species of Chaetodont at One Tree Reef, Southern Great Barrier Reef. Ph.D. thesis, University of Sydney, Sydney, Australia.

Gaston, K.J., 1994. *Rarity*. London: Chapman & Hall.

Gladfelter, W.B. & Gladfelter, E.H., 1978. Fish community structure as a function

of habitat structure on West Indian patch reefs. *Reviews in Biology of the Tropics*, **26**, 65–84.
Gladfelter, W.B. & Johnson, W.S., 1983. Feeding niche separation in a guild of tropical reef fishes (Holocentridae). *Ecology*, **64**, 552–63.
Gladfelter, W.B., Ogden, J.C. & Gladfelter, E.H., 1980. Similarity and diversity among coral reef fish communities: A comparison between tropical western Atlantic (Virgin Islands) and tropical central Pacific (Marshall Islands) patch reefs. *Ecology*, **61**, 1156–68.
Harmelin-Vivien, M.L., 1983. Etude comparative de l'ichthyofauna des herbiers de Phanerogames marines en milieu tropical et tempere. *Revue d'Ecologie: Terre et Vie*, **38**, 179–210.
Harmelin-Vivien, M.L., 1989. Reef fish community structure: An Indo-Pacific comparison. In *Vertebrates in Complex Tropical Systems*, ed. M.L. Harmelin-Vivien & F. Bourlière, pp. 21–60. New York: Springer-Verlag.
Harmelin-Vivien, M.L. & Bouchon-Navaro, Y., 1983. Feeding diets and significance of coral feeding among chaetodontid fishes in Moorea (French Polynesia). *Coral Reefs*, **2**, 119–27.
Hixon, M.A., 1980. Competitive interactions between Californian reef fishes of the genus *Embiotoca* (Embiotocidae). *Ecology*, **61**, 918–31.
Hixon, M.A., 1991. Predation as a process structuring coral reef communities. In *The Ecology of Fishes on Coral Reefs*, ed. P.F. Sale, pp. 475–508. San Diego, CA: Academic Press.
Hixon, M.A. & Beets, J.P., 1989. Shelter characteristics and Caribbean fish assemblages: Experiments with artificial reefs. *Bulletin of Marine Science*, **44**, 666–80.
Holbrook, S.J. & Schmitt, R.J., 1989. Resource overlap, prey dynamics, and the strength of competition. *Ecology*, **70**, 1943–53.
Hobson, E.S., 1984. The structure of reef fish communities in the Hawaiian archipelago. In *Proceedings of the Second Symposium on Resource Investigation of the Northwest Hawaiian Islands*, ed R.W. Grigg & K.Y. Tanoue, vol. **1**, pp. 101–22. Honolulu, HA: University of Hawaii.
Hourigan, T.F. & Reese, E.S., 1987. Mid-ocean isolation and the evolution of Hawaiian reef fishes. *Trends in Ecology and Evolution*, **2**, 187–91.
Hutchings, P. (ed.), 1992. *A Survey of Elizabeth & Middleton Reefs, South Pacific*. Canberra: Australian National Parks and Wildlife Service.
ICBP (International Council for Bird Preservation), 1992. *Putting Biodiversity on the Map: Priority Areas for Global Conservation*. Cambridge: ICBP.
Jackson, J.B.C., Winston, J.E. & Coates, A.G., 1985. Niche breadth, geographic range, and extinction of Caribbean reef-associated Cheilostome Bryozoa and Scleractinia. In *Proceedings of the Fifth International Coral Reef Symposium*, ed. C. Gabrie, B. Salvat, C. LaCroix & J.-L. Toffart, vol. **4**, pp. 151–8. Moorea, French Polynesia: Antenne Museum-EPHE.
Jokiel, P. & Martinelli, F.J., 1992. The vortex model of coral reef biogeography. *Journal of Biogeography*, **19**, 449–58.
Jones, G.P., 1986. Food availability affects growth in a coral reef fish. *Oecologia*, **70**, 136–9.
Jones, G.P., 1988. Ecology of rocky reef fish of north-eastern New Zealand: A review. *New Zealand Journal of Marine and Freshwater Research*, **22**, 445–62.
Jones, G.P., 1990. The importance of recruitment to the dynamics of a coral reef fish population. *Ecology*, **71**, 1691–8.
Jones, G.P., 1991. Postrecruitment processes in the ecology of coral reef fish

populations: A multifactorial perspective. In *The Ecology of Coral Reef Fishes*, ed. P.F. Sale, pp. 294–328. San Diego, CA: Academic Press.

Kami, H.T. & Ikehara, I.I., 1976. Notes on the annual juvenile siganid harvest in Guam. *Micronesica*, **12**, 323–5.

Kikkawa, J. & Green, A., 1997. Why are coral reefs and rainforests so diverse? In *International Workshop on Biodiversity: Its Complexity and Role*. Tusuka, Japan. (In press).

Kohn, A.J., 1985. Evolutionary ecology of *Conus* on Indo-Pacific coral reefs. *Proceedings of the Fifth International Coral Reef Symposium*, ed. C. Gabrie, B. Salvat, C. LaCroix & J.-L. Toffart, vol. **4**, pp. 139–44. Moorea, French Polynesia: Antenne Museum-EPHE.

Kulbicki, M., 1992. Distribution of the major life-history strategies of coral reef fishes across the Pacific Ocean. *Proceedings of the Seventh International Coral Reef Symposium*, ed. R.H. Richmond, vol. **2**, pp. 908–19. Mangilao, Guam: University of Guam Press.

Larson, R.J., 1980. Competition, habitat selection, and the bathymetric segregation of two rockfish (*Sebastes*) species. *Ecological Monographs*, **50**, 221–39.

Lawton, J.H., 1993. Range, population abundance and conservation. *Trends in Ecology and Evolution*, **8**, 409–13.

Luckhurst, B.E. & Luckhurst, K., 1978. Analysis of the influence of the substrate variables on coral reef fish communities. *Marine Biology*, **49**, 317–23.

May, R.M., 1984. An overview: Real and apparent patterns in community structure. In *Ecological Communities: Conceptual Issues and the Evidence*, ed. D.R. Strong Jr, D. Simberloff, L.G. Abele & A.B. Thistle, pp. 3–16. Princeton, NJ: Princeton University Press.

McAllister, D.E., 1991. What is the status of the world's coral reef fishes? *Sea Wind*, **5**, 14–18.

McAllister, D.E., Schueler, F.W., Roberts, C.M. & Hawkins, J.P., 1994. Mapping and GIS analysis of the global distribution of coral reef fishes on an equal-area grid. In *Advances in Mapping the Diversity of Nature*, ed. R. Miller, pp. 155–75. London: Chapman & Hall.

McArthur, R.H., 1972. *Geographical Ecology: Patterns in the Distribution of Species*. New York: Harper and Row.

McCoy, E.D. & Heck, K.L., 1976. Biogeography of corals, sea-grasses and mangroves: An alternative to the centre of origin concept. *Systematic Zoology*, **25**, 201–10.

McManus, J.W., 1985. Marine speciation, tectonics and sea-level changes in Southeast Asia. *Proceedings of the Fifth International Coral Reef Symposium*, ed. C. Gabrie, B. Salvat, C. LaCroix & J.-L. Toffart, vol. **4**, pp. 133–8. Moorea, French Polynesia: Antenne Museum-EPHE.

Milicich, M.J. & Doherty, P.J., 1994. Larval supply of coral reef fish populations: Magnitude and synchrony of replenishment to Lizard Island, Great Barrier Reef. *Marine Ecology Progress Series*, **110**, 121–34.

Milicich, M.J., Meekan, M.G. & Doherty, P.J., 1992. Larval supply: A good predictor of recruitment of three species of reef fish (Pomacentridae). *Marine Ecology Progress Series*, **86**, 153–66.

Morse, D.R., Lawton, J.H., Dodson, M.M. & Williamson, M.H., 1985. Fractal dimension of vegetation and the distribution of arthropod body lengths. *Nature*, **314**, 731–2.

Myers, R.F., 1991. *Micronesian Reef Fishes*. Guam: Coral Graphics.

Myrberg, A.A. & Thresher, R.E., 1974. Interspecific aggression and its relevance to the concept of territoriality in reef fishes. *American Zoologist*, **14**, 81–96.

Nee, S. & May, R.M., 1992. Dynamics of metapopulations: Habitat destruction and competitive coexistence. *Journal of Animal Ecology*, **61**, 37–40.

Ormond, R.F.G., 1980. Aggressive mimicry and other interspecific feeding associations among Red Sea coral reef predators. *Journal of Zoology*, **191**, 247–62.

Ormond, R.F.G. & Edwards, A.J., 1987. Red Sea reef fishes. In *Red Sea*, ed. A.J. Edwards & S.M. Head, pp. 251–87. Oxford: Pergamon Press.

Ormond, R.F.G., Bradbury, R., Bainbridge, S., Fabricius, K., Keesing, J., De Vantier, L., Medley, P. & Steven, A., 1991. Test of a model of regulation of Crown-of-Thorns starfish by fish predators. In *Acanthaster and the Coral Reef: A Theoretical Perspective*, ed. R. Bradbury, pp. 189–207. Berlin: Springer-Verlag.

Ormond, R.F.G., Jan, R.Q. & Roberts, J.M., 1996. Behavioural differences in microhabitat use by damselfishes (Pomacentridae): Implications for reef fish biodiversity. *Journal of Experimental Marine Biology and Ecology*, **202**, 85–95.

Pandolfi, J.M., 1992. Successive isolation rather than evolutionary centres for the origination of Indo-Pacific reef corals. *Journal of Biogeography*, **19**, 593–609.

Paxton, J.R., Hoese, D.F., Allen, G.R. & Hanley, J.E., 1989. *Zoology Catalogue of Australia*, vol. **7**. *Pisces Petromyzontidae to Carangidae*. Canberra: Australian Government Publishing Service.

Pearson, D.L., 1977. A pantropical comparison of bird community structure on six lowland forest sites. *The Condor*, **79**, 232–44.

Peters, R.H., 1983. *The Ecological Implications of Body Size*. New York: Cambridge University Press.

Potts, D.C., 1985. Sea-level fluctuations and speciation in Scleractinia. *Proceedings of the Fifth International Coral Reef Symposium*, ed. C. Gabrie, B. Salvat, C. LaCroix & J.-L. Toffart, vol. **4**, pp. 127–32. Moorea, French Polynesia: Antenne Museum-EPHE.

Pulliam, H.R., 1988. Sources, sinks and population regulation. *American Naturalist*, **132**, 652–61.

Randall, J.E. & Cea Egaña, A., 1989. *Canthigaster cyanetron*, a new toby (Teleoistei: Tetraodontidae) from Easter Island. *Revue Française d'Aquariologie*, **15**, 93–6.

Righton, D., Kemp, J. & Ormond, R., 1996. Biogeography, community structure and diversity of Red Sea and western Indian Ocean butterflyfishes. *Journal of the Marine Biological Association, UK*, **76**, 223–8.

Risk, M.J., 1972. Fish diversity on a coral reef in the Virgin Islands. *Atoll Research Bulletin*, **153**, 1–6.

Roberts, C.M., 1991. Larval mortality and the composition of coral reef fish communities. *Trends in Ecology and Evolution*, **6**, 83–7.

Roberts, C.M., 1993. Coral reefs: Health, hazards and history. *Trends in Ecology and Evolution*, **8**, 425–7.

Roberts, C.M., 1995. Effects of fishing on coral reefs: A massive uncontrolled experiment with ecosystem structure. *Conservation Biology*, **9**, 988–95.

Roberts, C.M. & Ormond, R.F.G., 1987. Habitat complexity and coral reef fish diversity abundance on Red Sea fringing reefs. *Marine Ecology Progress Series*, **41**, 1–8.

Roberts, C.M., Ormond, R.F.G. & Shepherd, A.R.D., 1988. The usefulness of butterflyfishes as environmental indicators on coral reefs. *Proceedings of the Sixth International Coral Reef Symposium*, ed. J.H. Choat, D. Barnes, M.A. Borowitzka *et al.*, vol. **2**, pp. 331–6. Townsville, Queensland: Sixth International Coral Reef Symposium Executive Committee.

Roberts, C.M., Shepherd, A.R.D. & Ormond, R.F.G., 1992. Large-scale variation in assemblage structure of Red Sea butterflyfishes and angelfishes. *Journal of Biogeography*, **19**, 239–50.

Robertson, D.R., 1988a. Abundance of surgeonfishes on patch reefs in Caribbean Panama: Due to settlement or post-settlement events? *Marine Biology*, **97**, 495–501.

Robertson, D.R., 1988b. Extreme variation in settlement of the Caribbean triggerfish *Balistes vetula* in Panama. *Copeia*, **1988**, 698–703.

Robertson, D.R., 1988c. Settlement and population dynamics of *Abudefduf saxatilis* on patch reefs in Caribbean Panama. *Proceedings of the Sixth International Coral Reef Symposium*, ed. J.H. Choat, D. Barnes, M.A. Borowitzka *et al.*, vol. 2, pp. 839–43. Townsville, Queensland: Sixth International Coral Reef Symposium Executive Committee.

Robertson, D.R. & Gaines, S.D., 1986. Interference competition structures habitat use in local assemblage of coral reef surgeonfishes. *Ecology*, **67**, 1327–83.

Robertson, D.R. & Lassig, B., 1980. Spatial distribution patterns and coexistence of a group of territorial damselfishes from the Great Barrier Reef. *Bulletin of Marine Science*, **30**, 187–203.

Robertson, D.R., Polunin, N.V.C. & Leighton, K., 1979. The behavioural ecology of three Indian Ocean surgeon fishes (*Acanthurus lineatus*, *A. leucosternum* and *Zebrasoma scopas*): Their feeding strategies, and social and mating systems. *Environmental Biology of Fishes*, **4**, 125–70.

Robertson, D.R., Sweatman, H.P.A., Fletcher, E.A. & Cleland, M.G., 1976. Schooling as a mechanism for circumventing the territoriality of competitors. *Ecology*, **57**, 1208–20.

Rohde, K., 1992. Latitudinal gradients in species diversity: The search for the primary cause. *Oikos*, **65**, 514–27.

Rosen, B.R., 1984. Reef and coral biogeography and climate through the late Cainozoic: Just islands in the sun or a critical pattern of islands? In *Fossils and Climate*, ed. P. Brenchley, pp. 201–62. New York: John Wiley.

Roughgarden, J., 1974. Species packing and the competition function with illustrations from coral reef fish. *Theoretical Population Biology*, **5**, 163–86.

Russ, G.R., 1984a. Distribution and abundance of herbivorous grazing fishes in the central Great Barrier Reef. I. Levels of variability across the entire continental shelf. *Marine Ecology Progress Series*, **20**, 23–34.

Russ, G.R., 1984b. Distribution and abundance of herbivorous grazing fishes in the central Great Barrier Reef. II. Patterns of zonation of mid-shelf and outershelf reefs. *Marine Ecology Progress Series*, **20**, 35–44.

Russell, B.C., 1983. *Annotated Checklist of the Coral Reef Fishes in the Capricorn-Bunker Group, Great Barrier Reef, Australia*. Townsville, Queensland: Great Barrier Reef Marine Park Authority.

Russell, B.C., Talbot, F.H. & Domm, S., 1974. Patterns of colonisation of artificial reefs by coral reef fishes. *Proceedings of the Second International Coral Reef Symposium*, ed. A.M. Cameron, B.M. Campbell, A.B. Cribb *et al.*, vol. 1, pp. 207–15. Brisbane: The Great Barrier Reef Committee.

Sale, P.F., 1974. Mechanisms of co-existence in a guild of territorial fishes at Heron Island. *Proceedings of the Second International Coral Reef Symposium*, ed. A.M. Cameron, B.M. Campbell, A.B. Cribb *et al.*, vol. 1, pp. 193–206. Brisbane: The Great Barrier Reef Committee.

Sale, P.F., 1977. Maintenance of high diversity in coral reef fish communities. *American Naturalist*, **111**, 337–59.

Sale, P.F., 1978. Coexistence of coral reef fishes – a lottery for living space. *Environmental Biology of Fishes*, **3**, 85–102.

Sale, P.F., 1979. Habitat partitioning and competition in fish communities. In *Predator–Prey Systems in Fisheries Management*, ed. H.E. Clepper, pp. 323–31. Washington, DC: Sport Fishing Institute.

Sale, P.F., 1980a. The ecology of fishes on coral reefs. *Oceanography and Marine Biology*, **18**, 367–421.

Sale, P.F., 1980b. Assemblages of fish on patch reefs – predictable or unpredictable? *Environmental Biology of Fishes*, **5**, 243–9.

Sale, P.F., 1982. Stock-recruitment relationships and regional co-existence in a lottery competitive system: A simulation study. *American Naturalist*, **120**, 139–59.

Sale, P.F., 1984. The structure of communities on coral reefs, and the merits of an hypothesis-testing manipulative approach to ecology. In *Ecological Communities: Conceptual Issues and the Evidence*, ed. D.R. Strong Jr, D. Simberloff, L.G. Abele & A.B. Thistle, pp. 478–90. Princeton, NJ: Princeton University Press.

Sale, P.F., 1991. Reef fish communities: Open nonequilibrial systems; In *The Ecology of Coral Reef Fishes*, ed. P.F. Sale, pp. 564–98. San Diego, CA: Academic Press.

Sale, P.F. & Douglas, W.A., 1984. Temporal variability in the community structure of fish on coral patch reefs and the relation of community structure to reef structure. *Ecology*, **65**, 409–22.

Sale, P.F. & Dybdahl, R., 1975. Determinants of community structure for coral reef fishes in an experimental habitat. *Ecology*, **56**, 1343–55.

Sale, P.F. & Dybdahl, R., 1978. Determinants of community structure for coral reef fishes in isolated coral heads at lagoonal and reef slope sites. *Oecologia*, **34**, 57–74.

Sale, P.F. & Williams, D.McB., 1982. Community structure of coral reef fishes: Are the patterns more than those expected by chance? *American Naturalist*, **120**, 121–7.

Sheppard, C.R.C., 1987. Coral species of the Indian Ocean and adjacent seas: A synonymized compilation and some regional distributional patterns. *Atoll Research Bulletin*, **307**, 1–32.

Sheppard, C.R.C., Price, A.R.G. & Roberts, C.M., 1992. *Marine Ecology of the Arabian Region: Patterns and Processes in Extreme Tropical Environments.* London: Academic Press.

Shulman, M.J., Ogden, J.C., Ebersole, J.P., McFarland, W.N., Miller, S.L. & Wolf, N.G., 1983. Priority effects in the recruitment of juvenile coral reef fishes. *Ecology*, **64**, 1508–13.

Smith, C.L., 1978. Coral reef fish communities: A compromise view. *Environmental Biology of Fishes*, **3**, 109–28.

Smith, C.L. & Tyler, J.C., 1972. Space resource sharing in a coral reef fish community. *Bulletin of the Natural History Museum, Los Angeles County*, **14**, 125–70.

Smith, G.B., Saleh, M. & Sangoor, K., 1987. The reef ichthyofauna of Bahrain (Arabian Gulf) with comments on its zoogeographic affinities. *Arabian Gulf Journal of Scientific Research*, **B5**, 127–46.

Steele, J.H., 1984. Kinds of variability and uncertainty affecting fisheries. In *Exploitation of Marine Communities*, ed. R.M. May, pp. 245–62. Berlin: Springer-Verlag.

Stevens, G.C., 1989. The latitudinal gradient in geographical range: How so many species coexist in the tropics. *American Naturalist*, **133**, 240–56.

Talbot, F.H. & Goldman, G., 1972. A preliminary report on the diversity and feeding relationships of the reef fishes on One Tree Island, Great Barrier Reef system. *Proceedings of the Symposium on Corals and Coral Reefs*, ed. C. Mukundan & C.S. Gopinadha Pillai, vol. 1, pp. 425–40. Cochin, India: Marine Biological Association of India.

Talbot, F.H., Russell, B.C. & Anderson, G.R.V., 1978. Coral reef fish commentaries: Unstable high-diversity systems? *Ecological Monographs*, **48**, 425–40.

Thresher, R.E., 1983a. Environmental correlates of the distribution of planktivorous fishes in the One Tree Reef lagoon. *Marine Ecology Progress Series*, **10**, 137–45.

Thresher, R.E., 1983b. Habitat effects on reproductive success in the coral reef fish, *Acanthochromis polyacanthus* (Pomacentridae). *Ecology*, **64**, 1184–99.

Thresher, R.E., 1991. Geographic variability in the ecology of coral reef fishes: Evidence, evolution, and possible implications. In *The Ecology of Fishes of Coral Reefs*, ed. P.F. Sale, pp. 401–36. San Diego, CA: Academic Press.

UNEP/IUCN (United Nations Environment Programme/International Union for the Conservation of Nature), 1988a. *Coral Reefs of the World*. Vol. **1**: *Atlantic and Eastern Pacific*. UNEP Regional Seas Directories and Bibliographies. Cambridge/Nairobi: IUCN/UNEP.

UNEP/IUCN (United Nations Environment Programme/International Union for the Conservation of Nature), 1988b. *Coral Reefs of the World*. Vol. **2**: *Indian Ocean, Red Sea and Gulf*. UNEP Regional Seas Directories and Bibliographies. Cambridge/Nairobi: IUCN/UNEP.

UNEP/IUCN (United Nations Environment Programme/International Union for the Conservation of Nature), 1988c. *Coral Reefs of the World*. Vol. **3**: *Central and Western Pacific*. UNEP Regional Seas Directories and Bibliographies. Cambridge/Nairobi: IUCN/UNEP.

Veron, J.E.N., 1993. *A Biogeographic Database of Hermatypic Corals*. Townsville, Queensland: Australian Institute of Marine Science.

Victor, B.C., 1983. Recruitment and populations dynamics of a coral reef fish. *Science*, **219**, 419–20.

Victor, B.C., 1984. Coral reef fish larvae: Patch size estimation and mixing in the plankton. *Limnology and Oceanography*, **29**, 1116–19.

Victor, B.C., 1991. Settlement strategies and biogeography of reef fishes. In *The Ecology of Fishes on Coral Reefs*, ed. P.F. Sale, pp. 231–60. San Diego, CA: Academic Press.

Waldner, R.E. & Robertson, D.R., 1980. Patterns of habitat partitioning by eight species of Caribbean territorial damselfish (Pisces: Pomacentridae). *Bulletin of Marine Science*, **30**, 171–86.

Watson, M. & Ormond, R.F.G., 1994. Effect of an artisanal fishery on the fish and urchin populations of a Kenyan coral reef. *Marine Ecology Progress Series*, **109**, 115–29.

Wiens, J.A., 1984. On understanding a non-equilibrium world: myth and reality in community patterns and processes. In *Ecological Communities: Conceptual Issues and the Evidence*, ed. D.R. Strong Jr, D. Simerloff, L.G. Abele & A.B. Thistle, pp. 439–57. Princeton, NJ: Princeton University Press.

Wilkinson, C.R., 1992. Coral reefs of the world are facing widespread devastation: Can we prevent this through sustainable management practices? *Proceedings of the Seventh International Coral Reef Symposium*, ed. R.H. Richmond, vol. **1**, pp. 11–21. Mangilao, Guam: University of Guam Press.

Williams, D.McB., 1979. Factors Influencing the Distribution and Abundance of

Pomacentrids (Pisces: Pomacentridae) on Small Patch Reefs in the One Tree Lagoon (Great Barrier Reef). Ph.D. thesis, University of Sydney, Sydney, Australia.

Williams, D.McB., 1982. Patterns in the distribution of fish communities across the central Great Barrier Reef. *Coral Reefs*, **1**, 35–43.

Williams, D.McB., 1983. Longitudinal and latitudinal variation in the structure of reef fish communities. In *Proceedings, Inaugural Great Barrier Reef Conference*, ed. J.T. Baker, R.M. Carter, P.W. Sammarco & K.P. Stark, pp. 265–70. Townsville, Queensland: James Cook University Press.

Williams, D.McB., 1986. Temporal variation in the structure of reef slope fish communities (central Great Barrier Reef): Short-term effects of *Acanthaster planci* infestation. *Marine Ecology Progress Series*, **28**, 157–64.

Williams, D.McB., 1989. Geographical and environmental variation in shallow-water reef fishes of Vanuatu. In *Vanuatu Marine Resources Survey, March–April 1988*, ed. T.J. Done & K.F. Navin. Consultancy Report. Townsville, Queensland: Australian Institute of Marine Science.

Williams, D.McB., 1991. Patterns and processes in the distribution of coral reef fishes. In *The Ecology of Coral Reef Fishes*, ed. P.F. Sale, pp. 437–74. San Diego, CA: Academic Press.

Williams, D.McB. & Hatcher, A.I., 1983. Structure of fish communities on outer slopes of inshore, mid-shelf and outer shelf reefs of the Great Barrier Reef. *Marine Ecology Progress Series*, **10**, 239–50.

Woodland, D.J., 1983. Zoogeography of the Siganidae (Pisces): An interpretation of distribution and richness patterns. *Bulletin of Marine Science*, **33**, 713–17.

Chapter 11
The historical component of marine taxonomic diversity gradients

J. ALISTAIR CRAME and ANDREW CLARKE
British Antarctic Survey, High Cross, Madingley Road, Cambridge, CB3 0ET, UK

Abstract

In the past it has been widely assumed that latitudinal diversity gradients in the marine realm must be underpinned in some way by a cline in evolutionary rates. Although there is some theoretical justification for believing that speciation is enhanced in the tropics and extinction towards the poles, this has not been borne out so far by empirical studies based on the fossil record. A detailed comparison of ten tropical and ten temperate molluscan clades showed that, although the former were, on average, three times as diverse as the latter, they had not radiated at a significantly faster rate. Many temperate clades, in both the marine and terrestrial realms, appear to have evolved comparatively rapidly; they are less diverse at the present day simply because they are geologically younger. Available evidence from the marine realm suggests that extinction is no more likely in cold than in warm waters. Explanations of large-scale biodiversity patterns may now reside more in regional rather than local phenomena. In particular, extant latitudinal taxonomic diversity gradients may reflect as much range fluctuation in response to Cenozoic climatic cycles as to any intrinsic properties of either tropical or polar biotas.

11.1 Introduction

In recent years there has been a growing realisation that the patterns of biodiversity we see today may be the product of a substantial historical legacy. Despite a vast array of both theoretical and empirical ecological evidence that might be taken to the contrary, it is becoming increasingly apparent that community patterns cannot be controlled by local processes alone. The strong correlation of local with regional diversity in a number of case studies, and the lack of community convergence in similar but widely separated habitats, are but two of the lines of evidence that are starting to reshape our views on the origin and maintenance of biodiversity. The biological wheel is beginning to turn full circle, with regional and historical perspectives once more coming to the fore (Ricklefs, 1987; Ricklefs & Schluter, 1993).

Aspects of this new approach are being applied to the study of latitudinal diversity gradients, perhaps the most important of all the large-scale biodiversity patterns on the face of the earth. So much of our thinking to date in this field has been dominated by some form of equilibrium hypothesis, which permits greater co-existence in the tropical than temperate regions (Farrell & Mitter, 1993, and references therein). In recent years diversity gradients seem to have become almost the exclusive preserve of the ecologist, with the focus of attention being on the twin issues of resource partitioning and limiting similarity (Ricklefs & Schluter, 1993). Nevertheless, the regional processes that have traditionally been the province of the biogeographer and palaeontologist are now beginning to warrant serious attention (Ricklefs, 1987).

In this chapter we concentrate in particular on how the evolution of latitudinal gradients in the shallow marine realm may have been influenced by fluctuations in both speciation and extinction rates through time. To do this we commence with a short theoretical consideration of the sorts of biological parameters that are likely to underpin changes in evolutionary rates in the marine realm, and then move on to consider if there may be any simple relationship between rates and latitude *per se*. In the main part of this chapter we present an attempt to estimate evolutionary rates from a variety of extant molluscan taxa with extensive fossil records. It should be emphasised at the outset that this is very much a pilot study and the results can only be regarded as preliminary. If nothing else, we hope to be able to demonstrate the main types of historical and regional processes that may have been of fundamental importance in the generation of marine taxonomic diversity gradients.

11.2 Some theoretical considerations

There are at least two potential processes that might suggest that rate of evolution should vary with latitude: the first of these is the influence of metabolic rate on mutation, and the second is generation time.

It has been proposed from studies of mitochondrial DNA (specifically, the 16 S RNA gene) that rate of mutation might be correlated with metabolic rate. The suggested mechanism is that mitochondrial respiratory processes produce free oxygen radicals, which in turn cause mutation by oxidative damage (Rand, 1994). Such a mechanism would tie the speed of the molecular clock to metabolic rate, and it would seem intuitively reasonable that the rate of molecular evolution would therefore also vary. The link to morphological evolution, as witnessed on a palaeontological time-scale, is, of course, very indirect, but Bargelloni et al. (1994) have suggested that a slow rate of molecular evolution may be responsible for the very recent divergence times calculated for the Antarctic notothenioid fish radiation. If typical rates of mitochondrial DNA evolution and mammalian rates of the accumulation of transversion mutations are assumed, then molecular data for notothenioids suggest that the earliest cladogenic events occurred at only 7–15 million years (my) ago. This contrasts with a more traditional age, based on morphological studies and major tectonic events in the Southern Ocean, of about 38 my ago. Since a known correlation between environmental temperature and routine metabolic rate in marine ectotherms (Clarke, 1991, 1993) also extends to a low metabolic rate in notothenioid fish (Eastman, 1993), it is possible that the mis-match in estimated divergence times is related to a reduced rate of molecular evolution at polar temperatures.

The second factor that might cause evolutionary rate to vary with latitude is that of generation time. Most polar marine ectothermic organisms grow slowly and have extended lifetimes (Clarke, 1983). This means that, on average, generation times will vary inversely with temperature (and thus latitude). Since the genomic reorganisation that occurs in meiosis during the process of gamete maturation is critical to the generation of raw material for evolution, this would suggest that evolutionary processes should proceed more rapidly in warmer waters (see also Vermeij, 1987).

11.3 The canonical view of taxonomic diversity gradients in the marine realm

Our perceptions of large-scale diversity patterns in the shallow marine realm were profoundly influenced by the seminal studies of Stehli et al. (1967, 1969). Working with bivalves, these authors were able to demonstrate clear equatorial–polar gradients of decreasing diversity in each hemisphere; their tropical high diversity foci seemed to correspond exactly with those established for many other marine and terrestrial groups (Stehli *et al.*, 1967, fig. 3). In their later paper, Stehli *et al.* (1969) suggested that these patterns were caused by a fundamental difference in evolutionary rates; in some way speciation was enhanced in the low latitudes and extinction enhanced towards the poles. The concept of marine taxa proliferating from tropical centres of origin has received widespread support, especially within palaeontological circles (e.g. Stehli & Wells, 1971; Hecht & Agan, 1972; Briggs, 1987).

More recent research has suggested that biodiversity patterns in general, and marine ones in particular, may be considerably more complex than was envisaged originally (see Chapter 6). One particularly important feature to emerge is that southern hemisphere patterns are not necessarily a mirror-image of those seen in the north. It is becoming apparent too, that there may not be any simple correlation between latitude and evolutionary rate. Use of the reciprocals of taxonomic longevities as an indicator of evolutionary rates (*sensu* Stehli *et al.*, 1969) may be inappropriate (Stanley, 1978, 1979), and tropical concentrations of geologically young taxa might equally well be explained by some form of range-retraction hypothesis (Crame, 1997 and references therein). Younger taxa, having had less time to evolve broad environmental tolerances, may be restricted periodically to just the core regions of large marine provinces.

11.4 Variation of speciation rates with latitude

To reinvestigate rates of speciation in the marine realm we selected ten tropical and ten cold temperate molluscan taxa that range from shallow into deep (but not abyssal) waters (Table 11.1). These taxa are either of generic, subfamily or family status and are believed to be, in essence, monophyletic. For each clade we have then estimated both the number of living species that it contains, and its geological age. Following a technique pioneered by Stanley (1979, 1985), we calculated rates of clade diversification (r) from the log of the number of extant species divided by time since its origin: i.e. $r = \ln N/t$ (Table 11.1). Once r is known, it is possible to estimate the time taken for the number of species within a clade to double (t_2) from the following equation: $t_2 = \ln 2/r$ (Stanley, 1979).

It was established that the number of species in a clade (Table 11.1) varied significantly with age ($F = 8.73$, $p < 0.01$), and this relationship did not differ between tropical and temperate clades (analysis of covariance: $F = 0.67$, $p = 0.424$). This applies to the linear (untransformed)

Table 11.1. *A comparison of rates of adaptive radiation in ten tropical and ten temperate molluscan taxa*

	Tropical				Temperate		
	No. species	Age (my)	r (my^{-1})		No. species	Age (my)	r (my^{-1})
Conidae	500	55	0.113	Buccininae	87	29	0.154
Murex	26	23	0.142	*Neptunea*	25	30	0.107
Haustellum	17	23	0.123	Volutopsinae	25	20	0.161
Mitridae	375	80	0.074	Oenopotinae	78	23	0.189
Strombus	50	23	0.170	*Aforia*	20	38	0.079
Strombidae	75	55	0.078	*Pareuthria*	17	38	0.075
Babylonia	12	42	0.059	*Trophon*	29	55	0.061
Cypraea	200	97	0.055	Philobryidae	25	55	0.058
Harpa	10	33	0.069	*Fusitriton*	5	26	0.062
Tridacnidae	6	42	0.043	Neptuneinae	102	38	0.122
x values	127	47	0.0920		41	35	0.1068

Data sources as described in the legend for Fig. 11.1. The two mean rates ($r = 0.0926$ my^{-1}, $r = 0.1068$ my^{-1}) are not significantly different at the 95% level, using either a Student's t-test or a Mann–Whitney U-test.

data, which is biased by a few very species-rich clades. Use of logarithmic data instead (Fig. 11.1) still showed the number of species in a clade to vary significantly with age ($F = 11.02$, $p < 0.001$), and this relationship also did not differ between tropical and temperate clades (analysis of covariance: $F = 1.68$, $p = 0.213$).

A number of other results from this study (Fig. 11.1 and Table 11.1) are immediately apparent. First of all, the tropical clades are not only the most diverse, but also the oldest (Fig. 11.1). For example, the Cypraeidae (cowries), with some 200 living species, first diversified at the base of the Late Cretaceous period, although it should be pointed out that certain representatives may even go back to the base of the Early Cretaceous (Table 11.1). The Mitridae, with some 375 living representatives, can be traced back to the Late Cretaceous, and the Conidae, one of the most diverse of all living marine gastropod groups (500 living species), has an Early Eocene origin. There is a distinct impression here that time may be an important component of tropical marine diversity; we will return to this theme below.

The mean rate of diversification of the 20 clades investigated ($r = 0.0099$ my^{-1}; Fig. 11.1) is significantly higher than that calculated by

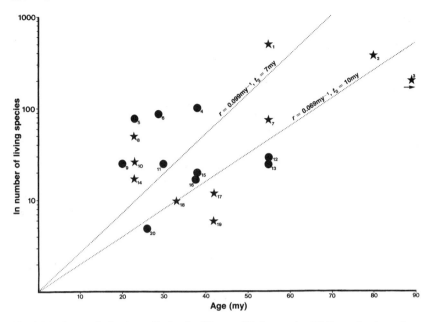

Fig. 11.1. Rates of adaptive radiation for 20 taxa of living marine Mollusca (gastropods and bivalves). For each taxon (which is of generic or higher rank), the logarithm of the number of living species (y axis) is plotted against its age (my; x axis). Further details about the calculation of mean rates of radiation (r) and doubling time (t_2) are given in the text. The upper dotted line is the mean rate of radiation for the 20 taxa investigated in this study; the lower line is the mean rate taken from Stanley & Newman (1980, fig. 2). The numerical key for each taxon and data source is as follows: 1, Conidae, Kohn (1990); 2, Mitridae, Cernohorsky (1976); 3, Cypraeidae (NB date of principal radiation taken as 97 my, but some forms may go back to Late Jurassic/Early Cretaceous), E.A. Kay (personal communication, 1994); 4, Neptuneinae, Golikov (1963), Nelson (1978), Titova (1994); 5, Oenopotinae, Bogdanov (1990); 6, Buccininae, Golikov (1980), Titova (1994); 7, Strombidae, Abbott (1960); 8, *Strombus*, Abbott (1960); 9, Volutopsinae, Kantor (1990), Titova (1994); 10, *Murex*, Ponder & Vokes (1988); 11, *Neptunea*, Golikov (1963), Nelson (1978), Titova (1994); 12, *Trophon*, Powell (1951), Oliver & Picken (1984), Crame (1996); 13, Philobryidae, Nicol (1966, 1970), Hayami & Kase (1993), Crame (1996); 14, *Haustellum*, Ponder & Vokes (1988); 15, *Aforia*, Sysoev & Kantor (1987), Dell (1990); 16, *Pareuthria*, Stillwell & Zinsmeister (1992), Crame (1996); 17, *Babylonia*, Altena & Gittenberger (1981); 18, *Harpa*, Rehder (1973); 19, Tridacnidae, Rosewater (1965); 20, *Fusitriton*, Smith (1970). ★, tropical; ●, cold temperate and polar.

Stanley & Newman (1980, fig. 2; $r = 0.069$ my^{-1}) in an earlier comparative study of molluscan taxa. However, that work encompassed a number of cosmopolitan groups that it would have been inappropriate to include here. Correspondingly, the mean doubling time (t_2) for a molluscan clade is reduced from 10 my to 7 my.

It can be seen that there is a genuine admixture of tropical and cold-temperate/polar clades both above and below the mean radiation rate (Fig. 11.1). Whereas some tropical taxa appear to have radiated comparatively slowly, some cold-temperate/polar forms have evolved rapidly. Examples within the former of these categories include the buccinid gastropod genus *Babylonia*, the volutid *Harpa*, and the bivalve family Tridacnidae (giant clams). Although giant clams occur on coral reefs throughout the Indo-West Pacific province, only six species are known (Table 11.1). Temperate clades that appear to have radiated much more rapidly include the buccinid subfamilies Buccininae and Neptuneinae, together with the turrid gastropod subfamily Oenopotinae. Indeed, the latter taxon gave the highest rate of clade diversification recorded in this study (Fig. 11.1; Table 11.1).

Perhaps the most striking feature to emerge from this investigation was that although tropical clades contain, on average, three times as many taxa as temperate ones (127 : 41), they have not radiated at significantly faster rates (Table 11.1). This, perhaps, can be taken as a further pointer towards the critical role of time. In some way that is not yet fully understood, taxa may have accumulated preferentially in the geologically oldest environments (see below). Of course, it must be emphasised that this study is by no means comprehensive, and the results can only be regarded as preliminary. It should also be appreciated that this method of determining rates rests heavily on the assumption that the adaptive radiation of a clade is exponential (Stanley, 1979, 1985; Kohn, 1990). Obviously, this will only be true for its initial stages, and there must come a point in its history when the rate of radiation will begin to level-off substantially. In addition, this technique is really only concerned with the two end points of a clade; the number of living species and its point of origin. No account is taken of any speciation events that may have occurred during its history, and consequently this method will always produce an underestimate of the true rate of speciation (Stanley, 1985; Kohn, 1990). Perhaps a better metric to use would be one that is based, in essence, on the number of speciation events in a clade's history, divided by time. In practice, a simple technique developed by Sepkoski (1978) and others is based on the following equation:

$r_s = 1/D \times S/\Delta t$,

Table 11.2. *A comparison of rates of speciation in tropical and temperate gastropods*

	Conidae (tropical)	Buccininae (temperate)
Miocene diversification	$r_s = 0.035$ my^{-1}	$r_s = 0.052$ my^{-1}
	Conidae (tropical)	Neptuneinae (temperate)
Pleistocene diversification	$r_s = 0.393$ my^{-1}	$r_s = 0.430$ my^{-1}

Calculations are based on the equation: $r_s = 1/D \times S/\Delta t$. See the text and Fig. 11.2 for further details.

where S is the number of first appearances (apparent speciations) within a specified time interval (Δt, in my), and D is the total number of species (apparent diversity) within that interval. Use of the reciprocal of D means that these are *per taxon* rates, but it should be noted that no distinction is made between phyletic and cladogenetic (or apparent and true) speciation events.

Using the very comprehensive data base developed by Kohn (1990), rates of speciation (r_s) for the gastropod family Conidae are replotted in Fig. 11.2. It is apparent that there are three very marked phase of diversification: one in the Miocene, one in the Pleistocene, and one in the Pleistocene–Recent. Because of possible taxonomic artefacts within the latter, only the first two of these phases will be considered here. Although we might expect, *a priori*, these Miocene and Pleistocene rates to be amongst the highest for any marine gastropod clade, it is interesting to note that, in both instances, they are exceeded by those obtained from two whelk (Buccinidae) subfamilies (Buccininae and Neptuneinae, Fig. 11.2 and Table 11.2).

Unfortunately, these are the only three molluscan groups to date for which speciation rates have been calculated in this way. Not enough information is yet available to check for statistical differences between tropical and temperate rates, and in any event it may be inappropriate to make rigid comparisons between different neogastropod superfamilies. Nevertheless, these are three of the commonest molluscan clades at the present day (Fig. 11.1), and the close similarity in rates (Fig. 11.2; Table 11.2) is extremely interesting. In a much larger study, using similar techniques and based on some 149 Neogene planktonic forminifera species and subspecies, Wei & Kennett (1986) were unable to detect

(a)

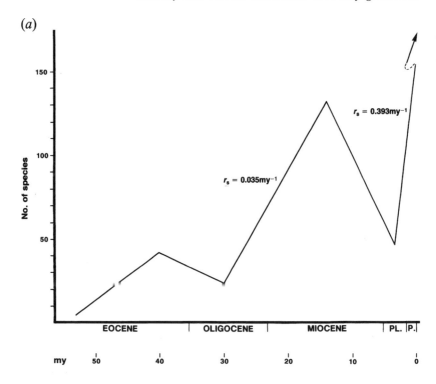

(b)

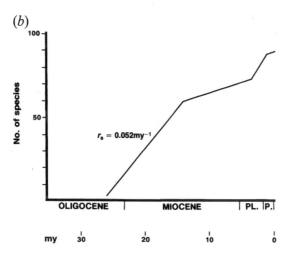

Fig. 11.2. A comparison of rates of speciation in tropical and temperate gastropods, based on the equation: $r_s = 1/D \times S/\Delta t$ (see the text for further details). Data sources as given in the legend to Fig. 11.1. (a) Conidae (tropical). (b) Buccininae (temperate). (c) Neptuneinae (temperate).

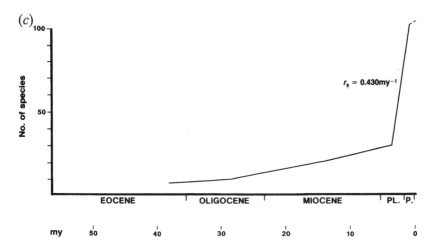

any latitudinal clines in evolutionary rates. Although numbers of both speciation and extinction events were highest in the tropics, there were no significant differences in either speciation or extinction *rates* between tropical, transitional and temperate regions (Wei & Kennett, 1986, table 2).

11.5 Variation of extinction rates with latitude

Despite the enormous amount of interest in recent years in the phenomenon of mass extinction, it is somewhat surprising to find that there have been very few systematic investigations of the variation in extinction rate with latitude (Crame, 1997). There is a widespread impression that reef-associated taxa are more susceptible to mass extinctions than corresponding level-bottom ones (Jablonski, 1991), especially at the Cretaceous–Tertiary boundary (e.g. Kauffman, 1984, figs 8-2 and 8-3). However, in a recent focused study of just one molluscan group (the bivalves) across this interval, Raup & Jablonski (1993) were unable to detect any form of global gradient in extinction intensity. The nature of faunal change across high-latitude Cretaceous–Tertiary boundary sections is currently a matter of intense debate (Keller *et al.*, 1993).

One palaeontological technique that has thrown some light on geographical variation in extinction rates, albeit in a rather circuitous fashion, is that of the construction of Lyellian Curves. Pioneered in their use by Stanley (1979, 1985), such curves represent plots of the percentage of extant species within Cenozoic molluscan faunas against their age.

Faunas dating back over approximately the last 15 my were found to form a reasonably tight curve against which anomalously low percentages (representing high extinction rates) could be judged. Initial studies indicated that endemism related to geographical configuration appeared to be a much stronger determinant of extinction rate than simply latitude on its own (Stanley et al., 1980, p. 425). Endemic Late Cenozoic molluscan faunas of the tropical Americas were much more prone to extinction than those of the tropical western Pacific.

Finally, reference can again be made to the fact that Wei & Kennett's (1986) comprehensive investigation of evolutionary rates in Neogene planktonic foraminifera found no statistical correlation between extinction rate and latitude.

11.6 Discussion

Despite a strong intuitive feeling, and some theoretical evidence to the contrary, it would appear that latitudinal diversity gradients are not necessarily underpinned by a cline in either speciation or extinction rates. Of course, we recognise that such a conclusion rests on a comparatively small data base, and the evolutionary history of many other prominent Cenozoic marine taxa has yet to be taken into consideration. There is considerable scope too, for improving the nature of analytical procedures on available data sets.

It would seem to be particularly important to apply some form of phylogenetic test to the study of latitudinal variations in evolutionary rates. Whilst no such tests are yet available in the marine realm, they have been applied recently to the diversification of phytophagous insect clades, many of which have clearly spread through the Cenozoic from tropical into temperate habitats (Farrell & Mitter, 1993). A statistical comparison of five sister group pairs showed no significant difference in tropical/temperate diversification rates, and it can be shown that a tropical portion of a clade is more diverse simply because it is older. Such a hypothesis is in turn open to testing since, if it is correct, there should be a greater proportion of higher taxa within the tropical portion of any clade. In other words, much of the greater diversity of the tropics should lie at higher taxonomic levels (Farrell & Mitter, 1993).

We may conclude then, albeit rather tentatively, that studies to date based on the fossil record indicate that there is not necessarily any greater rate of speciation in the tropics or extinction towards the poles. If such a conclusion is correct, it is important to assess what the implica-

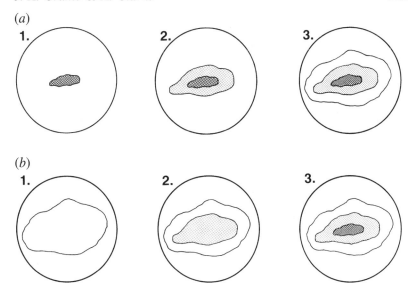

Fig. 11.3. Hypothetical evolution of taxonomic diversity gradients. (*a*) Three-stage development by radiation in the size of species' ranges from a point source (centre of origin). (*b*) Three-stage development by retraction in the size of species' range from a former widespread distribution. (*a*)-3 and (*b*)-3 are identical; both mechanisms produce latitudinal and longitudinal gradients.

tions might be for the formation of large-scale biodiversity patterns. To do this, and for the sake of simplicity, it can be envisaged that these patterns may form in one of two basic ways (Fig. 11.3): they may either represent an increase in the size of species' ranges, which thereby appear to radiate from a point source (centre of origin) (Fig. 11.3(*a*), 1–3), or conversely a decrease in range size, which generates a contraction from a former widespread distribution (Fig. 11.3(*b*), 1–3). If our conclusion about the absence of a latitudinal cline in evolutionary rates is correct, then we would have to accept that the sorts of processes depicted in Fig. 11.3 could, in theory, occur on any part of the Earth's surface. They need not necessarily be confined to any particular latitudinal (or longitudinal) zone.

If latitudinal diversity gradients are not the product of any intrinsic differences in evolutionary rates between the tropics and the poles, should we be looking instead to a greater role for extrinsic (i.e. regional) factors? One historical influence that may have been particularly important, especially in the generation of the sorts of sequential patterns depicted in Fig. 11.3, is that of climate change. We now know that, over at least the last 65–70 my, marine climates have fluctuated repeatedly and widely (Clarke & Crame, 1992, and references therein). Within the geological record, such fluctuations are detected on a scale ranging from tens of millions of years at one end to just a few thousand years at the other. It could well be that range expansions and contractions in response to these repeated climate changes were a greater determinant of large-scale diversity patterns than any intrinsic evolutionary properties of either the tropics or the poles. Climate-driven range fluctuations may be the key to a better understanding of the evolution of taxonomic diversity gradients in the marine realm (Crame, 1993; Valentine & Jablonski, 1993).

Acknowledgements

We are extremely grateful to Professors J.D. Taylor, E.A. Kay and A.J. Kohn for taxonomic and stratigraphical information that improved the quality of the analysis presented in Fig. 11.1 and Table 11.1.

References

Abbott, R.T., 1960. The genus *Strombus* in the Indo-Pacific. *Indo-Pacific Mollusca*, **1**, 33–146.

Altena, C.O. van R. & Gittenberger, E., 1981. The genus *Babylonia* (Prosobranchia, Buccinidae). *Zoologische Verhandelingen*, **188**, 1–57.

Bargelloni, L., Ritchie, P.A., Patarnello, T., Battaglia, B., Lambert, D.M. & Meyer, A., 1994. Molecular evolution at subzero temperatures: Mitochondrial and nuclear phylogenies of fishes from Antarctica (suborder Notothenioidei) and the evolution of antifreeze glycopeptides. *Molecular Biology and Evolution*, **11**, 854–63.

Bogdanov, I.P., 1990. *Fauna of USSR. Mollusks*, vol. **5**, no. 3. *Mollusks of Oenopotinae subfamily (Gastropoda, Pectinibranchia, Turridae) in the seas of the USSR*. Leningrad: 'Nauka' Publishing House. [In Russian].

Briggs, J.C., 1987. *Biogeography and Plate Tectonics*. Amsterdam: Elsevier.

Cernohorsky, W.O., 1976. The Mitridae of the world. *Indo-Pacific Mollusca*, **3**, 273–528.

Clarke, A., 1983. Life in cold water: The physiological ecology of polar marine ectotherms. *Annual Reviews of Oceanography and Marine Biology*, **21**, 341–453.

Clarke, A., 1991. What is cold adaptation and how should we measure it? *American Zoologist*, **31**, 81–92.

Clarke, A., 1993. Temperature and extinction in the sea: A physiologist's view. *Paleobiology*, **19**, 499–518.

Clarke, A. & Crame, J.A., 1992. The Southern Ocean benthic fauna and climate change: A historical perspective. *Philosophical Transactions of the Royal Society, London, Series B*, **338**, 299–309.

Crame, J.A., 1993. Latitudinal range fluctuations in the marine realm through geological time. *Trends in Ecology and Evolution*, **8**, 162–6.

Crame, J.A., 1996. Evolution of high-latitude molluscan faunas. In *Origin and Evolutionary Radiation of the Mollusca*, ed. J.D. Taylor, pp. 119–31. Oxford: Oxford University Press.

Crame, J.A., 1997. An evolutionary framework for the polar regions. *Journal of Biogeography*, **24**, 1–9.

Dell, R.K., 1990. Antarctic Mollusca. *Bulletin of the Royal Society of New Zealand*, **27**, 1–311.

Eastman, J.T., 1993. *Antarctic Fish Biology: Evolution in a Unique Environment*. San Diego, CA: Academic Press.

Farrell, B.D. & Mitter, C., 1993. Phylogenetic determinants of insect/plant community diversity. In *Species Diversity in Ecological Communities. Historical and Geographical Perspectives*, ed. R.E. Ricklefs & D. Schluter, pp. 253–66. Chicago: University of Chicago Press.

Golikov, A.N., 1963. *Fauna of USSR. Mollusks*, vol. **5**, no. 1, *Gastropod mollusks of the genus* Neptunea *Bolten*. Moscow & Leningrad: Akademi Nauk USSR. [In Russian].

Golikov, A.N. 1980. *Fauna of USSR Mollusks*, vol **5**, no. 2. *Buccininae mollusks of the World Ocean*. Leningrad: 'Nauka'. [In Russian].

Hayami, I. & Kase, T., 1993. Submarine cave Bivalvia from the Ryukyu Islands: Systematics and evolutionary significance. *Bulletin of the University Museum, University of Tokyo*, **35**, 1–33.

Hecht, A.D. & Agan, B., 1972. Diversity and age relationships in Recent and Miocene bivalves. *Systematic Zoology*, **21**, 308–12.

Jablonski, D., 1991. Extinctions: A paleontological perspective. *Science*, **253**, 754–7.

Kantor, Yu.I., 1990. *Gastropods of the subfamily Volutopsinae of the world ocean*. Moscow: 'Nauka'. [In Russian, with English abstract].

Kauffman, E.G., 1984. The fabric of Cretaceous marine extinctions. In *Catastrophes and Earth History*, ed. W.A. Berggren & J.A. van Couvering, pp. 151–246. Princeton, NJ: Princeton University Press.

Keller, G., Barrera, E., Schmitz, B. & Mattson, E., 1993. Gradual mass extinction, species survivorship, and long-term environmental changes across the Cretaceous–Tertiary boundary in high latitudes. *Bulletin of the Geological Society of America*, **105**, 979–97.

Kohn, A.J., 1990. Tempo and mode of evolution in Conidae. *Malacologia*, **32**, 55–67.

Nelson, C.M., 1978. *Neptunea* (Gastropoda: Buccinacea) in the Neogene of the North Pacific and adjacent Bering Sea. *The Veliger*, **21**, 203–15.

Nicol, D., 1966. Descriptions, ecology and geographic distribution of some antarctic pelecypods. *Bulletins of American Paleontology*, **51**, 1–102.

Nicol, D., 1970. Antarctic pelecypod faunal peculiarities. *Science*, **168**, 1248–9.

Oliver, P.G. & Picken, G.B., 1984. Prosobranch gastropods from Signy Island, Antarctica: Buccinacea and Muricacea. *British Antarctic Survey Bulletin*, **62**, 95–115.

Ponder, W.F. & Vokes, E.H., 1988. A revision of the Indo-West Pacific fossil and Recent species of *Murex* ss. and *Haustellum* (Mollusca: Gastropoda: Muricidae). *Records of the Australian Museum*, **Supplement 8**, 1–160.

Powell, A.W.B., 1951. Antarctic and subantarctic Mollusca: Pelecypoda and Gastropoda. *Discovery Reports*, **26**, 47–196.

Rand, D.M., 1994. Thermal habit, metabolic rate and the evolution of mitochondrial DNA. *Trends in Ecology and Evolution*, **9**, 125–31.

Raup, D.M. & Jablonski, D., 1993. Geography of end-Cretaceous marine bivalve extinctions. *Science*, **260**, 971–3.
Rehder, H.A., 1973. The family Harpidae of the world. *Indo-Pacific Mollusca*, **3**, 207–15.
Ricklefs, R.E., 1987. Community diversity: Relative roles of local and regional processes. *Science*, **235**, 167–71.
Ricklefs, R.E. & Schluter, D., 1993. Species diversity: Regional and historical influences. In *Species Diversity in Ecological Communities. Historical and Geographical Perspectives*, ed. R.E. Ricklefs & D. Schluter, pp. 350–63. Chicago: University of Chicago Press.
Rosewater, J., 1965. The family Tridacnidae in the Indo-Pacific. *Indo-Pacific Mollusca*, **1**, 347–93.
Sepkoski, J.J., Jr, 1978. A kinetic model of Phanerozoic taxonomic diversity. I. Analysis of marine orders. *Paleobiology*, **4**, 223–51.
Smith, J.T., 1970. Taxonomy, distribution and phylogeny of the cymatiid gastropods *Argobuccinum, Fusitriton, Mediargo* and *Priene*. *Bulletins of American Paleontology*, **56**, 443–573.
Stanley, S.M., 1978. Chronospecies longevities, the origin of genera, and the punctuational model of evolution. *Paleobiology*, **4**, 26–40.
Stanley, S.M., 1979. *Macroevolution. Pattern and process*. San Francisco: W.H. Freeman & Co
Stanley, S.M., 1985. Rates of evolution. *Paleobiology*, **11**, 13–26.
Stanley, S.M. & Newman, W.A., 1980. Competitive exclusion in evolutionary time: The case of the acorn barnacles. *Paleobiology*, **6**, 173–83.
Stanley, S.M., Addicott, W.O. & Chinzei, K., 1980. Lyellian curves in paleontology: Possibilities and limitations. *Geology*, **8**, 422–6.
Stehli, F.G. & Wells, J.W., 1971. Diversity and age patterns in hermatypic corals. *Systematic Zoology*, **20**, 115–26.
Stehli, F.G., McAlester, A.L. & Helsley, C.E., 1967. Taxonomic diversity of recent bivalves and some implications for geology. *Bulletin of the Geological Society of America*, **78**, 455–66.
Stehli, F.G., Douglas, R.G. & Newell, N.D., 1969. Generation and maintenance of gradients in taxonomic diversity. *Science*, **164**, 947–9.
Stilwell, J.D. & Zinsmeister, W.J., 1992. *Molluscan systematics and biostratigraphy. Lower Tertiary La Meseta Formation, Seymour Island, Antarctic Peninsula*. Antarctic Research Series, no. **55**, pp. 1–192. Washington, DC: American Geophysical Union.
Sysoev, A.V. & Kantor, Yu.I., 1987. Deep-sea gastropods of the genus *Aforia* (Turridae) of the Pacific: Species composition, systematics, and functional morphology of the digestive system. *The Veliger*, **30**, 105–26.
Titova, L.V., 1994. Cenozoic history of Turritelloidea and Buccinoidea (Mollusca: Gastropoda) in the North Pacific. *Palaeogeography, Palaeoclimatology and Palaeoecology*, **108**, 319–34.
Valentine, J.W. & Jablonski, D., 1993. Fossil communities: Compositional variation at many time scales. In *Species Diversity in Ecological Communities. Historical and Geographical Perspectives*, ed. R.E. Ricklefs & D. Schluter, pp. 341–9. Chicago: University of Chicago Press.
Vermeij, G.J., 1987. *Evolution and Escalation*. Princeton, NJ: Princeton University Press.
Wei, K.-Y. & Kennett, J.P., 1986. Taxonomic evolution of Neogene planktonic foraminifera and paleoceanographic relations. *Paleoceanography*, **1**, 67–84.

Note added in proof:

Since this text was completed in 1994, Flessa & Jablonski (1996) have undertaken a detailed study of faunal turnover in bivalve molluscs, which does find evidence for a small difference in turnover rates between tropical and temperate faunas. This is a subtly different measure than the speciation rates we have estimated, but it is indicative of geographical variation in some aspects of faunal differentiation.

Flessa, K.W. & Jablonski, D., 1996. Geography of evolutionary turnover. In *Evolutionary Paleobiology*, ed. D. Jablonski, D.H. Erwin & J. Lipps, pp. 176–397. Chicago: Chicago University Press.

Chapter 12
Population genetics and demography of marine species

JOSEPH E. NEIGEL
Department of Biology, The University of Southwestern Louisiana, Lafayette, Louisiana 70504, USA

Abstract

The life histories of many marine organisms are characterised by high fecundity, planktonic larvae and a remarkable capacity to colonise remote habitats. These species are commonly viewed as consisting of large, undifferentiated populations that are seldom limited by recruitment. This view leads to the expectation that substantial changes in marine populations generally must be brought about by forces acting over large geographical and temporal scales. However, recent surveys of molecular genetic variation suggest a very different perspective. Sharp geographical discontinuities have been found in the absence of any recognisable barriers to dispersal. These may be interpreted in terms of adaptation to local or regional environments. In many species, low levels of genetic variation or high levels of temporal variance in allele frequencies have indicated surprisingly small genetically effective population sizes. An important question raised by these findings is whether the stability of such populations may be critically dependent on the reproductive success of relatively few individuals. If so, then the potential consequences of habitat or range reduction may be more severe than previously anticipated.

12.1 Introduction

The concept of biodiversity unites the variety of life-forms that are the products of evolution with the diversity of habitats and ecosystems in which they are found. These two aspects of biodiversity have traditionally been the subjects of separate disciplines: biosystematics and ecology. However their union is now crucial. The rapid global decline of biological diversity at all levels represents a crisis that neither discipline, by itself, can adequately address. The developing science of biodiversity must relate ecological processes, which determine patterns of distribution and abundance, to the evolutionary histories of the organisms that undergo these processes. Only by understanding such relationships will it be possible to predict how changes in ecosystems and communities will impact specific taxa and, conversely, how losses or introductions of specific taxonomic groups will impact ecosystem or community function.

The marine realm has presented special problems for both systematists and ecologists (discussed by Ray, 1988). Extended, direct observation of marine organisms in their natural habitats is often difficult, and in some cases, nearly impossible. As a result, many marine environments have barely been explored, and historical information is available for only a few. Furthermore, because of the dynamic nature of oceanographic processes and the extensive movements of organisms in water, marine systems have no true boundaries. It is thus difficult to define small, easily managed units. These problems demand that marine biodiversity be studied and managed in ways that are fundamentally different from those developed for terrestrial systems. Because we know so much less about marine systems, we must find ways to extrapolate what we do know to places we have barely visited, and across time-scales much greater than our history of observation. Management strategies that are concerned with only the most visible (or most valuable) stage of a marine organism's life history are unlikely to succeed. Successful approaches to the study and management of marine biodiversity will probably be more conceptual, integrative and global than those developed for terrestrial systems.

It is to be expected that, in general, strategies for the conservation of biodiversity will be based on well established concepts and principles. My purpose in this chapter is to examine some new findings concerning the population genetics of marine organisms that do not fit within well established views. It is too early to predict whether these new findings will lead to a major shift in the way we think about the population biology of marine organisms. It is also true that much of what they imply has been proposed before, and may even represent views that have claimed adherents for some time. However, in the light of the current biodiversity crisis, the findings that are discussed below have particular significance because they suggest that populations of marine organisms may be far more delicate than is generally perceived.

12.2 The significance of population genetics for biodiversity studies

The importance of genetics to conservation biology has been much debated. Franklin (1980) and Soulé (1980) argued that loss of genetic variation in small refuge populations could threaten their viability and potential for evolutionary adaptation. Based on simple population genetic models, they proposed that a minimum population size of 50 was necessary to prevent inbreeding depression, and a minimum population size of 500 was necessary to prevent loss of evolutionary potential. The inadequacies of such generalisations have led to extensive criticism of genetic criteria for defining minimum viable population sizes (for a review, see Simberloff, 1988). Furthermore, as pointed out by Lande (1988), demographic factors may be more likely to cause the extinction of small populations than lack of genetic variation.

The debate over the role of genetic variation in determining the minimum viable size of a population has had the unfortunate effect of drawing attention away from other, less equivocal, applications of population genetics to conservation biology. Population genetics may be especially helpful in the study of marine biodiversity. A solid conceptual foundation for the study of marine biodiversity must somehow bridge the present gap between ecology and systematics. In principle, population genetics provides such a bridge. Population genetic processes are driven by events that take place in an ecological setting, but lead to evolutionary changes. These events: birth, death and migration, comprise the demography of a population. The evolutionary changes that follow can produce species, as well as higher levels of evolutionary diversity. In turn, adaptations that result from evolutionary change feed back to the ecological level by altering rates of birth, death and migration.

An optimistically long-term strategy for the preservation of biodiversity would strive to balance extinction with speciation. Geographical areas have been identified in which some taxa have recently undergone unusually high rates of speciation. It has been argued that these areas will continue to generate species and, on this basis, should be afforded special protection (Erwin, 1991). However, the conditions that are believed to lead to speciation are not inextricably linked to geography, however much they may be influenced by geographical circumstances. Precursors to speciation such as local adaptation and isolation are population genetic conditions, and as such may be altered by human activities, and are potential objects of management. However, although population genetic models do a reasonably good job of predicting how much genetic divergence will result from a given set of demographic conditions, they do not predict how much genetic divergence will lead to speciation. It is unlikely that a simple rule will be found that relates some standard measure of genetic divergence to speciation. The magnitude of genetic divergence between congeneric species appears to vary considerably (Avise & Aquadro, 1982). It may be difficult to develop a general understanding of speciation if reproductive isolation depends on specific and idiosyncratic genetic mechanisms, rather than on any overall level of genetic divergence.

The chain of causality can also be followed in the opposite direction, to predict the ecological effects of microevolutionary processes. Two major types of effects are expected: local adaptation and inbreeding depression. While both have been demonstrated in particular cases, it is difficult to reach any generalisation about their importance in natural populations. Theoretical models indicate a range of possibilities that depend on the particular characteristics of a species, its population biology and the nature of genetic variation (e.g. Wright, 1969). Empirical studies are generally biased towards species that are easy to maintain under controlled conditions. Some well entrenched generalisations are based on only a few studies of domesticated mammals and laboratory populations of *Drosophila* (Lande, 1988; Simberloff, 1988). Clearly, the importance of local adaptation and inbreeding depression in marine species needs to be explored.

12.3 Gene flow and local adaptation

Throughout its range, a species will experience different environmental conditions, constituting different regimes of natural selection. Transplantation of organisms to a new location may fail if they are not adapted to the conditions they encounter, even though the species may have once flourished at the same location. Furthermore, local adaptation requires that selection is strong enough to counter the homogenising effects of gene flow among populations (Ehrlich & Raven, 1969; Slatkin, 1985). Thus an increase in gene flow between populations may reverse the process of local adaptation, and cause populations to decline.

Natural selection is often viewed as a deterministic process, in which individual alleles are favoured in particular environments where they increase in frequency. This view is basically correct if the effect of each gene is largely independent of others. However, if there are significant interactions among genes, there may be many different ways for a population to respond to the same set of environmental conditions, with different combinations of genes that work together (Mayr, 1970; Wright, 1977; Hedrick et al., 1978). Which combination is used may depend on the initial genetic composition of the population, but may also be random to some degree. If local adaptation involves combinations of interacting genes, gene flow between populations that are adapted to the same conditions may nonetheless undo local adaptation by mixing incompatible combinations of genes. A consequent reduction in fitness would be considered a form of 'outbreeding depression' (Endler, 1977).

There is no simple test for local adaptation. It is not sufficient to demonstrate spatial or geographical variation in adaptive traits, because most complex traits are subject to direct environmental influence, as well as genetic control. Some form of experimental manipulation is required to control for environmental influences. Typically, a 'common garden' experiment is performed, in which stocks from different sources are reared in a common environment for several generations. However, for practical reasons, this approach can only be used for a few marine species.

The tide-pool copepod, *Tigriopus californicus*, is exceptionally well suited to experimental manipulation. Populations occur in high intertidal rock pools along the west coast of North America. These pools are relatively free of predators, but subject to extremes of temperature and salinity during periods of warm weather and low tide. Burton and co-workers have studied these populations extensively over the past 15 years. Allozyme allele frequencies in tide-pool populations were surveyed on several geographical scales (Burton et al., 1979; Burton & Feldman, 1981). Among a group of tide-pools located on a single rock outgroup, allele frequencies were generally uniform. However, between outcrops, sharp differences in allele frequencies were observed, even in cases where outcrops were separated by less than 500 m. Allele frequencies were also stable through several years of observation, although individual tide-pool populations cycled through extinction and recolonisation. These results indicated that tide-pools within a rock outcrop were linked by gene flow, but little gene flow occurred between tide-pools on different outcrops. This interpretation was confirmed by experiments in which individuals with characteristic allozyme alleles were transferred between populations (Burton & Swisher, 1984). Differences in allele frequencies between outcrops did not appear to constitute any large-scale geographical patterns; along the Californian coast, regional averages of allozyme frequencies were very similar.

By itself microgeographical variation in allozyme frequencies indicates that gene flow is sufficiently restricted to provide the opportunity for local adaptation, but does not imply that local adaptation has actually occurred. Differences in allozyme frequencies can be easily explained as the result of genetic drift, and may have no adaptive significance. Fortunately, the natural habitat of *T. californicus* can be approximated in the laboratory. Matings can be arranged, and progeny reared in small containers. Burton (1987, 1990) crossed animals from either the same population or from populations on different outcrops, raised F_1 progeny, and then intercrossed these F_1 to produce an F_2 generation. In an inter-population cross, haploid combinations of genes derived from each source population are passed on intact to F_1 progeny. However F_2 progeny inherit recombinant genotypes, in which the original haploid combinations of genes have been broken apart. To assess the fitness effects of these crosses, Burton measured the development time of progeny, and rates of mortality under osmotic stress. In a continuously breeding organism such as *T. californicus*, development time should be inversely correlated with fitness. The development times of F_1 progeny were generally between 15 and 17 days, regardless of whether they were the progeny of inter-population or intra-population crosses. In contrast, development times for F_2 progeny of inter-population crosses were typically much longer, indicating a substantial reduction in fitness. Furthermore, in crosses between two of the four populations (SD and AB), most of the F_2 progeny failed to even survive. These declines in fitness did not appear to be the effects of prolonged culture; development times for F_2 progeny of intra-population crosses were either about the same or only slightly longer than their F_1 parents. Thus it appears that interactions between genetic loci play a large role in local adaptation for *T. californicus*.

Tigriopus californicus is an excellent model organism for genetic studies, nearly a 'marine *Drosophila*'. However, its tide-pool habitat is unusual, and so we need to consider gene flow and local adaptation in other, more 'typical' marine species. A commonly held view is that for most marine species, dispersal in water provides high levels of gene flow, and, therefore, inter-population divergence and local adaptation are unlikely (Crisp, 1978; Hedgecock, 1986; Utter & Ryman, 1993). Allozyme surveys of both marine fish (Gyllensten, 1985) and invertebrates (Hedgecock, 1986) with nektonic or extended planktonic life-stages have typically found little genetic divergence among populations, a finding consistent with high rates of gene flow. Exceptions have been found (Burton, 1983) but have not altered the general view, that typical marine species consist of very large, panmictic (randomly breeding) populations, from becoming an established paradigm. This has been especially frustrating for fisheries biologists, who hoped to use genetic markers to identify natural stocks that could be managed separately.

12.4 Molecular markers and unexpected population structure

Recently, surveys of DNA sequence variation in marine animals, along with some new methods for the analysis of allozyme data, have led to some fairly radical reassessments of the established paradigm. Whether or not these new findings prove to be of general significance remains to be seen. To some extent, they represent not so much new ideas as revivals of old ones, such as the idea that the frequencies of allozyme alleles are controlled by natural selection. What is new is that recent technological developments should allow these ideas to be tested with greater rigour than ever before. These studies thus have the potential to force a major reassessment of the prevailing view that marine populations are subject only to large-scale events and processes.

Some of the most interesting new findings concern allozyme variation in the American oyster, *Crassostrea virginica*. This species is of considerable commercial importance, and has been the subject of extensive genetic and physiological studies. Its life history is typical of benthic marine invertebrates. Individual females live for up to 20 years, producing from 10–114 million eggs in a single reproductive cycle (Galtsoff, 1964). A high potential for dispersal is provided by a free-swimming planktonic veliger stage, which drifts in the plankton for 2–3 weeks.

An allozyme survey of American oyster populations along the Atlantic and Gulf of Mexico coasts of North America yielded fairly typical results (Buroker, 1983). Allozyme allele frequencies at most loci were generally uniform over a large geographical range, although 2 of the 21 polymorphic loci appeared to vary with latitude. These findings fit well with the accepted paradigm. Uniformity of allozyme allele frequencies is an expected consequence of high levels of gene flow. Unusual allele frequencies in one population, near Brownsville, Texas, were interpreted as evidence for isolation of this population. Only one locus (*Lap-2*), with allele frequencies that varied with latitude, was proposed to be under selection.

A survey of mitochondrial DNA (mtDNA) variation for essentially the same series of populations yielded a pattern strikingly at odds with the allozyme data (Reeb & Avise, 1990). Two very divergent groups of mtDNA sequences were found to distinguish Atlantic populations from those in the Gulf of Mexico. The location of the break was on the east coast of Florida, near a known biogeographical boundary. This difference is clearly not due to the greater variability of mtDNA as compared to allozyme loci. Even if detection of mtDNA variation had been limited to the two major forms, the pattern would still be very clear. Several possible explanations for this discrepancy were considered. The difference could be due to the maternal inheritance of mitochondria; the observed pattern would be expected if gene flow between Atlantic and Gulf of Mexico populations was mediated by sperm, but not eggs or larvae. Another possibility was that natural selection, acting on mtDNA, had overcome what appeared to be the substantial gene flow that maintained uniform allozyme frequencies. A third possibility required that Atlantic and Gulf of Mexico populations have been separated for a long time. To achieve the observed level of divergence with typical rates of mtDNA evolution, roughly 1 million years would be required. During that time, selection could have maintained uniform allozyme frequencies, while mitochondrial genomes steadily diverged. Support for the latter explanation came from studies on nuclear DNA polymorphisms. Karl & Avise (1992) examined variation in 'anonymous' nuclear single copy DNA sequences, which for the most part do not appear to encode proteins, and would therefore not be subject to the same forms of selection as protein-encoding allozyme genes. Although these sequences were only moderately polymorphic, they clearly separated Gulf of Mexico and Atlantic populations and therefore corroborated the mtDNA pattern. Thus it appears that in American oysters, selection may have maintained similar allozyme allele frequencies in populations that had been separated for roughly a million years.

The idea that allozyme alleles are subject to selection is hardly novel or obscure. Indeed, from Buroker's (1983) allozyme study of oysters, it was concluded that selection was probably acting at one or two loci. Yet the conclusions reached by Karl & Avise (1992) represent a sharp departure from conventional modes of interpretation. Firstly, while it may be acknowledged that some allozyme loci are under selection, it is often assumed that most are not. Thus, when a number of allozyme loci are surveyed, and most exhibit a similar pattern (i.e. geographical uniformity), selection is not usually invoked to explain this pattern. Only loci that depart from a general pattern are considered candidates for selection. This idea has even been used as the basis for statistical tests of selection on allozyme loci (eg. Lewontin & Krakauer, 1973). Thus while Buroker (1983) argued that the enzyme locus, *Lap-2*, was under selection that led to geographical variation, Karl & Avise (1992) suggested that nearly all of the other loci surveyed by Buroker were subject to selection that maintained uniformity. Under this interpretation, *Lap-2* may be the only locus that is not under selection.

12.5 Geographical and temporal scales

General invalidation of the 'majority rule' principle for deciding which allozyme loci are not under selection would have broad implications for the interpretation of numerous allozymes surveys (over 1000) that have been conducted. This principle is often used to justify the use of allozyme data to estimate levels of gene flow between populations (Slatkin & Barton, 1989) and to infer historical relationships among populations. Certainly our view of marine populations has been strongly influenced by this principle.

Buroker's (1983) interpretation of allozyme variation in the American oyster is consistent with the paradigm of large, panmictic marine populations. However, the distributions of variation in mitochondrial (Reeb & Avise, 1990) and nuclear DNA (Karl & Avise, 1992) imply the presence of an impenetrable wall across what appears to be a nearly continuous habitat. The completeness of this boundary (which is most apparent in the mitochondrial data) implies that whenever oyster larvae are carried across it, they fail to become established. Even if it is argued that two cryptic species are involved, a high degree of local adaptation is implied.

The paradigm of large, panmictic marine populations, implies that regional patterns of variation must be shaped by major oceanographical events that isolate populations for long periods of time. Population genetics theory shows that the rate of genetic drift should be inversely proportional to the breeding size of a population (Fisher, 1930; Wright, 1931). It follows that larger populations require more time to diverge by genetic drift (Crow & Aoki, 1984), and require less inter-population migration to prevent divergence (Wright, 1965). Thus it might be expected that regional patterns of allozyme diversity reflect major events that have taken place in the Pleistocene, or earlier. Such events would provide the degree of isolation and the amount of time required to generate major regional patterns.

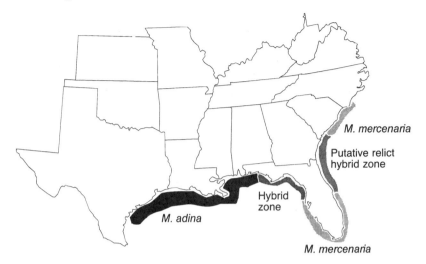

Fig. 12.1. Geographical ranges of *Menippe adina* and *M. mercenaria*, and two putative hybrid zones.

12.6 Historical versus ecological explanations for geographical variation in stone crabs

This historical view guided an initial investigation of regional patterns of diversity in the stone crab, *Menippe* (Schneider-Broussard, 1993; Schneider-Broussard & Neigel, 1996). Two species of stone crab are recognised from the Atlantic and Gulf of Mexico coasts of North America (Fig. 12.1). *Menippe mercenaria* ranges from North Carolina south to peninsular Florida. *Menippe adina*, recently recognised as a distinct species, occurs in the western Gulf of Mexico (Williams & Felder, 1986). These two species differ in a number of characteristics, including frequencies of allozyme alleles, colouration patterns and physiology (Bert, 1986; Williams & Felder, 1986). However, crabs with intermediate characteristics are found along the western coast of Florida, in what is considered to be a broad hybrid zone (Bert, 1986; Williams & Felder, 1986; Bert & Harrison). This distribution can be explained as the result of allopatric speciation and secondary contact at the hybrid zone. However, a second apparent hybrid zone on the Atlantic coast is more difficult to explain, because it is completely embedded within the range of *M. mercenaria* (Bert, 1986). Allozyme alleles and phenotypic characteristics of *M. adina* appear to have introgressed into *M. mercenaria* without any obvious direct route. This pattern suggests that some ancient, large-scale event must have allowed entry of *M. adina* into the Atlantic. Bert & Harrison (1988) proposed that this event was the opening of the Suwanee Straits, a waterway across northern Florida that most probably appeared during times of extremely high sea level in the Miocene (12 million years (my) before present (bp)) and Pliocene (3.5 my bp).

We used mtDNA to estimate the time of separation between *M. mercenaria* and *M. adina* (Schneider-Broussard, 1993; Schneider-Broussard & Neigel, 1996). In many taxa, mtDNA evolves more rapidly than nuclear DNA (Vawter & Brown, 1986), and in some instances rates of mitochondrial DNA evolution have been calibrated with paleontological data (Avise *et al.*, 1987). We chose to examine sequence variation in the mitochondrial large subunit ribosomal RNA gene (often referred to as the '16 S' gene), because rates of evolution in this sequence have been calibrated for other decapod crustaceans (Cunningham *et al.*, 1992)

In comparing mtDNA sequences from two species, it is possible that some of the divergence seen actually accumulated before speciation, as genetic variation within the ancestral species (Neigel & Avise, 1986). Thus estimates of divergence time should be based on the most similar pairs of sequences from two species. In our study, we included samples of *M. mercenaria*, *M. adina* and putative Atlantic hybrids. We expected to see evidence for two types of relationship between sequences (Fig. 12.2). Firstly, we expected to find sequences that were separated at or prior to the time of speciation. The most similar of these would be used to estimate the date of the speciation. In addition, we expected Atlantic hybrids to have sequences that were originally derived from *M. adina*, but somewhat divergent from modern *M. adina* sequences because of the long isolation of these relict hybrids from *M. adina* in the Gulf of Mexico. These sequences would be used to estimate the date of hybridisation.

The results of our mitochondrial DNA studies are illustrated in Fig. 12.3. Although we found some differences among DNA sequences, the most similar sequences from *M. adina* and *M. mercenaria* were identical. The same sequence was also found in individuals from the presumed Atlantic hybrid zone. It appears that the present distributions of genetic variation in *Menippe* were not generated over millions of years, as had been supposed. The existence of an Atlantic hybrid zone was particularly difficult to explain, as there is no geological evidence of a recent opening of the Suwanee Straits.

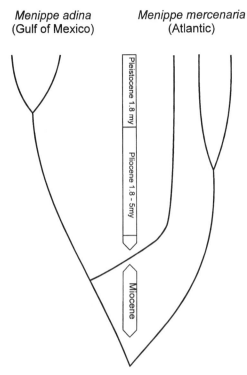

Fig. 12.2. Predicted relationships among ancestral and introgressed mitochondrial DNA sequences, based on secondary contact during a late Pliocene opening of the Suwanee straits.

The possibility of natural selection on allozyme alleles suggests a very different interpretation of genetic variation in *Menippe*. '*M. adina*-like' alleles may be maintained by contemporary conditions in populations along the Atlantic coast, rather than remaining there as historical relicts. Contemporary current patterns could transport larvae from *M. adina* to Atlantic populations at frequent intervals (Schneider-Broussard, 1993). Thus the observed geographical pattern may have no historical basis. This is not to say that historical events had no role in the initial separation of *M. mercenaria* and *M. adina*. A very recent separation might not have provided enough time for divergence of mtDNA sequences. However, long-distance dispersal of large numbers of larvae would tend to quickly erase historical patterns, once barriers to dispersal were removed. In contrast, the constant availability of a genetically diverse pool of larvae would allow selection to generate new, and potentially more complex, geographical distributions.

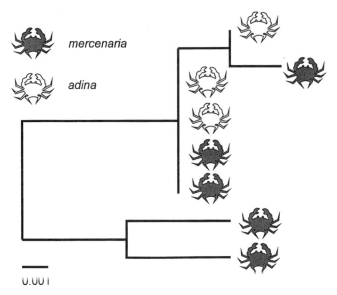

Fig. 12.3. Modified Kimura distance phenogram of *Menippe* mitochondrial large subunit ribosomal DNA sequences. The bar represents 0.001 units.

The idea that highly fecund organisms could be subject to intense selection was proposed by Williams (1975). In his 'Elm–Oyster' model, enormous numbers of genetically variable progeny allowed selection to limit success to only those that were extremely well adapted to current environmental conditions. A contrasting view would be that larval success is mostly a matter of luck, of being carried along by currents to food but not predators, and, ultimately, to a suitable habitat. However, there may be room for both luck and genetics, if the number of larvae is large enough to sustain reductions by both. If this is the case, then large numbers of larvae do not represent a surplus, but rather the raw material for local adaptation.

12.7 Inbreeding and effective population size

The second microevolutionary process with direct ecological effects is inbreeding. Inbreeding may occur at several levels within a population. Many plants and lower animals are capable of a most extreme form of inbreeding, self-fertilisation. Self-fertilisation may occur under conditions in which mating with other individuals is not possible; for example, when population density is extremely low. A more general form of inbreeding occurs when matings occur between related individuals. Finally, genetic drift can be considered to be a form of inbreeding. Genetic drift is the random change in gene frequencies that occurs because each generation represents a limited statistical sample of the genetic diversity present in the preceding generation. As a result of genetic drift, a population's gene pool will tend to represent a decreasing portion of the original founders. This effect is most severe in small populations, in which the limited statistical sample represented by each generation is subject to large sampling errors. As genetic diversity is lost from a small population, the effect becomes the same as if individuals were mating between close relatives. In sufficiently large populations, genetic variation generated by mutation is sufficient to balance the loss of diversity from genetic drift.

A general consequence of all forms of inbreeding is an increase in the proportion of homozygous loci among inbred progeny. This increase in homozygosity can in turn cause a reduction in fitness, referred to as inbreeding depression. There are at least two genetic mechanisms that can explain inbreeding depression (reviewed by Charlesworth & Charlesworth, 1987). One is an increase in the proportion of genetic loci that are homozygous for rare, recessive alleles with deleterious effects. These alleles are not eliminated from large populations because they generally occur in heterozygous combinations with other alleles, which thus shield them from selection. If a population suddenly decreases in size, or if mating occurs between related individuals, these alleles will more frequently combine as homozygotes that express their deleterious effects. The other genetic mechanism that can explain inbreeding depression is a decrease in the proportion of genetic loci that are heterozygous for pairs of alleles that confer greater fitness as heterozygotes than either would as a homozygote. Sickle cell anaemia is a well-known example of heterozygote superiority. The sickle cell allele of the human haemoglobin locus confers resistance to malaria as a heterozygote, but severe anaemia as a homozygote. While the difference between these two mechanisms of inbreeding depression may appear subtle, they are quite different with respect to the potential for a population to resist or recover from inbreeding depression. If inbreeding occurs gradually, recessive deleterious alleles can be purged from a population by selection. Elimination of these deleterious alleles will both reduce the severity of inbreeding depression as well as make the population somewhat resistant to further inbreeding depression. However, if loss of heterozygosity is responsible for inbreeding depression,

288   *Population genetics and demography*

gradual inbreeding will not prevent the loss of alleles. Once allelic diversity is lost, it can only be restored by immigration or mutation.

One of the most common criticisms of the 50/500 rule advocated by Franklin (1980) and Soulé (1980) is that it must be applied to the genetically effective size of a population (denoted N_e), which determines the rate of genetic drift, rather than the census size, or total number of individuals (N). Only breeding individuals contribute to N_e, and if a few individuals are much more successful at breeding than others, N_e will be closer to their number. Thus in general, N_e tends to be smaller than N. For terrestrial vertebrates with limited fecundity, N_e and N are usually at least of the same order of magnitude, and may be nearly equal (e.g. Nunney & Elam, 1994). However, in highly fecund species, including many plants and marine organisms, N_e may be orders of magnitude lower than N (e.g. Orive, 1993).

The best way to estimate N_e, and thus the rate at which genetic variation is lost, is from detailed pedigree information (e.g. Wright, 1965). However, such information is generally only available for captive populations. An alternative approach is to observe some effect of genetic drift, and derive a value of N_e consistent with that observation. This approach is complicated by the fact that there are several distinct manifestations of genetic drift, and each corresponds to a different definition of N_e (Ewens, 1982).

The variance effective size of a population is based on the variance in allele frequencies over time that is caused by genetic drift. Thus by measuring fluctuations in allele frequencies over time, and assuming they are caused by drift, it is possible to estimate variance effective population size for the period of observation. Such data must be analysed and interpreted carefully. One approach is described by Waples (1989). Hedgecock *et al.* (1992) have applied this method to populations of several marine species, including American oysters. The results are quite startling. Estimates of N_e for natural populations of American oysters in the upper Chesapeake Bay were as low as 5–50. Taken at face value, this implies that a very large and geographically extensive population is sustained by relatively few individuals. The potential importance of this finding is apparent by asking the question: what if these individuals had been removed from the population; would others have taken their place, or would the population have declined?

Another approach to estimating N_e is based on the expected relationship between the amount of genetic variation in a population, and the product of effective population size and mutation rate: N_e (for a review, see Neigel, 1996). Using known mutation rates for mtDNA, it is possible to estimate N_e from a sample of mtDNA sequences taken from a population. Because this approach is based on levels of genetic variation that have developed over evolutionary time, the parameter estimated is sometimes referred to as 'evolutionary effective population size', although it may also be considered an estimate of inbreeding effective population size. Estimates of N_e based on mtDNA for American eels, *Anguilla rostrata*, and hardhead catfish, *Arius felis*, are two or three orders of magnitude lower than estimates of the current breeding size of their respective populations (Avise et al., 1988). One possible explanation for this discrepancy that populations of these species have only recently reached their present sizes, and thus estimates of N_e reflect a history of smaller populations. However, another possibility is that these populations are sustained by the progeny of relatively few individuals.

Whether it is luck or genetic superiority that limits reproductive success to a small proportion of a population, very small effective population sizes will lead to a loss of genetic variation. In large populations, the loss of genetic variation from genetic drift is compensated for by mutation. However, in small populations, mutations occur too infrequently to replenish variation, and most genetic variation will be lost within $4N_e$ generations. The detrimental effects of reduced genetic variation, inbreeding depression and loss of adaptive potential, are probably not serious for populations with effective sizes of several thousand (Lande, 1988). However, estimates of N_e for marine species suggest that corresponding total numbers of individuals must be much larger, perhaps by several orders of magnitude. Thus if the geographical range of a once widespread marine species becomes restricted to a few reserves or protected habitats, a rapid loss of genetic diversity may result.

A major question in conservation biology is: how much genetic diversity can a species lose, without risking extinction? In game populations that will be carefully managed in parks or zoos, slight inbreeding depression or loss of adaptive potential may be offset by the elimination of other threats, such as drought or predation (Simberloff, 1988). However, it is unlikely that many marine species, with their complex life histories and exacting requirements, will be managed in captivity. Can marine populations that are compromised by loss of genetic variation be sustained under natural conditions? If recent interpretations of allozyme studies are taken at face value, the potential for adaptation that is provided by a genetically diverse pool of planktonic larvae may be indispensable for survival.

References

Avise, J.C. & Aquadro, C.F., 1982. A comparative summary of genetic distances in the vertebrates. *Evolutionary Biology*, **15**, 151–85.

Avise, J.C., Ball, R.M. & Arnold, J., 1988. Current versus historical population sizes in vertebrate species with high gene flow: A comparison based on mitochondrial DNA lineages and inbreeding theory for neutral mutations. *Molecular Biology and Evolution*, **5**, 331–44.

Avise, J.C., Arnold, J., Ball, R.M., Bermingham, E., Lamb, T., Neigel, J.E., Reeb, C.A. & Saunders, N.C., 1987. Intraspecific phylogeography. The mitochondrial DNA bridge between population genetics and systematics. *Annual Review of Ecology and Systematics*, **18**, 489–522.

Bert, T.M., 1986. Speciation in western Atlantic stone crabs (genus *Menippe*): The role of geological processes and climatic events in the formation and distribution of species. *Marine Biology*, **93**, 157–70.

Bert, T.M. & Harrison, T.M., 1988. Hybridization in western Atlantic stone crabs (genus *Menippe*): Evolutionary history and ecological context influence complex species interactions. *Evolution*, **42**, 528–44.

Buroker, N.E., 1983. Population genetics of the American oyster *Crassostrea virginica* along the Atlantic coast and the Gulf of Mexico. *Marine Biology*, **75**, 99–112.

Burton, R.S., 1983. Protein polymorphisms and genetic differentiation of marine invertebrate populations. *Marine Biology Letters*, **4**, 193–206.

Burton, R.S., 1987. Differentiation and integration of the genome in populations of the marine copepod *Tigriopus californicus*. *Evolution*, **41**, 504–13.

Burton, R.S., 1990. Hybrid breakdown in developmental time in the copepod *Tigriopus californicus*. *Evolution*, **44**, 1814–22.

Burton, R.S. & Feldman, M.W., 1981. Population genetics of *Tigriopus californicus*. II. Differentiation among neighbouring populations. *Evolution*, **35**, 1192–205.

Burton, R.S. & Swisher, S.G., 1984. Population structure of the intertidal copepod *Tigriopus californicus* as revealed by field manipulation of allele frequencies. *Oecologia*, **65**, 108–11.

Burton, R.S., Feldman, M.W. & Curtsinger, J.W., 1979. Population genetics of *Tigriopus californicus* (Copepoda: Harpacticoida): I. Population structure along the central California coast. *Marine Ecology Progress Series*, **1**, 29–39.

Charlesworth, D. & Charlesworth, B., 1987. Inbreeding depression and its evolutionary consequences. *Annual Review of Ecology and Systematics*, **18**, 237–68.

Crisp, J.D., 1978. Genetic consequences of different reproductive strategies in marine invertebrates. In *Marine Organisms: Genetics, Ecology and Evolution*, ed. B. Battaglia & J.A. Beardmore, pp. 257–73. New York: Plenum Press.

Crow, J.F. & Aoki, K., 1984. Group selection for a polygenic behavioral trait: Estimating the degree of population subdivision. *Proceedings of the National Academy of Sciences of the USA*, **81**, 6073–7.

Cunningham, C.W., Blackstone, N.W. & Buss, L.W., 1992. Evolution of king crabs from hermit crab ancestors. *Nature*, **355**, 539–42.

Ehrlich, P.R. & Raven, P.H., 1969. Differentiation of populations. *Science*, **165**, 1228–31.

Endler, J.A., 1977. *Geographic Variation, Speciation, and Clines*. Princeton, NJ: Princeton University Press.

Erwin, T.L., 1991. An evolutionary basis for conservation strategies. *Science*, **253**, 750–2.
Ewens, W.J., 1982. On the concept of effective population size. *Theoretical Population Biology*, **21**, 373–8.
Fisher, R.A., 1930. *The Genetical Theory of Natural Selection*. Oxford: Clarendon Press.
Franklin, I.R., 1980. Evolutionary change in small populations. In *Conservation Biology: An Evolutionary-Ecological Perspective*, ed. M.E. Soulé & B.A. Wilcox, pp. 135–50. Sunderland, MA: Sinauer Associates.
Galtsoff, P.S., 1964. The American oyster, *Crassostrea virginica* Gmelin. Fishery Bulletin no. **64**. Washington, DC: USA. Fisheries and Wildlife Service. U.S. Government Printing Office.
Gyllensten, U., 1985. The genetic structure of fish: Differences in the intraspecific distribution of biochemical genetic variation between marine, anadromous and freshwater species. *Journal of Fisheries Biology*, **26**, 691–9.
Hedgecock, D., 1986. Is gene flow from pelagic larval dispersal important in the adaptation and evolution of marine invertebrates? *Bulletin of Marine Science*, **39**, 550–64.
Hedgecock, D., Chow, V. & Waples, R.S., 1992. Effective population numbers of shellfish broodstocks estimated from temporal variance in alleleic frequencies. *Aquaculture*, **108**, 215–32.
Hedrick, P., Jain, S. & Holden, L., 1978. Multilocus systems in evolution. *Evolutionary Biology*, **11**, 101–84.
Karl, S.A. & Avise, J.C., 1992. Balancing selection at allozyme loci in oysters: Implications from nuclear RFLPs. *Science*, **256**, 100–2.
Lande, R., 1988. Genetics and demography in biological conservation. *Science*, **241**, 1455–60.
Lewontin, R.C. & Krakauer, J., 1973. Distribution of gene frequency as a test of the theory of the selective neutrality of polymorphisms. *Genetics*, **74**, 175–95.
Mayr, E., 1970. *Populations, Species and Evolution*. Cambridge, MA: Harvard University Press.
Neigel, J.E., 1996. Estimation of effective population size and migration parameters from genetic data. In *Molecular Genetic Approaches in Conservation*, ed. T. Smith & R.K. Wayne, pp. 329–46. Oxford: Oxford University Press.
Neigel, J.E. & Avise, J.C., 1986. Phylogenetic relationships of mitochondrial DNA under various demographic models of speciation. In *Evolutionary Processes and Theory*, ed. E. Nevo & S. Karlin, pp. 515–34. New York: Academic Press.
Nunney, L. & Elam, D.R., 1994. Estimating the effective population size of conserved populations. *Conservation Biology*, **8**, 175–84.
Orive, M.E., 1993. Effective population size in organisms with complex life histories. *Theoretical Population Biology*, **44**, 316–40.
Ray, C.G., 1988. Ecological diversity in coastal zones and oceans. In *Biodiversity*, ed. E.O. Wilson & F.M. Peter, pp. 36–50. Washington, DC: National Academy Press.
Reeb, C.A. & Avise, J.C., 1990. A genetic discontinuity in a continuously distributed species: Mitochondrial DNA in the American Oyster, *Crassostrea virginica*. *Genetics*, **124**, 397–406.
Schneider-Broussard, R.M., 1993. Composition, variation and divergence in the nuclear and mitochondrial genomes of the stone crabs *Menippe adina* and *M. mercenaria*. Ph.D. thesis. University of Southwestern Louisiana.
Schneider-Broussard, R.M. & Neigel, J.E., 1996. A large subunit mitochondrial

ribosomal DNA sequence translocated to the nuclear genome of two stone crabs (*Menippe*). *Molecular Biology and Evolution*, **14**, 156–65.
Simberloff, D., 1988. The contribution of population and community biology to conservation science. *Annual Review of Ecology and Systematics*, **19**, 473–511.
Slatkin, M., 1985. Gene flow in natural populations. *Annual Review of Ecology and Systematics*, **16**, 393–430.
Slatkin, M. & Barton, N.H., 1989. A comparison of three indirect methods for estimating average levels of gene flow. *Evolution*, **43**, 1349–68.
Soulé, M.E., 1980. Thresholds for survival: Maintaining fitness and evolutionary potential. In *Conservation Biology: An Evolutionary-Ecological Perspective*, ed. M.E. Soulé & B.A. Wilcox, pp. 111–24. Sunderland, MA: Sinauer Associates.
Utter, F.M. & Ryman, N., 1993. Genetic markers and mixed stock fisheries. *Fisheries*, **18**, 11–21.
Vawter, L. & Brown, W.M., 1986. Nuclear and mitochondrial DNA comparisons reveal extreme rate variation in the molecular clock. *Science*, **234**, 194–6.
Waples, R.S., 1989. A generalized approach for estimating effective population size from temporal changes in allele frequency. *Genetics*, **121**, 379–91.
Williams, A.B. & Felder, D.L., 1986. Analysis of stone crabs: *Menippe mercenaria* (Say), restricted, and a previously unrecognized species described (Decapoda: Xanthidae). *Proceedings of the Biological Society of Washington*, **99**, 517–43.
Williams, G.C., 1975. *Sex and Evolution*. Princeton, NJ: Princeton University Press.
Wright, S., 1931. Evolution in Mendelian populations. *Genetics*, **16**, 97–159.
Wright, S., 1965. The interpretations of population structure by F-statistics with special regard to systems of mating. *Evolution*, **19**, 395–420.
Wright, S., 1969. *Evolution and the Genetics of Populations*. Vol. **2**. *The Theory of Gene Frequencies*. Chicago: University of Chicago Press.
Wright, S., 1977. *Evolution and the Genetics of Populations*. Vol. **3**. *Experimental Results and Evolutionary Deductions*. Chicago: University of Chicago Press.

Chapter 13
Discovering unrecognised diversity among marine molluscs

J. GRAHAME†, S.L. HULL‡, P.J. MILL† and
R. HEMINGWAY†
†*Department of Biology, The University of Leeds, Leeds, LS2 9JT, UK*
‡*Science Section, University College Scarborough, Filey Road, Scarborough, YO11 3AZ, UK*

Abstract

The biology of closely related and recently separated species is of special interest in the study of speciation. Recognition of the status of such populations presents problems – it is not always easy to discriminate which populations are fully reproductively isolated from those where some hybridisation between 'species' occurs. In our chosen study organism, *Littorina saxatilis* in the prosobranch subgenus *Neritrema*, there is still considerable controversy over which populations are in a state of gene flow with which others. Recent opinion holds that there are three differentiated sibling species: *L. saxatilis*, *L. arcana* and *L. nigrolineata*, with the status of the small barnacle-dwelling *L. neglecta* still controversial. In this chapter, we question the genetic unity of *L. saxatilis sensu stricto*, and discuss the relative importance of study of shell morphology, enzyme polymorphisms and DNA polymorphisms. We conclude that in this organism, enzyme polymorphisms are a weak tool for the study of sibling species.

13.1 Introduction

A 'species' is often held to comprise a group of populations that freely exchange genes among themselves, but not with other groups. The idea of species is, for some workers, fundamental in biology and studies of biodiversity (e.g. Futuyma, 1987), and indeed inasmuch as species are genetically circumscribed units, this has force. Our opening sentence is an expression of the Biological Species Concept (BSC) especially associated with Mayr (e.g. Mayr, 1963) – there are other species concepts. Their application to marine organisms has been recently reviewed by Knowlton (1993), Gosling (1994) and Palumbi (1994); it is not our purpose to re-work this ground here, but nevertheless some initial consideration is necessary. This is because the terminology used sets a framework that affects the argument. While some maintain that the BSC is superior to alternatives (e.g. Coyne, 1994), others find an objectionable circularity in the BSC (and indeed in all concepts with interbreeding criteria), considering that they confound cause and effect (Mallet, 1995).

Whatever concept is favoured, there is increasing recognition of one problem: more and more instances are being shown of closely related species that show hybridisation. Maintenance of largely separate gene pools in the face of limited gene flow suggests an equilibrium between homogeneity on the one hand and complete isolation on the other, how is this maintained? Moreover, if we knew the nature of the equilibrium (is it static? – or shifting, favouring either introgression or divergence?), this would lead to insights concerning the often held universality of allopatric speciation and the possible role of ecology (as well as, or instead of, biogeography) in speciation.

To address these questions, we need to know what is the status quo for as wide a sample of organisms as possible – and this will mean working with populations whose members do not necessarily make tractable laboratory subjects. The basic fundamental is to assess the degree of isolation between component populations – we need to recognise their status in terms of gene flow, so as to decide whether we are dealing with fully isolated species, or the polymorphic members of a panmictic unit, or some intermediate situation between these possibilities. A favourite approach is to use the behaviour inferred genetic loci, studied by electrophoretic differentiation of enzymes. With the tools of an overall measure of genetic identity (e.g. Nei, 1972) and one of the many analytic and sorting approaches (e.g. 'Phylip': Felsenstein, 1989; see also Sundberg *et al.*, 1990) a variety of expressions of genetic relatedness among populations or subpopulations is possible.

Haylor *et al.* (1984) have suggested that the green morph of the sea anemone *Actinia equina* L. is a separate species, for which they proposed the name *Actinia prasina*, using as trivial the varietal name used by Gosse (1860). At 0.91, the genetic identity (I) between the two is high, but besides differences in biology the authors reported that there were significant differences in allele frequencies at four loci, indicating a reproductive barrier between the two. A similar state of affairs is reported for the anemones *Urticina felina* (L.) and *U. eques* (Gosse), with an I value of 0.907 (Solé-Cava *et al.*, 1985). Solé-Cava & Thorpe (1992) reconsidered *Actinia*, reporting I values of between 0.687 and 0.983 for various pairs of populations in three groupings, each of which they concluded were reproductively isolated. These authors pointed out that high values of I between different species are unusual but not unknown. Elsewhere among Cnidaria, Van Veghel & Bak (1993) considered three morphotypes of *Montastraea annularis* (Ellis & Solander) to be conspecific (I values ranging from 0.86 to 0.93 between selected pairs of morphotypes) while Knowlton *et al.* (1992) considered that they were reproductively isolated (therefore three species), with I values as high as 0.942 between supposed species.

Among molluscs, the mussel *Mytilus* has been much studied. Skibinski *et al.* (1980) found that between *M. edulis* L. and *M. galloprovincialis* Lmk. genetic identity was 0.847, this was in contrast with values of ≈0.2 between these two 'species' and the mytilid *Modiolus modiolus* L. On the basis of enzyme polymorphism studies, Blot *et al.* (1988) concluded that *M. edulis*, *M. galloprovincialis* and *M. desolationis* Lamy formed a 'super species' with actual gene flow between *M. edulis* and *M. galloprovincialis* (principal distributions in the North Atlantic and Mediterranean respectively), and at least the potential for gene flow between these populations and those of *M. desolationis* in the southern Indian Ocean (Kerguelen Islands). Gosling (1994) took the same view of this group, including the form *M. trossulus* Gould in the super species. This genus is interesting in that it has also been studied using mitochondrial DNA genotypes by using restriction enzymes to reveal patterns characteristic of groups of individuals – such studies confirm the hybridisation of *M. edulis* and *M. galloprovincialis* where they are sympatric in Britain (Edwards & Skibinski, 1987). Net gene flow was from *M. edulis* to *M. galloprovincialis*, inferred to be caused by larval transport, with larvae taking *M. edulis* genes into *M. galloprovincialis* populations. This was at variance with allozyme studies, which failed to show the predicted concomitant flow of nuclear genes. Varvio *et al.* (1988) agree that it is impossible to regard the *Mytilus* super species members as fully differentiated species, but maintained that *M. edulis*, *M. trossulus* and *M. galloprovincialis* do represent three separate evolutionary lineages. In the venerid bivalves *Chamelea gallina* (L.) and *C. striatula* (Da Costa), Backeljau *et al.* (1994) concluded that the populations represent separate species, with a low genetic identity (0.327).

Beaumont & Wei (1991) reported on the Antarctic limpet *Nacella concinna* (Strebel), which shows distinct and consistent differences between littoral and sublittoral populations in both morphology and behaviour. I values between the populations they sampled were ≥ 0.949, and it was concluded that the forms were phenotypic variants within one panmictic species. In *Nucella lapillus* (L.) there is no evidence from enzyme polymorphism of anything other than one species being present, yet Pascoe & Dixon (1994), after considering chromosome polymorphisms, speculated that some populations may be reproductively isolated.

Thus a sample of the various genetic studies on 'difficult' groups in the Cnidaria and Mollusca shows instances where slight differences in morphology between populations are evidently indicative of species differentiation when genetic data are considered (anemones, venerid clams), or of interbreeding even in partially distinct populations (mytilids, Antarctic limpets, dog-whelks), with the studies on the coral *Montastraea annularis* providing controversy.

We turn now to the 'rough periwinkles', in the subgenus *Neritrema* (see Reid, 1989) of *Littorina*. The constituent species are well studied, abundant and accessible shore animals. Reid (1996) presents a full review of the systematics, we shall sketch the principal views only as far as is necessary for the context of this chapter.

Quite apart from problems over synonymy and priority, there has been and remains considerable discussion over where polymorphism ends and species differences begin. The doubt and confusion over the relationships among rough periwinkles was expressed more than a century ago by Forbes & Hanley (1853): 'although the typical forms of *rudis*, *tenebrosa*, *patula*, and *saxatilis*, are so very unlike, certain aberrant individuals almost indicate, that they form but one species'. We draw attention to this to emphasise what seems a very real and contemporary problem: while the shell characters of the various 'forms' may be fairly distinct at any one location where two or more co-occur, samples from other locations quickly erode any distinctness the investigator might initially have appreciated. In the absence of clear differences in the biology of the animals, the tendency is naturally to treat such 'forms' as members of a polymorphic species.

James (1968) referred to *Littorina saxatilis* (Olivi) as 'the most widely distributed intertidal mollusc in Britain'. Considering the range of morphology and habitat, occurrence of 'intermediates' and the apparent likelihood of interbreeding, he concluded that the populations represented a single, variable species. He divided the species into subspecies and varieties, noting that the variety '*nigrolineata*' within the subspecies '*rudis*' might 'revert to oviparity' on a shore at Roscoff, on the Brittany coast. Seshappa (1947) had previously reported gelatinous masses of gastropod eggs in rock crevices high on the shore at Cullercoats, on the north-east coast of England, concluding that they had been spawned by *L. saxatilis*. His account describes the '*L. saxatilis*' population as dimorphic: as well as 'some of the individuals [being] of the viviparous *rudis* type', most of the animals were smaller, with a thinner shell, larger operculum and squatter spire. They were found to have a pallial oviduct typical of egg-layers (by comparison with *L. obtusata* (L.)) rather than brooders, and when kept in the laboratory laid eggs in masses like those found on the shore. This form was referred to the taxa *rudissima* or *patula* (these names have been used for both varieties and species – *L. rudissima* Johnston and *L. patula* Thorpe). Interestingly, Seshappa (1948, 1976) later reported that some of the thin-shelled *rudissima* forms were ovoviviparous, but he remained of the opinion that these forms were subspecies within *L. saxatilis*.

Heller (1975) stated firmly that there were four species in the group (*L. rudis* (Maton), *L. patula*, *L. nigrolineata* Gray* and *L. neglecta* Bean), using evidence from the morphology of shell, radula and penis, from esterase phenotypes and from shore distributions. One species (*L. nigrolineata*) was oviparous. Hannaford Ellis (1979) again claimed four species to be present, two ovoviviparous (*L. saxatilis* and *L. neglecta*) and two oviparous (*L. nigrolineata*, and a new species, *L. arcana* Hannaford Ellis). Given its position on the shore, the thin shell and shape of shell, *L. arcana* no doubt accounts for some, though not all, of the *rudissima* forms seen by Seshappa.

This is more or less where matters rest, except for continuing controversy over the status of *L. neglecta* for which there are conflicting accounts of its genetic distinctness from *L. saxatilis* (compare Wilkins & O'Regan, 1980, with Johannesson & Johannesson, 1990). We consider that in Britain *L. neglecta* is distinct in shape (Caley *et al.*, 1995; Grahame *et al.*, 1995), and we concur with Wilkins & O'Regan (1980) that it is genetically distinct (S.L. Hull, J. Grahame & P.J. Mill, unpublished results), at least in the British Isles. However the status of this taxon is not settled, as Johannesson *et al.* (1993) maintain that there is gene flow between '*L. saxatilis*' populations in Spain, where one of the populations has the appearance and habits of *L. neglecta*.

Among large brooding periwinkles showing obvious phenotypic and habitat differences, recent authors find good evidence that *L. saxatilis* is a single, variable and polymorphic species. Thus, the estuarine form

---

* This species' name is being changed to *L. compressa* Jeffreys on grounds of priority (Reid, 1996); however for simplicity we retain the more familiar name in this chapter.

L. tenebrosa Montagu has been ascribed to *L. saxatilis* on grounds of genetic similarity and the occurrence of phenotypic intermediates in large numbers (Janson & Ward, 1985). Again, the occurrence of exposed and sheltered phenotypes in Sweden was interpreted as evidence of intraspecific polymorphism (Janson & Ward, 1984) – importantly here the intermediates were reported to be abundant in several sites. In these cases, genetic identities are very high – 0.996 and 0.990 between *L. saxatilis* and *L. tenebrosa* in Scotland and Sweden, respectively, and ranging between 0.960–0.994 between pairs of exposed and sheltered subpopulations. These identities certainly are such as to be expected of conspecific populations (see Thorpe, 1982). They may be contrasted with an I value of 0.537 between *L. sitkana* Philippi and *Littorina* sp. in the western north Pacific, where these are considered to be sibling species (Zaslavskaya, 1995). The identity between the sibling bivalves *Chamelea gallina* and *C. striatula* is even lower, being 0.327 (Backeljau et al., 1994).

The status of *L. saxatilis*, *L. nigrolineata* and *L. arcana* as reproductively isolated species is generally accepted, and is supported by the extensive studies of Janson, Ward and their collaborators on polymorphic enzymes (e.g. Ward & Janson, 1985; Knight & Ward, 1991). Here genetic identities are again very high, being 0.946 between *L. saxatilis* and *L. nigrolineata* and 0.996 between *L. saxatilis* and *L. arcana* (Knight & Ward, 1991). Nevertheless, the clustering of putatively conspecific populations from geographically separate locations supports the accepted taxonomy, and Knight & Ward (1991) considered that their genetic data did not suggest that further species were 'hidden' in the complex.

However, these species are so closely related that investigators perhaps may not be able to recognise species differentiation on the basis of genetic data alone, and indeed they have sometimes failed to do so (e.g. see Wilkins et al., 1978). Having not initially taken into account the species *L. nigrolineata*, these authors considered (*a posteriori*) that it was very likely that they had included specimens of it with some of their *L. saxatilis*, but yet there was no sign from the genetic data that this might have happened. Reid & Golikov (1991) suggested that enzyme electrophoresis studies confirmed species status where it had already been suspected on the basis of morphological work. The corollary of this is that the primary sorting of the material is of paramount importance (see Caley et al., 1995). If there is no morphological 'signal' from the material – or if the signal is missed or misconstrued – then it will be easy to overlook small genetic differences. The trees separating *L. saxatilis*, *L. arcana* and

L. nigrolineata (e.g. Knight & Ward, 1991) are generated *a posteriori*, after careful morphological sorting of material prior to genetic analysis. Suppose the diagnostic morphological differences between these species (misinterpreted for many decades) were still unclear?

The problem is made more difficult because genetic markers for phylogenetic inference ideally should be neutral. This was stated to be the case in the study of Johannesson & Johannesson (1990), where no substantial differences could be found between *L. saxatilis* and *L. neglecta*. If selection is supposed to be involved, similar shifts of allele frequency can be explained without inferring gene flow, as in the case of aspartate amino transferase in *L. saxatilis* and *L. arcana* (Grahame *et al.*, 1992; see also Grahame *et al.*, 1993; Johannesson & Johannesson, 1993). The problem is, are the allozyme polymorphisms neutral? Karl & Avise (1992) have made striking observations in *Crassostrea virginica* (Gmelin): DNA polymorphisms show that a variety of putatively neutral allozymes seem in fact to be under selection. This must caution us against making inferences of gene flow based on assumptions of allozyme neutrality.

Using data from morphological, enzyme polymorphism and random amplification of polymorphic DNAs (RAPDs) analyses we offer evidence that in one restricted location that we have investigated '*L. saxatilis*' does not represent one fully panmictic population. Two forms that differ in shape show evidence for at least a partial reproductive barrier between them (Hull *et al.*, 1996).

13.2 Methods

Samples were collected from two sites on the east coast of England, namely Old Peak (British Grid reference NZ/984021) and Filey Brigg (TA/137815). At both sites '*L. saxatilis*' were collected from boulders of sandstone resting on bedrock between the upper region of the *Semibalanus balanoides* (L.) zone to about 1 m above the zone of *Pelvetia canaliculata* Dcne. & Thur. We designated the former as midshore (M) and the latter as high-shore (H). *L. arcana* occurs at both sites, usually on the H boulders along with the *L. saxatilis* H form.

Collections were kept separate, and sorted in the laboratory using shell characters. Animals were characterised using shell colour and sculpture, thus avoiding the circularity of sorting by shape and then analysing shape. Shape was measured as described by Grahame & Mill (1989). Shape data have been analysed using principal components analysis and canonical variate analysis. For these analyses, linear measurements were transformed to base 10 logarithms. A geometric mean size transform was also used to minimise the effect of size (Reist, 1985; Caley et al., 1995).

The polymorphic enzyme loci glucose 6 phosphate isomerase (Gpi: EC 5.3.1.9), phosphogluconate dehydrogenase (Pgdh: EC 1.1.1.44), phosphoglucomutase (Pgm-1: EC 5.4.2.2), glycerol 3 phosphate dehydrogenase (G3pdh: EC 1.1.1.8) and superoxide dismutase (Sod-1: EC 1.15.1.1) were investigated using polyacrylamide gel electrophoresis. Bodies of unparasitised females were homogenised after removing the brood pouch or jelly gland.

Material for DNA extraction was collected from the Old Peak site and returned to the laboratory. Animals were held in the laboratory without food for a week, after which the animals were homogenised and DNA extracted. Again only females were used and the few parasitised individuals found were discarded. Care was taken to avoid contamination of individuals with tissue or mucus from the preceding individual, with forceps and dishes washed between handling each animal. After cracking the shell of an individual in a pestle and mortar the body (without brood pouch) was homogenised and extraction carried out following the method described by Crossland et al. (1993). Polymerase chain reaction (PCR) amplification was also carried out as described by Crossland et al. (1993), with the modifications of using a 'hot start' (primer and template DNA introduced to the reaction tubes during an initial phase at 95°C) and the total number of cycles reduced to 35. PCR product was subject to electrophoresis in a 3% (w/v) agarose gel with ethidium bromide, and inspected under ultra-violet light. Ten primers were screened; two were found that suggest population markers. These are Operon Technologies OPY-01, sequence 5' GTG GCA TCT C 3', and RAPD-h (A. Corner, personal communication), sequence 5' GCC GTG GTT A 3'.

Further analysis of the populations involved PCR amplification from individuals from several field samples using RAPD-h, after which the bands generated from each individual were scored. Migration distances on 3% agarose gels were used to calculate fragment sizes. Bands from different individuals that were within plus or minus 3 base-pairs of one another were assumed to be the same product and treated as such. The resulting presence/absence data matrix was analysed using the BOOT procedure from Phylip 3.4 (Felsenstein, 1989).

13.3 Results

Fig. 13.1 shows a map of the sampling sites, with a profile sketch of the shores together with outline drawings of the *L. saxatilis* H and M form shells. On both shores larger boulders occur near high water (H), close to the foot of a cliff; smaller boulders occur at about mid-tide level (M). The two forms of *L. saxatilis* were usually well sorted onto their respective boulders, but occasional individuals of the 'wrong' sort were found out of their respective habitats, and after rough weather the whole population was found to be rather mixed, with many individuals of H in M locations and *vice versa*. As well as the H and M forms there were occasional individuals of a form we designated as intermediate, or I, usually found on the M boulders with that population. The shells of *L. saxatilis* H are very thin, with pronounced ridging, and almost black in colour. Those of *L. saxatilis* M are light brown, thick and heavy, and with a very fine sculpture of spiral lines. I shells are like M shells but thinner, while *L. arcana* shells are like those of *L. saxatilis* H but lighter in colour and with no ridging.

At both locations, the two *L. saxatilis* populations evidently meet and show evidence of hybridisation. At Filey Brigg, M individuals commonly inhabit the lower portions of the boulders that are used in their upper portions by H individuals. They are probably best thought of as being parapatric.

Fig. 13.2 shows the shells of *L. arcana* and *L. saxatilis* H, M and I forms plotted on the first three principal component vectors from a principal components analysis (PCA). Since this is a PCA, there is no question of groups being imposed by the investigators – the scatter of points on the graph arises from an analysis that is 'blind' to any external grouping. The eigenvalues and coefficients for each variable for the first three components are shown in Table 13.1. These components are all associated with eigenvalues >1. There is no evidence that size as such is contributing disproportionately to component 1, the coefficient values ranging from -0.44 to 0.48 (see Sundberg, 1988; Caley *et al.*, 1995, for a discussion).

The forms H and M of *L. saxatilis* fall into two quite discrete groups of points, with *L. arcana* more like H than M. Thus, the shells initially discriminated on characters of sculpture and colour show comparable separation in the PCA. The I forms of *L. saxatilis* are scattered within both H and M. The same symbol is used for specimens from both sites on the plot, in fact the specimens from Old Peak are like the M form,

302   *Discovering diversity among marine molluscs*

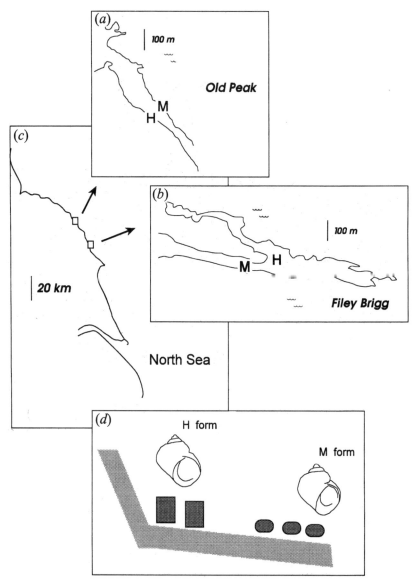

Fig. 13.1. Sketch maps showing the shores at (*a*) Old Peak and (*b*) Filey Brigg, and locating these on the east coast of Britain. (*c*) The shores are delimited by lines of high and low water spring tides. (*d*) Shore profile is sketched to indicate the form and distribution of boulders (stippled areas) on which the H and M forms of *L. saxatilis* occur, these are illustrated by outlines taken from camera lucida drawings.

J. Grahame et al.

Table 13.1. *Vectors of coefficients forming the first three principal components in a principal components analysis of shell measurements*

Vector 1 3.65		Vector 2 2.15		Vector 3 1.16	
Lip length	−0.44	Apical angle	−0.55	Aperture width	−0.26
Whorl width 2	−0.35	Lip length	−0.32	Aperture length	−0.26
Columella length	0.08	Aperture length	−0.02	Columella length	−0.23
Whorl width 0	0.23	Whorl width 0	−0.02	Lip length	0.06
Apical angle	0.26	Aperture width	0.00	Apical angle	0.07
Whorl width 1	0.29	Whorl width 1	0.36	Whorl width 2	0.09
Aperture width	0.48	Whorl width 2	0.38	Whorl width 1	0.45
Aperture length	0.48	Columella length	0.57	Whorl width 0	0.77

The eigenvalue for each vector is given.

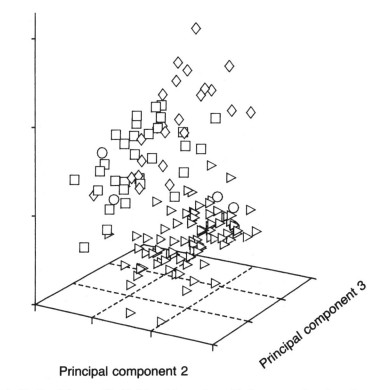

Fig. 13.2. Shells of *L. saxatilis* H, M and I together with *L. arcana* plotted on the first three vectors from a principal components analysis. Squares, *L. saxatilis* H; flags, *L. saxatilis* M; circles, *L. saxatilis* I; diamonds, *L. arcana*.

Table 13.2. Mahalanobis distance estimates from a canonical variates analysis of shell measurement data

	L. arcana, Old Peak	L. arcana, Filey Brigg	L. saxatilis H, Old Peak	L. saxatilis M, Old Peak	L. saxatilis H, Filey Brigg	L. saxatilis M, Filey Brigg	L. saxatilis I, Old Peak	L. saxatilis I, Filey Brigg
L. arcana, Old Peak	0	13.7	8.6	65.7	22.6	60.8	55.1	23.4
L. arcana, Filey Brigg		0	6.2	36.5	3.9	28.9	34	3.9
L. saxatilis H, Old Peak			0	45.5	9.2	39.6	40.2	9.3
L. saxatilis M, Old Peak				0	29.4	2.7	2.8	37.8
L. saxatilis H, Filey Brigg					0	22.5	31	0.7
L. saxatilis M, Filey Brigg						0	4.7	29.1
L. saxatilis I, Old Peak							0	39.3
L. saxatilis I, Filey Brigg								0

J. Grahame et al. 305

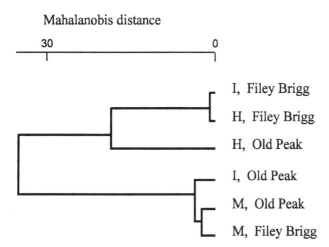

Fig. 13.3. Dendrogram expressing the Mahalanobis distance measures between the *L. saxatilis* groups.

while the specimens from Filey Brigg are like the H form. Individuals of form I are scarce, and so few have been available for measurement that we cannot be certain of their morphometric affinities, but the available evidence suggests that they resemble one form or the other. Table 13.2 shows the matrix of Mahalanobis distance estimates from a canonical variates analysis for the samples within and between shores. It is clear that the forms of *L. saxatilis* are much more closely related between the shores (H to H, M to M) than on either of the shores (H to M). This is clearly seen in Fig. 13.3, where the distance measures for just the *L. saxatilis* samples have been expressed as a dendrogram. The information for *L. arcana*, included for completeness in Table 13.2, has been omitted in the Figure, since we are principally concerned with relationships in *L. saxatilis*.

We turn now from consideration of shape to polymorphic enzymes. Tables 13.3 and 13.4 show the results of tests of genetic homogeneity between subpopulations at Filey Brigg (Table 13.3) and for selected comparisons between Filey Brigg and Old Peak (Table 13.4). There is no comparison within *L. saxatilis* H at Filey Brigg because one of the two subpopulations had only two animals in the sample. At Filey Brigg, there are only two instances of significant difference: between subpopulations of *L. saxatilis* M, and between *L. saxatilis* H and M. Both involve the locus G3pdh. Considering differences between shores (Table 13.4), there is about the same level of heterogeneity between the populations

Table 13.3. *Genetic homogeneity between subpopulations at Filey Brigg – sample sizes for the comparisons are given*

	L. arcana 11, 6		*L. saxatilis* M 7, 18, 12		*L. saxatilis* H:M 19, 37	
	χ^2	p	χ^2	p	χ^2	p
Gpi	1.19	0.779	3.55	0.78	5.58	0.931
Pgdh	1.57	0.301	1.48	0.821	6.83	0.548
Pgm-1	3.39	0.278	6.09	0.183	12.48	0.131
G3pdh	2.21	0.412	14.18	**0.004**	18.13	**0.019**
Sod-1	0.46	0.705	1.70	0.448	4.53	0.361

Significant values of χ^2 after Bonferroni correction are shown in bold type. Because of small sample sizes, χ^2 calculations have been carried out as described by Zaykin & Pudovkin (1993).

Table 13.4. *Genetic homogeneity between populations at Filey Brigg and Old Peak, calculated as for Table 13.3*

	L. arcana 17, ≥ 48		*L. saxatilis* H 19, ≥ 48		*L. saxatilis* M 37, ≥ 38	
	χ^2	p	χ^2	p	χ^2	p
Gpi	26.42	0.029	7.91	0.566	16.73	0.173
Pgdh	32.74	**0.001**	6.80	0.532	10.97	0.190
Pgm-1	16.92	0.083	1.94	0.919	12.80	0.126
G3pdh	32.74	**0.000**	14.23	0.031	22.59	**0.004**
Sod-1	5.03	0.422	19.93	**0.000**	3.87	0.424

Significant values of χ^2 after Bonferroni correction are shown in bold type. Sample sizes are indicated, the second value is for Old Peak and is given as a minimum since not all loci were scored in these samples.

on the different shores. Thus, the analysis does not suggest any greater heterogeneity between the shores within the groups, as opposed to between the groups within a shore.

The genetic relationships between the samples are expressed as an unrooted tree drawn using the continuous maximum likelihood routine (CONTML) in Phylip 3.4 (Felsenstein, 1989; Fig. 13.4). Confidence limits are not shown on the tree, they are typically about twice the branch lengths, and little confidence can be placed in this particular

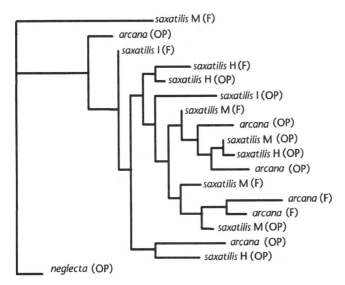

Fig. 13.4. Continuous maximum likelihood tree relating the populations of *L. saxatilis* and *L. arcana* at Old Peak (OP) and Filey Brigg (F). Data for *L. neglecta* have been used as an outgroup.

arrangement being 'correct'. The tree suggests no distinct ordering of the samples by either taxon or shore.

Finally we have used random amplification from total DNA extracts to search for possible diagnostic characters. Trials with 10 random primers have shown two primer sequences that cause products to be synthesised in samples from one population and not the other, suggesting the possibility of locally useful markers. Thus:

1. Primer OPY-01, product of interest 400 bases, band present in product from all of four *L. saxatilis* M, absent from all of four *L. saxatilis* H.
2. Primer RAPD-h, product of interest 700 bases, band present in product from all of ten *L. saxatilis* M, absent from all of eight *L. saxatilis* H, and present in all of three intermediate *L. saxatilis* including one specimen taken from the high shore site.

In each case the product of interest is represented by a brightly fluorescing band found in individuals from one population, this band being absent in individuals from the other. The sample sizes are very small; these findings are the result of a preliminary trial in which the object was to screen a number of primers to discover if there were potentially

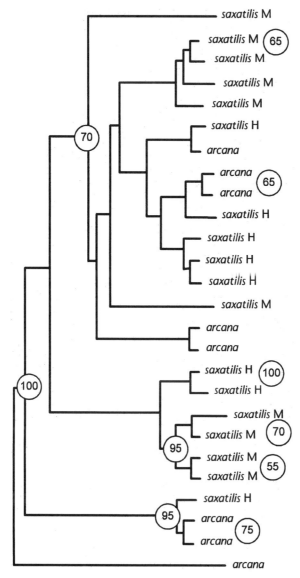

Fig. 13.5. Bootstrapped mixed parsimony algorithm (Wagner parsimony) tree relating 26 individuals of *L. saxatilis* and *L. arcana* from Old Peak only, using presence/absence data of bands generated by RAPD-h (see the text).

useful primer sequences. To date we have investigated only populations at Old Peak.

This investigation has been extended by using one primer (RAPD-h) in closely standardised conditions to amplify from total DNA extracts taken from 26 individuals of *L. saxatilis* and *L. arcana* from Old Peak. A total of 106 bands were resolved, the maximum number of individuals sharing a band was 20. Seven bands occurred in ≥10 individuals; 13 bands in between 5 and 9 individuals, and 86 bands in 4 or fewer individuals. All the data were used to construct an unrooted bootstrapped tree shown in Fig. 13.5. Because of time constraints, only 20 bootstraps were carried out, the tree shows the bootstrap values as percentages where these were ≥50%. The Figure shows a definite tendency for individuals of the recognised species *L. arcana* and *L. saxatilis* (H) to group together, *L. saxatilis* M groups separately. The analysis is weak in that many of the branches are poorly supported but, overall, there is no more or less support for differentiation of any one of the three groups than for any other. Notably, there is evidence of similarity between *L. saxatilis* H and *L. arcana*.

13.3 Discussion

13.3.1 Phenotypic differences

Shell morphology shows by both inspection and analysis as discrete and recognisable differentiation into two forms of *L. saxatilis* as one might expect to find in rough periwinkles (Figs 13.1 and 13.2). *L. arcana* is shown as being very like the H form of *L. saxatilis*, while *L. saxatilis* M is different from both of these. Notably the variation between H and M appears to be discontinuous - the forms do not merge into one another. It is true that there is some overlap where individual shells occur in the 'wrong' portion of Fig. 13.2, but there is a distinct space between the two clouds of symbols. The form we designate as 'intermediate' in fact generally more closely resembles the M form at Old Peak, and the H form at Filey Brigg, in neither place does it represent a continuously varying part of the population spanning the difference between H and M. In this respect, our 'intermediates' appear to be very different from those described by Janson & Sundberg (1983), which were numerous in the population and formed a continuum between the two morphs they designated as being characteristic of exposed and sheltered shores, respectively, at sites in Sweden. In the situation we describe at Old Peak and at Filey Brigg, differentiation into two morphs occurs on the same shore, between boulders at most only a few tens of metres apart, or even on portions of the same boulder. In fact, the situation is just that referred to by (for example) Mayr: there is a 'definite gap' between two forms (Mayr, 1942). It is important to note that the shape differences found on either of the shores are consistent between the shores, separated by some 20 km (Fig. 13.1). Admittedly the scale is local, but this itself is important: in rough periwinkles, discontinuities may be most evident locally, being blurred or disappearing altogether as the range of samples increases (Grahame & Mill, 1989, 1992). Such observations have of course been reported before (see Coyne, 1994), but we believe that this phenomenon – simple and obvious as it is – may not be borne sufficiently in mind by biologists surveying variable material from a range of sites.

How should we interpret these forms? There are several possibilities: the forms might be ecotypes, the variation representing phenotypic expression of the same range of genotypes in different environments. Thus, they might represent members of a genetically polymorphic population that, by habitat selection perhaps combined with differential mortality, come to be distributed mainly in their different habitats, but nevertheless interbreed. Considering phenotypic difference alone, this sort of explanation is attractive – *L. saxatilis* H and M live in very different microhabitats on the shore, and they are probably exposed to very different regimes of crab predation, among many environmental variables altering up and down the shore. Perhaps the similarity of *L. saxatilis* H to *L. arcana*, and the difference of *L. saxatilis* M from both of these, is a reflection of microhabitat maintained polymorphism in *L. saxatilis*. This sort of explanation is favoured for other rough periwinkle assemblages (Johannesson & Johannesson, 1990), perhaps associated with some degree of reproductive isolation as well as observed differences of form and distribution (Johannesson *et al.*, 1993; Rolán-Alvarez & Rolán, 1995).

However, we have observed that on several shores over some 60 km on the coast of Yorkshire, H animals have larger and fewer young in the brood pouch than do M forms (Hull *et al.*, 1996). This finding that reproductive traits differ in the females of forms H and M is, at first sight, no different from previously reported differences in such traits (e.g. Hart & Begon, 1982). What is novel, however, is that the H and M forms referred to here are behaving like populations showing some reproductive isolation: there is a rare intermediate form that is evidently a reproductively handicapped hybrid. Qualitative observations suggest that similar phenomena may be widespread in British *L. saxatilis* – are we seeing speciation?

13.3.2 Genotypic differences

In rejecting essentialist (typological) thinking for population thinking (Mayr, 1982), biologists focus on the extent and meaning of variation. In the present instance there is a pronounced discontinuity in the spectrum of variation between *L. saxatilis* H and M forms. Further progress requires knowledge of gene flow. As a way forward, the first response is often to use polymorphic enzymes. As so often in studies of rough periwinkles, the data are equivocal: the distribution of polymorphic enzyme alleles does not support the idea of much genetic differentiation, and what there is seems to be disordered. Johannesson and co-workers (e.g. Johannesson *et al.*, 1993) have found that morphologically very divergent populations of '*L. saxatilis*' on the coast of Galicia, northern Spain, seem to be related in such a way that intraspecific polymorphism is indicated (Fig. 13.6, model 1). Their data indicate that very divergent morphs are in fact more genetically related on a single shore than between comparable shores. From this, they infer that there is gene flow between the divergent morphs, albeit perhaps across a partial reproductive barrier. The Spanish forms are very distinct in shell character, shape and size, to the extent that the smaller low-shore form in Spain closely resembles what in Britain would be characterised as *L. neglecta* Bean. Elsewhere we deal with British populations of *L. neglecta*, showing that it is morphologically (Caley *et al.*, 1995) and genetically (S.L. Hull, J. Grahame & P.J. Mill, unpublished results) distinct, and therefore in Britain probably deserves full specific status.

Considering the two morphologically distinct *L. saxatilis* forms at Old Peak and Filey Brigg, we conclude that the situation is more like that of model 2 in Fig. 13.6. We cannot infer any particular pattern of gene flow between the forms on or between shores as being stronger or weaker, either on the basis of genetic heterogeneity (Tables 13.3 and 13.4) or maximum likelihood analysis (Fig. 13.4). Therefore a rather conventional interpretation is favoured: that because phenotypes between the shores are similar, there is presumably at least potential gene flow between them (they are one species), leaving open the question of gene flow between the divergent phenotypes on the same shore.

However, these divergent phenotypes on the Yorkshire coast belong

(a) Location 1   Location 2

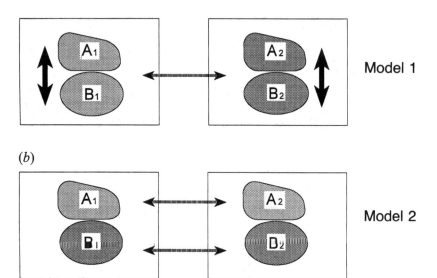

Model 1

(b)

Model 2

Fig. 13.6. Two parapatric populations, A and B, occur in different locations. They show phenotypic differences, indicated here by different shapes. (*a*) In Model 1, gene flow is inferred to occur between A_1 and B_1, and between A_2 and B_2 (heavy arrows, similar shading of the A/B pairs), so that these pairs are more connected than through relatively slight gene flow between locations (thin arrow). (*b*) In Model 2, there is no evidence of gene flow between A and B in either location, and the phenotypic differences are taken to reflect substantial or even complete reproductive isolation. Some gene flow presumably occurs between A_1, A_2, and B_1, B_2.

to one putative species – *L. saxatilis*. The suggestion that there may be a hitherto unsuspected reproductive barrier within this 'species' rests first on observation of the breeding biology of the animals. In the face of uncertainty about gene flow from polymorphic enzyme data, we have begun searching for DNA polymorphisms using RAPDs. It is fairly easy to show that there are such polymorphisms, evidently distinguishing one population from another. This of course does not justify specific status for either, but we note that the 700 base-pair band generated by RAPD-h occurs in *L.saxatilis* M and also in *L. saxatilis* I (the supposed hybrid) but not in *L. saxatilis* H. This is precisely what might be expected of a non-co-dominant marker in the absence of gene flow from M to H – it leaves open the possibility of gene flow from H to M, since we have not yet found a marker for H. It is apparently easy to generate RAPDs markers that will differentiate sets of individuals within a population at virtually any level of discrimination. This is seen in the clonal bivalve *Lasaea*, where RAPDs has been used to demonstrate variation in apparently homogeneous clones (Tyler-Walters & Hawkings, 1995). Therefore the presence of such a marker in *L. saxatilis* M currently has the status of an apparent DNA polymorphism that, because its distribution in the population is evidently limited by a reproductive barrier, merits further attention. Strikingly, it does not behave as do the putative allozyme markers (see also Karl & Avise, 1992).

A bootstrapped tree based on all the bands generated by RAPD-h (Fig. 13.5) provides no conclusive evidence concerning possible relationships for animals from Old Peak. Many branches are poorly supported, but it is interesting that *L. saxatilis* H shows at least as great a relationship to *L. arcana* as it does to *L. saxatilis* M. This is unlikely to have been due to cross-contamination, as strenuous efforts were taken to avoid this, and the specimens were not processed in any particular order, as would have to have been the case to generate the patterns seen. Supposing the pattern is real, as *L. saxatilis* H and *L. arcana* share the same boulder microhabitat, the observation raises the possibility of gene flow between them. Warwick *et al.* (1990) reported that in the laboratory these two species appeared to be interfertile, but they did not detect evidence of hybridisation in the field. Perhaps they are hybridising in the field, and perhaps DNA techniques will be more sensitive in picking this up (see also Edwards & Skibinski, 1987).

13.3.3 Summary and conclusions

In the rough periwinkles (subgenus *Neritrema*) the presently understood taxonomy of three or at most four species may be inadequate. Not only is *L. neglecta* morphologically distinct and reproductively isolated in Britain (Caley *et al.*, 1995; S.L. Hull, J. Grahame & P.J. Mill, unpublished results), but now there is evidence of a further reproductive barrier within *L. saxatilis sensu stricto*. This suggestion rests first on morphological evidence, and is best supported by DNA polymorphisms that indicate differences between the populations. These observations are consistent with the interpretation of the I form as a hybrid between H and M, and suggest interrupted gene flow.

We are impressed by how unhelpful enzyme polymorphism studies may be in *Neritrema*: if it were not for the striking differences in morphology and reproductive behaviour, with the occurrence of an evidently unfit form that we interpret as a hybrid, we would not make the suggestion of further species differentiation. While it is possible using enzyme polymorphism data to differentiate forms already acknowledged as species (e.g. Knight & Ward, 1991), it is difficult or impossible to use this approach *ab initio* in detecting sibling species in *Neritrema*. In other words, the findings may depend upon the initial recognition of discontinuous variation in morphology. In this context, we note that there is a further 'form' of rough periwinkle not discussed above that shares the barnacle-shell habitat at Old Peak with *L. neglecta*, breeds at a similar small size, and yet is quite distinct from *L. neglecta*. There is no evidence of a reproductive barrier in this case (the animal is not in contact with the larger, higher shore populations of *L. saxatilis*) but is to some extent ecologically and physiologically distinct (Grahame *et al.*, 1995), and presents another population evidently worth investigating with DNA-based techniques.

Our caveat about enzyme polymorphism studies is clearly not to be taken as universal – see for example the bivalve *Chamelea* (Backeljau *et al.*, 1994). Yet controversy often still abounds, as in the case of *Montastraea annularis*. Such controversy is also the case in recent studies of rough periwinkles.

Knowlton (1993) has suggested that sibling species complexes may be more numerous in the sea than on land, and has pointed out that such complexes are of enormous potential value for understanding evolution. We agree – but of course the complexes must be identified and understood for progress to be made. That the situation in *Neritrema* matters in the context of general biodiversity considerations rests on two suggestions. The first is methodological: that in sibling species, close inspection of morphology, biogeography and determination of breeding behaviour, may be a better guide to likely reproductive barriers and existence of species than are studies of polymorphic enzymes. Study of DNA polymorphisms is promising, but yet to be fully exploited. The second suggestion concerns the future research programme: it is that in this group, we have a valuable tool for investigating speciation (see also Rolán-Alvarez & Rolán, 1995). We suggest that the group seems likely to be more species-rich than hitherto suspected. Considering this possibility with the complex biogeography and relationship to the shore environment gradient, there is a set of populations and species with which to study the generation of biodiversity.

Acknowledgements

S.L.H. was supported by a studentship from the Natural Environment Research Council and by a grant from the Research Council, University College Scarborough. We are grateful for the help of Dr David Coates and the Molecular Biology Unit of the Department of Pure and Applied Biology, University of Leeds. Drs M.J. McPherson and A. Corner of the Centre for Plant Biochemistry and Biotechnology, University of Leeds, supplied primers.

References

Backeljau, T., Bouchet, P., Gofas, S. & de Bruyn, L., 1994. Genetic variation, systematics and distribution of the venerid clam *Chamelea gallina*. *Journal of the Marine Biological Association of the United Kingdom*, **74**, 211–23.

Beaumont, A.R. & Wei, J.H.C., 1991. Morphological and genetic variation in the Antarctic limpet *Nacella concinna* (Strebel, 1908). *Journal of Molluscan Studies*, **57**, 443–50.

Blot, M., Thiriot-Quiévreux, C. & Soyer, J., 1988. Genetic relationships among populations of *Mytilus desolationis* from Kerguelen, *M. edulis* from the North Atlantic and *M. galloprovincialis* from the Mediterranean. *Marine Ecology Progress Series*, **44**, 239–47.

Caley, K.J., Grahame, J. & Mill, P.J., 1995. A geographically-based study of shell shape in small rough periwinkles. *Hydrobiologia*, **309**, 181–93.

Coyne, J.A., 1994. Ernst Mayr and the origin of species. *Evolution*, **48**, 19–30.

Crossland, S., Coates, D., Grahame, J. & Mill, P.J., 1993. The use of random amplified polymorphic DNAs (RAPDs) in separating two sibling species of *Littorina*. *Marine Ecology Progress Series*, **96**, 301–5.

Edwards, C.A. & Skibinski, D.O.F., 1987. Genetic variation of mitochondrial DNA in mussel (*Mytilus edulis* and *M. galloprovincialis*) populations from south west England and south Wales. *Marine Biology*, **94**, 547–56.

Felsenstein, J., 1989. *Phylip 3.2 Manual*. Berkeley, CA: University of California Herbarium.

Forbes, E. & Hanley, S., 1853. *A History of British Mollusca, and their Shells*, vol. III. London: John Van Voorst.

Futuyuma, D.J., 1987. On the role of species in anagenesis. *American Naturalist*, **130**, 465–73.

Gosling, E.M., 1994. Speciation and wide-scale genetic differentiation. In *Genetics and Evolution of Aquatic Organisms*, ed. A.R. Beaumont, pp. 1–14. London: Chapman & Hall.

Gosse, P.H., 1860. *A History of the British Sea Anemones and Corals*. London: John Van Voorst.

Grahame, J. & Mill, P.J., 1989. Shell shape variation in *Littorina saxatilis* and *L. arcana*: A case of character displacement? *Journal of the Marine Biological Association of the United Kingdom*, **69**, 837–55.

Grahame, J. & Mill, P.J., 1992. Local and regional variation in shell shape of rough periwinkles in southern Britain. In *Proceedings of the Third International*

Symposium on Littorinid Biology ed. J. Grahame, P.J. Mill & D.G. Reid, pp. 99–106. London: The Malacological Society.

Grahame, J., Mill, P.J., Double, M. & Hull, S.L., 1992. Patterns of variation in Aat-1 allele frequencies in rough periwinkles (*Littorina*) suggest similar selection regimes rather than conspecificity. *Journal of the Marine Biological Association of the United Kingdom*, **72**, 499–502.

Grahame, J., Mill, P.J., Double, M. & Hull, S.L., 1993. Reply to Johannesson and Johannesson. *Journal of the Marine Biological Association of the United Kingdom*, **73**, 250.

Grahame, J., Mill, P.J., Hull, S.L. & Caley, K.J., 1995. *Littorina neglecta* Bean: Ecotype or species? *Journal of Natural History*, **29**, 887–99.

Hannaford Ellis, C.J., 1979. Morphology of the oviparous rough winkle *Littorina arcana* Hannaford Ellis, 1978, with notes on the taxonomy of the *L. saxatilis* species-complex (Prosobranchia: Littorinidae). *Journal of Conchology*, **30**, 43–56.

Hart, A. & Begon, M., 1982. The status of general reproductive strategy theories, illustrated in winkles. *Oecologia*, **52**, 37–42.

Haylor, G.S., Thorpe, J.P. & Carter, M.A., 1984. Genetic and ecological differentiation between sympatric colour morphs of the common intertidal sea anemone *Actinia equina*. *Marine Ecology Progress Series*, **16**, 281–9.

Heller, J., 1975. The taxonomy of some British *Littorina* species with notes on their reproduction (Mollusca: Prosobranchia). *Zoological Journal of the Linnean Society*, **56**, 131–51.

Hull, S.L., Grahame, J. & Mill, P.J., 1996. Morphological divergence and evidence for reproductive isolation in *Littorina saxatilis* (Olivi) in northeast England. *Journal of Molluscan Studies*, **62**, 89–99.

James, B.L., 1968. The characters and distribution of the subspecies and varieties of *Littorina saxatilis* (Olivi 1792) in Britain. *Cahiers de Biologie Marine*, **9**, 143–65.

Janson, K. & Sundberg, P., 1983. Multivariate morphometric analysis of two varieties of *Littorina saxatilis* from the Swedish coast. *Marine Biology*, **74**, 49–53.

Janson, K. & Ward, R.D., 1984. Microgeographic variation in allozyme and shell characters in *Littorina saxatilis* Olivi (Prosobranchia: Littorinidae). *Biological Journal of the Linnean Society*, **22**, 289–307.

Janson, K. & Ward, R.D., 1985. The taxonomic status of *Littorina tenebrosa* Montagu as assessed by morphological and genetic analysis. *Journal of Conchology*, **32**, 9–15.

Johannesson, K. & Johannesson, B., 1990. Genetic variation within *Littorina saxatilis* (Olivi) and *Littorina neglecta* Bean: Is *L. neglecta* a good species? *Hydrobiologia*, **193**, 89–97.

Johannesson, K. & Johannesson, B., 1993. The taxonomic status of *Littorina neglecta*: A comment to Grahame, Mill, Double and Hull. *Journal of the Marine Biological Association of the United Kingdom*, **73**, 249.

Johannesson, K., Johannesson, B. & Rolán-Alvarez, E., 1993. Morphological differentiation and genetic cohesiveness over a microenvironmental gradient in the marine snail *Littorina saxatilis*. *Evolution*, **47**, 1770–87.

Karl, S.L. & Avise, J.C., 1992. Balancing selection at allozyme loci in oysters: Implications from nuclear RFLPs. *Science*, **256**, 100–2.

Knight, A.J. & Ward, A.R.D., 1991. The genetic relationships of three taxa in the *Littorina saxatilis* species complex (Prosobranchia: Littorinidae). *Journal of Molluscan Studies*, **57**, 81–91.

Knowlton, N., 1993. Sibling species in the sea. *Annual Review of Ecology and Systematics*, **24**, 189–216.
Knowlton, N., Weil, E., Wcigt, L.A. & Guzman, H.M., 1992. Sibling species in *Montastraea annularis*, coral bleaching, and the coral climate record. *Science*, **255**, 330–3.
Mallet, J., 1995. A species definition for the modern synthesis. *Trends in Ecology and Evolution*, **10**, 294–9.
Mayr, E., 1942. *Systematics and the Origin of Species from the Viewpoint of a Zoologist*. New York: Columbia University Press.
Mayr, E., 1963. *Animal Species and Evolution*. Cambridge, MA: Harvard University Press.
Mayr, E., 1982. *The Growth of Biological Thought: Diversity, Evolution and Inheritance*. Cambridge, MA: Harvard University Press.
Mill, P.J. & Grahame, J., 1995. Shape variation in the rough periwinkle *Littorina saxatilis* on the west and south coasts of Britain. *Hydrobiologia*, **309**, 61–72.
Nei, M., 1972. Genetic distance between populations. *American Naturalist*, **106**, 283–91.
Palumbi, S.R., 1994. Genetic divergence, reproductive isolation, and marine speciation. *Annual Review of Ecology and Systematics*, **25**, 547–72.
Pascoe, P.L. & Dixon, D.R., 1994. Structural chromosomal polymorphism in the dog-whelk *Nucella lapillus* (Mollusca: Neogastropoda). *Marine Biology*, **118**, 247–53.
Reid, D.G., 1989. The comparative morphology, phylogeny and evolution of the gastropod family Littorinidae. *Philosophical Transactions of the Royal Society of London, Series B*, **324**, 1–110.
Reid, D.G., 1996. *Systematics and Evolution of* Littorina. London: The Ray Society.
Reid, D.G. & Golikov, A.N., 1991. *Littorina naticoides*, new species, with notes on the other smooth-shelled *Littorina* species from the northwestern Pacific. *The Nautilus*, **105**, 7–15.
Reist, J.D., 1985. An empirical evaluation of several univariate methods that adjust for size variation in morphometric data. *Canadian Journal of Zoology*, **63**, 1429–39.
Rolán-Alvarez, E. & Rolán, E., 1995. Applicability of the biological species concept in taxonomy: The example of *Littorina saxatilis* (Gastropoda: Prosobranchia). *Argonauta*, **9**, 1–8.
Seshappa, G., 1947. Oviparity in *Littorina saxatilis* (Olivi). *Nature*, **160**, 335–6.
Seshappa, G., 1948. Nomenclature of the British Littorinidae. *Nature*, **162**, 702–3.
Seshappa, G., 1976. Reproduction in *Littorina saxatilis rudissima* (Bean) and two other subspecies of *Littorina saxatilis* (Olivi) at Cullercoats, England. *Indian Journal of Marine Sciences*, **5**, 222–9.
Skibinski, D.O.F., Cross, T.F. & Ahmad, M., 1980. Electrophoretic investigation of systematic relationships in the marine mussels *Modiolus modiolus* L., *Mytilus edulis* L., and *Mytilus galloprovincialis* Lmk. (Mytilidae; Mollusca). *Biological Journal of the Linnean Society*, **13**, 65–73.
Solé-Cava, A.M. & Thorpe, J.P., 1992. Genetic divergence between colour morphs in populations of the common intertidal sea anemones *Actinia equina* and *A. prosina* (Anthozoa: Actiniaria) in the Isle of Man. *Marine Biology*, **112**, 243–52.
Solé-Cava, A.M., Thorpe, J.P. & Kaye, J.G., 1985. Reproductive isolation with little genetic divergence between *Urticina* (= *Tealia*) *felina* and *U. eques* (Anthozoa: Actiniaria). *Marine Biology*, **85**, 279–84.

Sundberg, P., 1988. Microgeographic variation in shell characters of *Littorina saxatilis* Olivi – a question mainly of size? *Biological Journal of the Linnean Society*, **35**, 169–84.
Sundberg, P., Knight, A.J., Ward, R.D. & Johannesson, K., 1990. Estimating the phylogeny in mollusc *Littorina saxatilis* (Olivi) from enzyme data: Methodological considerations. *Hydrobiologia*, **193**, 29–40.
Thorpe, J.P., 1982. The molecular clock hypothesis: Biochemical evolution, genetic differentiation and systematics. *Annual Review of Ecology and Systematics*, **13**, 139–68.
Tyler-Walters, H. & Hawkins, A.R., 1995. The application of RAPD markers to the study of the bivalve mollusc *Lasaea rubra*. *Journal of the Marine Biological Association of the United Kingdom*, **75**, 563–9.
Van Veghel, M.L.J. & Bak, R.P.M., 1993. Intraspecific variation of a dominant Caribbean reef building coral *Montastrea annularis*: Genetic, behavioral and morphometric aspects. *Marine Ecology Progress Series*, **92**, 255–65.
Varvio, S.L., Koehn, R.K. & Väinölä, R., 1988. Evolutionary genetics of the *Mytilus edulis* complex in the North Atlantic region. *Marine Biology*, **98**, 51–60.
Ward, R.D. & Janson, K., 1985. A genetic analysis of sympatric subpopulations of the sibling species *Littorina saxatilis* (Olivi) and *Littorina arcana* Hannaford Ellis. *Journal of Molluscan Studies*, **51**, 86–94.
Warwick, T., Knight, A.J. & Ward, R.D., 1990. Hybridisation in the *Littorina saxatilis* species complex. *Hydrobiologia*, **193**, 109–16.
Wilkins, N.P. & O'Regan, D., 1980. Generic variation in sympatric sibling species of Littorina. *The Veliger*, **22**, 355–9.
Wilkins, N.P., O'Regan, D. & Moynihan, E., 1978. Electrophoretic variability and temperature sensitivity of phosphoglucose isomerase and phosphoglucomutase in littorinids and other marine molluscs. In *Marine Organisms: Genetics, Ecology and Evolution*, ed. B. Battaglia & J.A. Beardmore, pp. 141–55. New York and London: Plenum Press.
Zaslavskaya, N.I., 1995. Allozyme comparison of four littorinid species morphologically similar to *Littorina sitkana*. *Hydrobiologia*, **309**, 123–8.
Zaykin, D.V. & Pudovkin, A.I., 1993. Two programs to estimate significance of χ^2 values using pseudo-probability tests. *Journal of Heredity*, **84**, 152.

Chapter 14
Ecosystem function at low biodiversity – the Baltic example

RAGNAR ELMGREN† and CATHY HILL‡
†*Department of Systems Ecology, Stockholm University, S-106 91 Stockholm, Sweden*
‡*Swedish Environmental Protection Agency, S-106 48 Stockholm, Sweden*

Abstract

For a marine area, the Baltic Sea has a uniquely low species diversity, due to its low salinity, recent geological origin and harsh climate. Its flora and fauna consist of a mixture of marine, freshwater and brackish-water species. The three major basins of the Baltic Sea, from south to north, have quite stable surface salinities of 6–8 psu in the Baltic proper, 5–6 psu in the Bothnian Sea and 2–4 psu in the Bothnian Bay.

A comparison of Steele's (1974) carbon flow model of the high diversity North Sea ecosystem, with Elmgren's (1984) carbon flow calculations for the Baltic Sea shows great similarity for the Baltic proper, and similar, but proportionally reduced, flows in the Bothnian Sea. Only the Bothnian Bay has a markedly different carbon flow pattern, due to a near-total absence of benthic filter-feeders. Only when a major functional group was lost, was there a drastic alteration of ecosystem function. The Baltic example suggests we should not claim without good proof that high biodiversity is needed to 'maintain ecosystem function'.

The possibility remains that low diversity may destabilise ecosystem function, even without the loss of major functional groups. With fewer species in each functional group, there is a greater risk that the loss or drastic reduction of a single species may affect ecosystem processes. Published data on ecosystem variability over time and on ecosystem resilience under experimental stress are not yet sufficient for testing this hypothesis in the Baltic Sea.

The Baltic also has fewer species of commercial interest than the North Sea, with a lower average value per unit of catch weight. The recent catastrophic decline in Baltic cod catches vividly illustrates the vulnerability of a fishery with few alternative target species.

14.1 Introduction

The discussion of biodiversity has to a large extent centred on areas of particularly high biodiversity, on threats to that biodiversity, and on efforts at protecting it. Yet it is often studies of simple systems that help us understand complex phenomena. Thus, to understand the causes and effects of biodiversity and its loss, it may be useful to also study areas of intermediate and low natural biodiversity.

The Baltic Sea is an area of exceedingly low natural biodiversity. In the following, we discuss mainly species diversity, but also touch upon genetic diversity and diversity at high taxon level. We first present the Baltic biotia, and the causes of its poverty, and then discuss how its low biodiversity influences the way in which the Baltic ecosystem functions, as exemplified by an overview of its major carbon flows and its fisheries yield. Finally, we ask what light studies of the Baltic Sea ecosystem may throw on the general biodiversity debate.

14.2 Abiotic conditions in the Baltic Sea

The Baltic Sea (see map, Fig. 14.1(a)), excluding the Kattegat and the Danish Sounds, covers 373 000 km^2 and has a volume of 20 500 km^3. It is a semi-enclosed sea, that is connected to the Kattegat, and indirectly to the North Sea, only through the narrow Danish straits. Some 500 km^3 of seawater enters the Baltic each year through these straits, and about the same volume of freshwater reaches the Baltic each year from land, and as excess of precipitation over evaporation. The sum, roughly 1000 km^3 of brackish water, leaves each year, giving a water residence time on the order of 20 years.

The Baltic may be viewed as a huge, atidal estuary, with the salinity of the surface waters varying from 10 psu in the Arkona Sea, just inside the Sound, to about 7 psu in the northern Baltic proper, 5–6 psu in the intermediate Bothnian Sea and only 2–4 psu in the northernmost basin, the Bothnian Bay. (psu is the practical salinity unit, based on seawater conductivity, and has approximately the same value as the older, weight-based unit ‰S). The salinity of the surface water varies relatively little over time. During intermittent inflows of North Sea water the salinity of the bottom water in the Arkona Basin may reach over 20 psu (Voipio, 1981).

The inflowing seawater is denser than the brackish surface water of the Baltic, creating permanently stratified conditions in the Baltic proper.

In the main basin, a primary halocline is found at 60–80 m, and there are secondary ones at greater depths, where the bottom water is exchanged only intermittently, following large inflows of seawater of particularly high density. In recent decades, the deep, more saline waters of the Baltic proper have become largely devoid of oxygen, as a result of a combination of long periods without deep-water exchange and an increasing rate of oxygen consumption due to eutrophication (Elmgren, 1989).

14.3 Baltic biodiversity: Patterns and causes

It has long been known that the Baltic has fewer plant and animal species than fully marine areas. Lovén (1864) identified low salinity as the primary cause of this biological poverty. The effects of salinity on the Baltic biota were extensively reviewed by Remane & Schlieper (1971), who also described some of the proximate, physiological reasons why so many marine and freshwater species are excluded from the intermediate salinities of the Baltic Sea. Dahl (1956) discussed the main ultimate (evolutionary) cause of low biological diversity in brackish waters, i.e. that such waters depend on a delicate balance between the supply of seawater and freshwater, a balance unlikely to remain relatively constant long enough to allow a distinct, species-rich flora and fauna to evolve.

Pleistocene history has exacerbated the species poverty of the Baltic Sea. Glaciations forced the European brackish-water fauna through a series of long-distance migrations up and down the European coastline (Wolff, 1973), with a gradual loss of species as a result. Since the last glaciation, the Baltic has gone through several freshwater and marine stages (the freshwater Baltic Ice-Lake, the Yoldia Sea, the freshwater Ancylus Lake and the Littorina Sea). Each meant a drastic reorganisation of the biota. The last change from lake to sea occurred only about 7000 years ago, and the present brackish-water stage is only 2–3000 years old (Voipio, 1981).

The climate of the northern part of the Baltic Sea area is harsh. The Bothnian Bay is covered with a thick layer of ice for almost half the year, and in particularly cold years the entire Baltic Sea may be ice-covered. The severe climate is thought to exclude some brackish-water animals from the northern Baltic (Segerstråle, 1972). Scouring by ice prevents perennial plants from becoming established in shallow waters (Wallentinus, 1991).

The Baltic fauna and flora consist of a mixture of marine, brackish-

322   *Ecosystem function at low biodiversity in the Baltic*

Fig. 14.1. The Baltic Sea, showing (*a*) major subdivisions and (*b*) isohalines of the surface water and distribution limits for some common marine species in the Baltic Sea: A, the bivalve *Macoma balthica*; B, the blue mussel, *Mytilus edulis*; C, the isopod *Idotea balthica* (Pallas); D, the red alga *Furcellaria lumbricalis* (Huds.) Lamour.; E, bladderwrack, *Fucus vesiculosus*; F, plaice, *Pleuronectes platessa*; G, serrated wrack, *Fucus serratus* L.; H, the kelp *Laminaria saccharina* (L.) Lamour., I, shore crab, *Carcinus maenas* L.; J, periwinkle, *Littorina littorea* (L.); K, starfish, *Asterias rubens* L.; L, sea urchin *Strongylocentrotus droebachiensis* Müller; M, limpet, *Patella vulgata* L. From Wallentinus (1991).

water and freshwater species (Segerstråle, 1957; Wallentinus, 1991). Freshwater fish such as European perch, *Perca fluviatilis* L., can be caught in the same net as marine species such as cod, *Gadus morhua* L. (Fig. 14.2). The distribution of Baltic organisms reflects the north–south salinity gradient, with marine species gradually disappearing as one

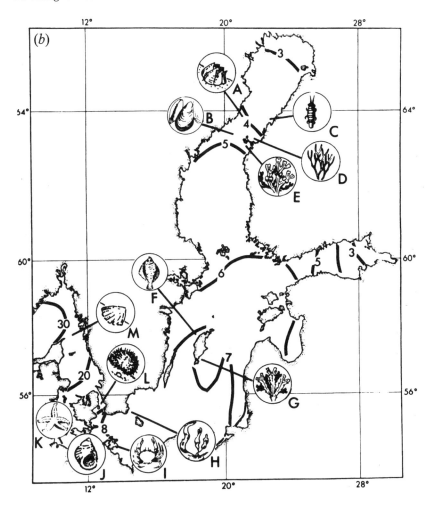

moves from the Swedish west coast to the southern Baltic and then northwards to the Bothnian Bay (Fig. 14.1(b)). Major marine groups such as coralline algae, kelps, radiolarians, echinoderms and elasmobranchs are absent from most of the Baltic Sea.

Freshwater taxa, such as insects, show a reduction in the opposite direction, from north to south, and their diversity in the Bothnian Bay, at 2–4 psu, is already much lower than in lakes. A few species typical of brackish water are common at intermediate salinities (Segerstråle, 1957). The resulting species diversity is low in comparison with fully marine or freshwater areas, as is generally the case in estuarine and brackish waters (Remane, 1934). The low diversity is evident both within

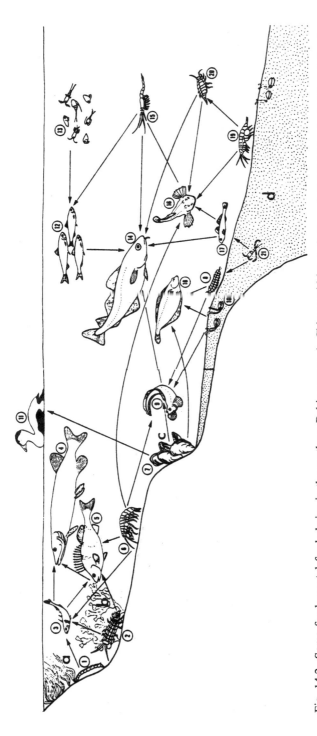

Fig. 14.2. Some fundamental food chains in the northern Baltic proper. 1, Chironomid larva; 2, Isopod; 3, Stickleback; 4, Pike; 5, Perch; 6, Amphipod; 7, Blue mussel; 8, Polynoid polychaete; 9, Viviparous blenny; 10, Baltic clam; 11, Eider duck; 12, Herring; 13, Plankton; 14, Cod; 15, Opossum shrimp; 16, Four horned sculpin; 17, Goby; 18, Flounder 19, Amphipod; 20, Isopod, *Saduria*; 21, Meiofauna; a, *Cladophora* belt; b, *Fucus* belt; c, Blue mussel belt; d, Soft bottoms. From Jansson (1972).

and between habitats, with Baltic organisms generally having wider realised niches than is normal in fully marine areas (Remane & Schlieper, 1971). There is a tendency for the reduction in species richness to be more drastic for macroscopic organisms, compared to microscopic organisms such as phytoplankton (Bernes, 1994) and benthic meiofauna.

Some 240 taxa of phytoplankton have been recorded from the southern Baltic and the Baltic proper (Willén, 1995), which is as high as about 6% of the total marine species pool of the world (approximately 3500–4500 species: Sournia *et al.*, 1991). The phytoplankton diversity declines to about 160 taxa in the Bothnian Bay.

The zooplankton communities in the Baltic consist of less than about 50 species, most of which are smaller than 1 mm in size (Voipio, 1981). Subsets of about 10 species account for most of the biomass and production in the Baltic proper and the Bothnian Bay (Hernroth & Ackefors, 1979; Kankaala, 1987).

Of about 180 species of macroalgae found in the Kattegat, less than half enter the Baltic proper, and only about a dozen reach the Bothnian Bay (Wallentinus, 1991). In most of the Baltic, the only large, belt-forming species is bladderwrack, *Fucus vesiculosus* L., which extends as far north as the northern Bothnian Sea. Eelgrass, *Zostera marina* L., is not found north of the Baltic proper. The Baltic wrack-belts and eelgrass beds support a comparatively high diversity of macroscopic animals. In the Bothnian Bay, most macrophytes are either annual freshwater phanerogams or stoneworts (Characeae).

The number of benthic macrofaunal species declines drastically from south to north in the Baltic, from about 1500 in the Skagerrak off the Swedish west coast, to about 150 in the Arkona area of the southern Baltic, 80 in the waters off Gotland and 50 in the northern Baltic proper (Voipio, 1981). On hard substrates in the Baltic proper, the filter-feeding blue mussel, *Mytilus edulis* L., forms large banks down to about 30 m depth, dominating the animal biomass (Kautsky, 1981). In the Bothnian Bay, where the salinity is as low as 2–4 psu, there are no mussels or other major filter feeders (Kautsky *et al.*, 1981; Elmgren *et al.*, 1984).

On soft-sediment bottoms, the species diversity of the benthic macrofauna is low. A single grab sample of 0.1 m^2, taken below the thermocline, normally contains about 6–7 species of macrobenthos in the northern Baltic proper, and only 1–2 in the Bothnian Bay (Elmgren *et al.*, 1984), whereas 40–50 species are commonly found in similar grab samples from the North Sea. The species diversity of the macrofauna on soft-sediment bottoms declines with depth in the Baltic proper (Ankar

& Elmgren, 1976), in contrast to the general pattern of an increase in diversity with depth in fully marine areas (Grassle & Maciolek, 1992). The diversity of meiofauna (animals less than 1 mm in size) is higher than that of the macrofauna, with up to 20 or more species in a few cubic centimetres of Baltic sediment (Elmgren, 1978, and unpublished results), but it is still lower than that in similar North Sea samples, where there may be more than 100 species of meiofauna.

The diversity of marine fishes decreases from 120 species in the North Sea to 69 in the Kiel Bay, 41 in the southern and central Baltic Sea, 20 in the Gulf of Bothnia and 6–10 in the Bothnian Bay (Voipio, 1981). Herring, *Clupea harengus* L. and sprat, *Sprattus sprattus* (L.), are found in the entire Baltic. Although cod can live in the Bothnian Bay, spawning is restricted to the deep, salty basins of the Baltic proper. Of the freshwater species, salmon, *Salmo salar* L., migrates from about 20 rivers along the whole coast to feed in the Baltic. The proportion of freshwater species is greatest in coastal areas, where perch and roach, *Rutilus rutilus* (L), are common, and in the Bothnian Bay (Voipio, 1981).

14.4 Does the Baltic Sea have endemic species?

Beginning with Linnaeus, a number of plant and animal species were first described from the Baltic Sea, but the overwhelming majority were later also found in other areas. A few are still known only from the Baltic Sea area, but those are from little-studied or taxonomically difficult groups, and are likely to represent poor knowledge rather than true endemism (Elmgren, 1984).

There are, however, several instances where genetically differentiated populations, or even physiological races, seem to be found in the Baltic (Wallentinus, 1991). Examples include bladderwrack (Bäck *et al.*, 1992), cod (Mork *et al.*, 1985), the bivalve *Macoma balthica* (L.) (Väinölä & Varvio, 1989), and the blue mussel (Johannesson *et al.*, 1990). The blue mussels of the Baltic have also been proposed to belong to a separate brackish-water species, *Mytilus trossulus* Gould, with wide distribution in brackish areas (McDonald & Koehn, 1988). On the other hand, Baltic herring, which on phenotypic grounds was long thought to be a distinct subspecies, *Clupea harengus harengus*, shows little real genetic differentiation from the Atlantic herring *Clupea harengus* (Ryman *et al.*, 1984).

14.5 Species gained and lost

The species diversity of the Baltic has increased in recent times, due to the successful invasion of foreign species from distant waters, probably in all cases helped by man. At least 30 immigrant species have become established (Leppäkoski, 1991, 1994; Jansson, 1994).

This influx may have started as far back as the Vikings (Petersen et al., 1992) and some immigrants have become important community members. Examples include the bivalve *Mya arenaria* L., the barnacle *Balanus improvisus* Darw., and the New Zealand snail *Potamopyrgus jenkinsi* (Smith). Some freshwater species are important in marginal water bodies of very low salinity, for instance the aquarium weed *Elodea canadensis* Michx., the zebra mussel, *Dreissena polymorpha* (Pall.), the American crayfish, *Orconectes limosus* (Raf.), and the freshwater mysid *Hemimysis anomala* G.O. Sars, which now dominates the food of fishes in the Couronian lagoon in Lithuania. Some are typically found in polluted conditions, e.g. the polychaete *Polydora redeki* Horst., and the bryozoan *Victorella pavida* Kent.

Most of the immigrants are found in shallow water, and only a few have established themselves below the thermocline of the Baltic, e.g. the north American polychaete *Marenzelleria viridis* Verril, which now dominates the benthos of the Vistula lagoon in Poland, and is spreading in the Baltic proper (Jansson, 1994).

Invasion by further new species is a real threat to the Baltic ecosystem as we know it today. There is a fear of large-scale changes should the filter-feeding zebra mussel, or its close relative the quagga mussel, *Dreissena bugensis* Andrusov, find its way to the Bothnian Bay, or should the American comb jelly, *Mnemiopsis leidyi* (A. Agassiz), spread via ballast water to the Baltic. In the Black Sea this comb jelly now occurs in enormous numbers and feeds on zooplankton and fish larvae (Mee, 1992). In contrast, few species have been totally lost to the Baltic ecosystem in historical times, but the sturgeon, *Acipenser sturio* L., may soon be gone (Witkowski, 1992).

14.6 Comparison of energy flows in the Baltic and North Seas

Elmgren (1984) summarised the patterns of energy flow in the Baltic Sea and its three main basins (the Baltic proper, the Bothnian Sea and the Bothnian Bay). Pelagic primary production is the main energy source, but riverine input of organic matter is almost as large in the oligotrophic

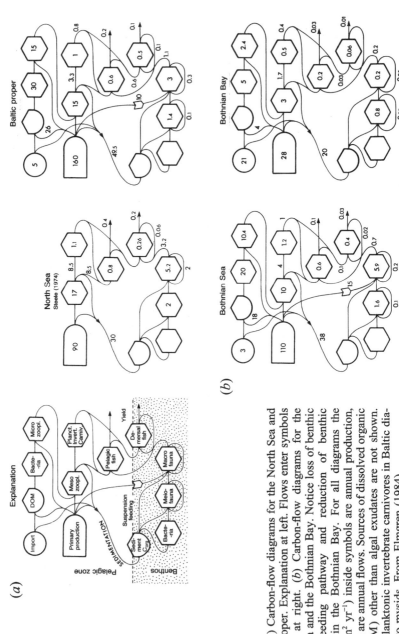

Fig. 14.3. (a) Carbon-flow diagrams for the North Sea and the Baltic proper. Explanation at left. Flows enter symbols at left, exit at right. (b) Carbon-flow diagrams for the Bothnian Sea and the Bothnian Bay. Notice loss of benthic suspension-feeding pathway and reduction of benthic macrofauna in the Bothnian Bay. For all diagrams the figures (gC/m² yr⁻¹) inside symbols are annual production, those outside are annual flows. Sources of dissolved organic matter (DOM) other than algal exudates are not shown. Figures for planktonic invertebrate carnivores in Baltic diagrams refer to mysids. From Elmgren (1984).

Bothnian Bay. Locally, in the coastal zone, benthic primary producers may also be important (Wallentinus, 1991). Primary production decreases along the salinity gradient from the Baltic proper to the Bothnian Bay, by a factor of about 4–6 (Elmgren, 1978).

There are four major routes for the organic carbon produced by phytoplankton: release of algal exudates that become available to bacteria; grazing by zooplankton; filter-feeding by zoobenthic animals such as mussels; and sedimentation to the benthic zone. Most of the carbon input is used up by the pelagic system, and the food chain based on algal exudates, pelagic bacteria and microzooplankton (e.g. ciliates) processes more carbon than that based on grazing of phytoplankton by larger zooplankton, even though the latter food chain leads more directly to pelagic fish.

Macrofauna account for most of the benthic biomass in the Baltic Sea, except in the Bothnian Bay, where meiofauna dominate the benthic biomass, in the absence of filter-feeding bivalves. Even though the Baltic is poor in species, biotic interactions can be quite complex (Ankar, 1977; Bonsdorff & Blomqvist, 1993). By aggregating species into functional groups, it is still possible to give an overview of carbon and energy flows.

Elmgren (1984) compared his estimated carbon flows for the three main basins of the Baltic Sea with Steele's (1974) model of carbon flows in the high-diversity North Sea ecosystem, and found striking similarity between the largest and southernmost Baltic basin, the Baltic proper and the North Sea (Fig. 14.3(a)). There were, however, also some differences, partly due to knowledge acquired since Steele formulated his model, and partly due to real differences between the systems. The North Sea model did not include the pelagic microheterotrophic food chain, which was insufficiently known in the 1970s. The Baltic calculations included not only open sea, but also coastal areas, where benthic suspension-feeders derive substantial amounts of energy from phytoplankton, a food chain link not included by Steele. Steele's estimate of phytoplankton primary production is a good deal lower than the estimate for the Baltic proper, but recent studies indicate that Steele's primary production estimate was too low (North Sea Task Force, 1993). In the North Sea, benthic invertebrate predators, such as echinoderms and large decapods, form a predation loop within the benthos, that is not present to the same degree in the Baltic.

Going north to the Bothnian Sea, there were still no principal differences in carbon flow from the North Sea model. The Bothnian Sea carbon flow chart indicated that the production of most trophic groups had diminished in rough proportionality to the lower phytoplankton primary production in this basin (Fig. 14.3(b)). The importance of benthic fauna had, however, increased, since, unlike the Baltic proper, the Bothnian Sea lacks extensive areas with oxygen deficiency.

In contrast, the carbon flow chart for the Bothnian Bay showed a radical re-structuring of the ecosystem, due mainly to the lack of a whole functional group, the benthic filter-feeders (Fig. 14.3(b)). In the absence of filter-feeding bivalves (e.g. blue mussels), there is almost no direct utilisation of phytoplankton by benthic filter-feeders, and a much smaller proportion of the primary production is used by macrobenthos. This leads to lower macrobenthos biomass and a biomass dominance by meiobenthos.

The Baltic Sea example thus suggests that as long as major functional groups are present, a simple, low-diversity ecosystem may be functionally almost equivalent to a high-diversity ecosystem, in terms of flows of organic carbon. Only the near-total loss of a major functional group resulted in drastic alterations of ecosystem processes. The Baltic/North Sea comparison is, however, as rough as the model calculations on which it is based, calculations that to some extent have been superseded by new research, and which do not take into account dynamic changes with time, the effect of fisheries, the eutrophication of the Baltic, fluctuations in climate and other complicating factors.

14.7 Effects of low biodiversity on natural resource

The low diversity of fish species in the Baltic affects the fishery and its commercial value. Fewer species are landed commercially in the Baltic, compared to the west coast. The main commercial species in the Baltic are herring, sprat, cod, eel, *Anguilla anguilla* (L.), salmon and freshwater fish, such as vendace, *Coregonus albula* L.

On the west coast, many more species are fished. Commercially important species include flatfishes such as witch flounder, *Glyptocephalus cynoglossus* (L.), common sole, *Solea solea* (L.) and plaice, *Pleuronectes platessa* L.; gadoids such as saithe, *Pollachius virens* (L.) and haddock, *Melanogrammus aeglefinus* (L.); and mackerel, *Scomber scrombus* L. There are also crustaceans of high commercial value: deepwater prawns, *Pandalus borealis* Krøyer, and Norway lobster, *Nephrops norvegicus* (L.).

The landed weight of the west-coast fishery in 1993 was about 38%

of the total Swedish catch of 336 kilotonnes, while that from the coasts off the Sound and the Baltic constituted about 12% (Anon., 1994). However, half of the Swedish fisheries yield was landed in Denmark, and consisted overwhelmingly of industrial fish (mostly herring and sprat) from the Baltic. This means that the actual yield of the Swedish fishery in the Sound and the Baltic in 1993 was larger than that of the west coast.

The greater diversity of the marine fish, crustaceans and molluscs on the Swedish west coast supports a fishery with almost double the value per unit catch. For instance, crustaceans and molluscs from the west coast, which constituted only 1% of the yield, accounted for 15% of the total commercial value of the Swedish fishery in 1993 (Anon., 1994). The recent catastrophic decline in Baltic cod stocks (Hammer et al., 1993) illustrates the vulnerability of a fishery with few alternative target species.

14.8 Does high diversity stabilise ecosystem function?

There are two opposing views about the importance of species diversity for ecosystem function. The first view holds that species in areas of high diversity are highly specialised and often co-evolved (Jackson, 1994) and that each species is of importance for ecosystem function. Every loss of a species is thought to affect ecosystem function, making it in some way less effective (Ehrlich & Wilson, 1991). According to the second view, many of the species present in high-diversity ecosystems may be redundant, as they carry out the same function, and the loss of some species is unimportant for ecosystem processes (Lawton, 1991; Walker, 1992).

The Baltic example suggests that we should not claim without good proof that high biodiversity is needed to maintain ecosystem function, even if it may be a boon to fisheries. The simple Baltic ecosystem, with its mixed origin and short history, is clearly not a co-evolved unit (cf. Jackson, 1994), yet functions similarly to the more diverse and co-evolved North Sea ecosystem. There is no scientific consensus on the relationship between the diversity of species within an ecosystem and its 'stability' (Pimm, 1984), and theoretical studies have even suggested that ecosystems with more species and a greater complexity should be less stable (May, 1972).

Nevertheless, the possibility remains that lowered diversity may destabilise ecosystem function, even before the loss of major functional groups. From an intuitive standpoint, one could expect that several species within each functional group will do a more reliable and less variable job than a single species. The risk of a single species being knocked out by a pathogen or an unusual climatic event must be larger than for a whole guild of species of similar function and different origins. Thus, we may perhaps expect more diverse systems to function in a less variable manner, given similar environmental variability, and to be more functionally resilient to perturbations (cf. Pimm, 1984).

For terrestrial ecosystems, both field studies and controlled experiments are now at hand in support of such a view. Experimental manipulations of species diversity in terrestrial ecosystems have shown that reducing species numbers affected ecosystem functions: species-poor ecosystems had a lower primary productivity, were more vulnerable to perturbation (drought) and took longer to recover (Naeem et al., 1994; Tilman & Downing, 1994). The simple ecosystems of the Baltic Sea would seem to be well suited for similar experiments in the aquatic environment, but this potential has not yet been exploited.

Can we judge whether the simple Baltic ecosystem is more variable than the more diverse North Sea system? Both areas have long series of observations, but in most cases they are not compatible, and lack information on external forcing functions that govern the recorded variability. At least one study reports co-variation between Baltic and North Sea communities, and indicates large-scale climatic cycles as a cause (Gray & Christie, 1983). In the Bothnian Bay there are regular fluctuations in populations of the benthic amphipod *Monoporeia affinis* (Lindström) (Andersin et al., 1978). However, there are also many reports of fluctuations in North Sea systems, e.g. the so-called Russell cycle in the abundance and species composition of zooplankton and pelagic fish in the English Channel, and changes in the relative abundances of plankton, fish and seabirds in the North Sea (Southward, 1980; Aebisher et al., 1990; Steele, 1991). Published analyses of natural ecosystem variability over time do not allow us to conclude whether the ecosystem of the Baltic is more or less variable than that of the North Sea.

As the Baltic is one of the most polluted seas in the world (Fitzmaurice, 1993), it is natural to wonder whether its low biodiversity influences its sensitivity to anthropogenic perturbations. Jernelöv & Rosenberg (1976) advanced the theory that Baltic ecosystems should be less sensitive to such stress than more diverse ecosystems on the Swedish west coast. Their main argument was that many of the dominant species in the Baltic are hardy, estuarine species that on the west coast are among the last to survive under polluted conditions. Elmgren (1984) countered by pointing out that at least some species in the Baltic, the so-called glacial-relict crustaceans, are in fact known to be quite sensitive to various pollutants. In addition, Tedengren & Kautsky (1986) argued that marine species in the Baltic are under greater osmotic stress than in the North Sea, due to the low salinity, and that this increases their susceptibility to other perturbations. Finally, as mentioned above, functional redundancy is low in the Baltic – loss or drastic reduction of a single species may mean the loss of a whole functional group, and thus greatly change ecosystem processes.

References

Aebischer, N.J., Coulson, J.C. & Colebrook, J.M., 1990. Parallel long-term trends across four marine trophic levels and weather. *Nature*, **347**, 753–5.

Andersin, A.-B., Lassig, J., Parkonen, L. & Sandler, H., 1978. Long-term fluctuations of the soft bottom macrofauna in the deep areas of the Gulf of Bothnia 1954–1974; with special reference to *Pontoporeia affinis* Lindström (Amphipoda). *Finnish Marine Research*, **244**, 137–44.

Ankar, S., 1977. The soft-bottom ecosystem of the northern Baltic proper, with special reference to the macrofauna. *Contributions from the Askö Laboratory, University of Stockholm*, **19**, 1–62.

Ankar, S. & Elmgren, R., 1976. The benthic macro- and meiofauna of the Askö-Landsort area. *Contributions from the Askö Laboratory, University of Stockholm*, **11**, 1–115.

Anon., 1994. Fiske 1993 – En översikt, Definitiva uppgifter. (*Swedish Sea-fisheries in 1993 – An Overview. Definitive Data*). Statistical Report, no. J55 SM 9402. Statistics Sweden, Publication Services, S-701 89 Örebro, Sweden. (In Swedish).

Bäck, S., Collins, J.C. & Russell, G., 1992. Effects of salinity on growth of Baltic and Atlantic *Fucus vesiculosus*. *British Phycological Journal*, **27**, 39–47.

Bernes, C. (ed.), 1994. Biological diversity in Sweden. A land study. *Swedish Environmental Protection Board, Monitor*, **14**, 1–280.

Bonsdorff, E. & Blomqvist, E.M., 1993. Biotic couplings on shallow water soft bottoms – examples from the northern Baltic Sea. *Oceanography and Marine Biology Annual Review*, **31**, 153–76.

Dahl, E., 1956. Ecological salinity boundaries in poikilohaline waters. *Oikos*, **7**, 1–21.

Elmgren, R., 1978. Structure and dynamics of Baltic benthos communities, with particular reference to the relationship between macro- and meiofauna. *Kieler Meeresforschungen*, **4**, 1–22.

Elmgren, R., 1984. Trophic dynamics in the enclosed, brackish Baltic Sea. *Rapports et Procés-verbaux des Réunions. Conseil International pour l'Exploration de la Mer*, **183**, 152–69.

Elmgren, R., 1989. Man's impact on the ecosystem of the Baltic Sea: Energy flows today and at the turn of the century. *Ambio*, **81**, 326–32.

Elmgren, R., Rosenberg, R., Andersin, A.-B., Evans, S., Kangas, P., Lassig, J., Leppäkoski, E. & Varmo, R., 1984. Benthic macro- and meiofauna in the Gulf of Bothnia. *Finnish Marine Research*, **250**, 3–18.
Ehrlich, P.R. & Wilson, E.O., 1991. Biodiversity studies: Science and policy. *Science*, **253**, 758–62.
Fitzmaurice, M., 1993. The new Helsinki convention on the protection of the marine environment of the Baltic Sea area. *Marine Pollution Bulletin*, **26**, 64–7.
Grassle, J.F. & Maciolek, N.J., 1992. Deep-sea species richness: Regional and local diversity estimates from quantitative bottom samples. *American Naturalist*, **139**, 313–41.
Gray, J.S. & Christie, H., 1983. Predicting long-term changes in marine benthic communities. *Marine Ecology Progress Series*, **13**, 87–94.
Hammer, M., Jansson, A.-M. & Jansson, B.-O., 1993. Diversity change and sustainability: Implications for fisheries. *Ambio*, **22**, 97–105.
Hernroth, L. & Ackefors, H., 1979. The zooplankton of the Baltic proper. A long-term investigation of the fauna, its biology and ecology. *Report from the Fisheries Board of Sweden, Institute of Marine Research*, **2**, 1–60.
Jackson, J.B.C., 1994. Community unity? *Science*, **264**, 1412–13.
Jansson, B.-O., 1972. Ecosystem approach to the Baltic problem. *Swedish Natural Science Research Council, Bulletins of the Ecological Research Committee*, **16**, 1–82.
Jansson, K., 1994. *Alien Species in the Marine Environment. Introductions to the Baltic Sea and the Swedish West Coast*. Report no. 4357. Solna: Swedish Environmental Protection Agency. (In Swedish).
Jernelöv, A. & Rosenberg, R., 1976. Stress tolerance of ecosystems. *Environmental Conservation*, **3**, 43–6.
Johannesson, K., Kautsky, N. & Tedengren, M., 1990. Genotypic and phenotypic differences between Baltic and North Sea populations of *Mytilus edulis* evaluated through reciprocal transplantations. II. Genetic variation. *Marine Ecology Progress Series*, **59**, 211–19.
Kankaala, P., 1987. Structure, dynamics and production of the mesozooplankton community in the Bothnian Bay, related to environmental factors. *International Revue der gesamten Hydrobiologie*, **72**, 121–46.
Kautsky, H., Widbom, B. & Wulff, F., 1981. Vegetation, macrofauna and benthic meiofauna in the phytal zone of the archipelago of Luleå – Bothnian Bay. *Ophelia*, **20**, 53–77.
Kautsky, N., 1981. On the trophic role of the blue mussel (*Mytilus edulis* L.) in a Baltic coastal ecosystem and the fate of the organic matter produced by the mussels. *Kieler Meeresforschungen*, **5**, 454–61.
Lawton, J.H., 1991. Are species useful? *Oikos*, **62**, 3–4.
Leppäkoski, E., 1991. Introduced species – Resource or threat in brackish-water seas? Examples from the Baltic and the Black Sea. *Marine Pollution Bulletin*, **23**, 219–23.
Leppäkoski, E., 1994. Non-indigenous species in the Baltic Sea. In *Introduced Species in European Coastal Waters*, ed. C.F. Boudouresque, F. Briand & C. Nolan, pp. 67–111. Ecosystems Research Report no. 8. European Commission EUR 15309 EN. Brussels: European Commission.
Lovén, S., 1864. Om Östersjön. Föredrag Skandinaviska Naturforskarsällskapets första offentliga möte 9 juli 1863. (*On the Baltic Sea.* Lecture at the first Public Meeting of the Scandinavian Association of Natural Scientists, 9 July, 1863). Stockholm: Norstedt & Söner. (In Swedish).

May, R.M., 1972. Will a large complex system be stable? *Nature*, **238**, 413–14.
McDonald, J.H. & Koehn, R.K., 1988. The mussels *Mytilus galloprovincialis* and *M. trossulus* on the Pacific coast of North America. *Marine Biology*, **99**, 111–18.
Mee, L.D., 1992. The Black Sea in crisis: A need for concerted international action. *Ambio*, **21**, 278–86.
Mork, J., Reuterwall, C., Ryman, N. & Ståhl, G., 1985. Genetic variation in Atlantic cod (*Gadus morhua*) throughout its range. *Canadian Journal of Fisheries and Aquatic Sciences*, **42**, 1580–7.
Naeem, S., Thompson, L.J., Lawler, S.P., Lawton, J.H. & Woodfin, R.M., 1994. Declining biodiversity can alter the performance of ecosystems. *Nature*, **368**, 734–7.
North Sea Task Force, 1993. *North Sea Quality Status Report 1993*. Oslo and Paris Commissions, London. Fredensborg, Denmark: Olsen & Olsen.
Petersen, K.S., Rasmussen, K.L., Heinemeier, J. & Rud, R., 1992. Clams before Columbus? *Nature*, **35**, 679.
Pimm, S.L., 1984. The complexity and stability of ecosystems. *Nature*, **307**, 321–6.
Remane, A., 1934. Die Brackwasserfauna (Mit besonderer Beruecksichtigung der Ostsee). *Zoologischer Anzeiger*, **7 (Supplementband)**, 34–74.
Remane, A. & Schlieper, C., 1971. *Biology of Brackish Water*, 2nd edn. *Die Binnengewässer, no.* **25**. Stuttgart: E. Schweizerbart'sche Verlagsbuchhandlung (Nägele u. Obermiller).
Ryman, N., Lagercrantz, U., Andersson, L., Chakraborty, R. & Rosenberg, R., 1984. Lack of correspondence between genetic and morphologic variability patterns in Atlantic herring (*Clupea harengus*). *Heredity*, **53**, 687–704.
Segerstråle, S.G., 1957. Baltic Sea. *Memoirs of the Geological Society of America*, **67**, 751–800.
Segerstråle, S.G., 1972. The distribution of some malacostracan crustaceans in the Baltic Sea in relation to the temperature factor. *Merentutkimuslaitoksen Julkaisu*, **237**, 13–26.
Sournia, A., Chrétiennot-Dinet, M.-J. & Ricard, M., 1991. Marine phytoplankton: How many species in the world ocean? *Journal of Plankton Research*, **13**, 1093–9.
Southward, A.J., 1980. The western English channel: An inconstant ecosystem? *Nature*, **285**, 361–6.
Steele, J.H., 1974. *The Structure of Marine Ecosystems*. Oxford: Blackwell Scientific Publications.
Steele, J.H., 1991. Marine functional diversity. *BioScience*, **41**, 470–4.
Tedengren, M. & Kautsky, N., 1986. Comparative study of the physiology and its probable effect on size in blue mussels (*Mytilus edulis* L.) from the North Sea and the northern Baltic proper. *Ophelia*, **25**, 147–55.
Tilman, D. & Downing, J.A., 1994. Biodiversity and stability in grasslands. *Nature*, **367**, 363–5.
Väinölä, R. & Varvio, S.-L., 1989. Biosystematics of *Macoma balthica* in northwestern Europe. In *Reproduction, Genetics and Distribution of Marine Organisms. Proceedings of the 23rd European Marine Biology Symposium*, ed. J.S. Ryland & P.A. Tyler, pp. 309–16. Fredensborg, Denmark: Olsen & Olsen.
Voipio, A. (ed.), 1981. *The Baltic Sea*. Amsterdam: Elsevier.
Walker, B.H., 1992. Biodiversity and ecological redundancy. *Conservation Biology*, **6**, 18–23.
Wallentinus, I., 1991. The Baltic Sea gradient. In *Intertidal and Littoral Ecosystems*

of the World, ed. A.C. Mathiesen & P.H. Nienhuis, pp. 83–108. Amsterdam: Elsevier.

Willén, T., 1995. Växtplankton i Östersjön 1979–1988. PMK – utsjöprogrammet. (Phytoplankton in the Baltic Sea 1979–1988. The Swedish Environmental Monitoring Programme – The Open Sea Programme). Report no. 4288. Solna: Swedish Environmental Protection Agency. (In Swedish).

Witkowski, A., 1992. Threats and protection of freshwater fishes in Poland. Netherlands Journal of Zoology, 42, 243–59.

Wolff, W.J., 1973. The estuary as habitat. An analysis of data on the soft-bottom macrofauna of the estuarine area of the rivers Rhine, Meuse and Scheldt. Zoologische Verhandelingen, 126, 1–242.

Chapter 15
Land–seascape diversity of the USA East Coast coastal zone with particular reference to estuaries

G.C. RAY, B.P. HAYDEN, M.G. McCORMICK-RAY and T.M. SMITH
Department of Environmental Sciences, University of Virginia, Charlottesville, VA 22903, USA

Abstract

We propose a landscape-pattern approach for interpreting the ecological function of coastal-zone biodiversity. Our model incorporates pattern from the ecosystem to the species levels and predicts that the decline or removal of a species or a set of species will not only directly influence the total biological diversity of the system, but, by removing ecological feedbacks, will also influence environmental conditions. That is, predicting biodiversity and faunal dynamics both requires that the response of the organism to the environment, and the modifications the organism may make to the environment, be made explicit at appropriate scales. The implication is that by removing the possible feedbacks between these two environmental conditions, it is possible that functions of the entire coastal-zone ecosystem will be altered in fundamental ways.

The coastal zone may be subdivided into a hierarchical structure that is reflected by faunal pattern. The largest spatial scale we will consider is that of the coastal zone as transitional between land and sea on a continental scale. Smaller scales, respectively, are coastal provinces, estuarine systems and the mosaic of habitats within those systems. We identify three benthic invertebrate assemblages and four estuarine-dependent fish assemblages by means of principal components analysis (PCA); these assemblages generally reflect the boundary conditions of the Virginian and Carolinian biogeographical provinces. We then describe estuarine function in terms of five principal components; these are assumed to play fundamental roles in determining the occurrence and position of various habitats, including oyster reefs. Oyster reefs provide 'keystone' functions and provide habitat for a large array of coastal fauna; also, we hypothesise, they exert major influences on estuarine function and coastal fauna.

'Ecosystem management' has been made urgent by the failures of single- or even multiple-species models and practices. For the coastal zone, the natural histories of species provide the functional linkages among differing scales of ecological interaction. Thus, the minimal additional data sets and models that are necessary are descriptions of the species' life histories in terms of their environmental relationships, determination of controls on boundary conditions, and quantitative, landscape-level descriptions of estuarine habitat. Ecosystem management may be unattainable until and unless these conditions are met.

15.1 Introduction

The principal purposes of this paper are to explore patterns of hierarchically scaled components of the USA East Coast coastal zone, and to relate these to functional relationships. We first examine faunal patterns of the shelf and estuaries. Next, we examine the morphometrics of estuaries and their possible influences on biotic communities. Finally, we describe the biogenic structures provided by oyster reefs and how these reefs act as habitats for portions of the estuarine biota and play an important role in the dynamics of estuaries. Much of our analysis for analysing the hierarchical structure and functional properties of the coastal zone depends on the multivariate statistical tool called principal component analysis (PCA). This is a useful method for providing insights into the structure of data, thus reaching conclusions that are not immediately obvious (Zitko, 1994).

The coastal zone includes the continental plains and shelves of the world. It may roughly be delimited by the rise and fall of sea level from the Pleistocene to the present (Fig. 15.1). This zone, which forms an ecotone between land and sea, may be conceived as a nested hierarchical system. The regional scale is that of biogeographical and physiological provinces (Hayden et al., 1984), distinguished by strong spatial and temporal gradients both along and across this zone (Ray & Hayden, 1992) and within the water column (Holligan & Boois, 1993). The mesoscale, where land and sea intimately interact, is represented by watersheds, estuaries, coastal islands, and lagoons. The smallest scale that we will consider here is that of interacting mosaics of habitats, for example, wetlands, hard and soft bottoms, and the water column; in our case, it is that of the oyster reef.

Our study concerns the USA East Coast Virginian and Carolinian provinces, within which occur among the world's most extensive estuar-

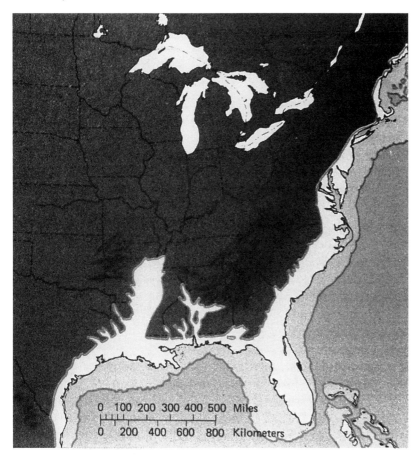

Fig. 15.1. The USA East and Gulf Coast coastal zone (courtesy Robert Dolan, University of Virginia). The dark grey, ocean tone represents areas deeper than 130 m that were unaffected by Pleistocene ice volumes. The light grey tone shows presently submerged areas that were exposed during low stands of the sea that accompanied maximum advances of the Pleistocene ice sheet. The lightest tone indicates land areas that would be submerged if the present continental ice sheets in Greenland and Antarctica were to melt, thereby raising sea level by about 65 m. The coastal-zone biota has adapted to these fluctuations (Ray, 1991).

ine systems (Fig. 15.2). The traditionally accepted boundaries for the Virginian Province are Cape Cod to Cape Hatteras and, for the Carolinian Province, Cape Hatteras to Cape Canaveral. These capes are significant points of deflection for major ocean currents, principally the warm, north-flowing Gulf Stream and the cold, south-flowing Labrador Current. At these capes, dramatic changes in such coastal characteristics as water temperatures and circulation patterns occur, and these physical features,

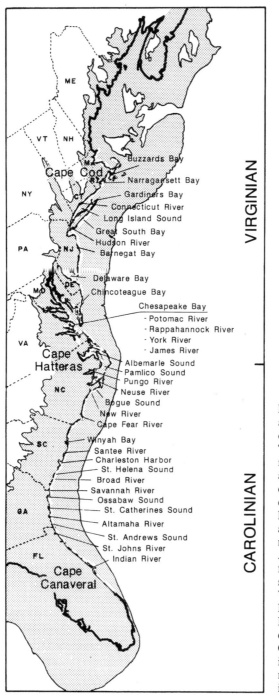

Fig. 15.2. The Virginian and Carolinian provinces of the USA East Coast. The capes that form major faunal boundaries are indicated, as are the estuaries that are considered in this chapter. From Cape Cod and the adjacent Georges Bank northward is the boreal Acadian Province, and from Cape Canaveral southward is the tropical West Indian Province. Note that Chesapeake Bay, by far the largest estuary on the USA East Coast, is not considered in its entirety; four of its major subestuaries are included in the analysis.

among others, play major roles in determining the ranges of the biota.

Despite these boundary conditions, the Virginian and Carolinian provinces share many features. They are, in common with temperate environments world-wide, transitional between high-latitude boreal and low-latitude tropical provinces. They possess relatively wide continental plains and shelves, and their coastal, sedimentary-plain environments are remarkable for the fact that 80–90% of their linear extents consist of estuaries and lagoons, some of which are among the world's largest (Emery, 1967). The major differences between these two provinces lie in climatic and oceanographical features, the sizes, shapes and hydrological features of their estuaries and lagoons, the relative proportions of habitat types, and their biotic patterns. Thus, these provinces are ideally suited for comparative study.

15.2 Concept

Our hypothetical model is presented in Fig. 15.3. This model expresses our concept that understanding biodiversity as a function of environmental conditions, as well as assessing faunal dynamics, requires that two components be explicitly considered: (a) the response of species to prevailing environmental conditions, including other members of the biotic community, and (b) how species may modify those conditions. This model inherently predicts that decline or removal of a species (or a set of species) will not only directly influence total biological diversity of the system, but also, by removing possible feedbacks, will influence environmental conditions. Such feedbacks have the potential to influence the ecological relationships of the coastal-zone system as a whole.

As Patrick (1983) has put it: 'The word *diversity* denotes variety or multiformity, a condition of being different in character and quality ... Diversity is the resultant of actions of the biota and the environment, and a cause of community properties'. As has repeatedly been observed, diversity exists at many scales, from genetic, to species, to population, and finally to ecosystems. We posit that analysis of the land- and seascape is a useful surrogate for all of these properties, and that coastal-zone biodiversity must be understood in the context of both large-scale, regional controls on biotic distributions and ecological feedback mechanisms. That is, the ability of a location to provide suitable habitat for a diversity of species is dependent on an array of physical and biotic features that vary at different temporal and spatial scales. At the largest scale, the limits of a species' range result from large-scale environmental

Levels of habitat description

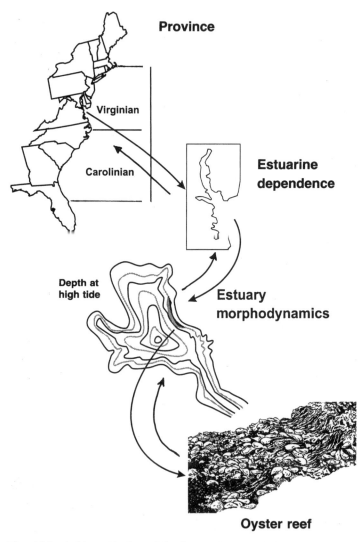

Fig. 15.3. A hierarchical model of coastal-zone relationships. See the text for explanation.

attributes, such as water-body characteristics, currents, and climate. At smaller scales, each species is dependent on the presence of particular habitat types, depending on the species' particular life history.

This concept is illustrated by the best-known of taxa, namely birds. Birds have lately been a special focus of attention due to the revelation that 71% of eastern neotropical migrants declined dramatically from 1978 to 1987 (Robbins *et al.*, 1989). Several inter-related factors have been advanced to explain these declines, among which are fragmentation of North American breeding habitat and loss of Central American wintering grounds (Terborgh, 1989). Determining the causes of the declines is inhibited by difficulties in relating bird habitat across broad geographical scales. These difficulties may be overcome by developing simulation models to integrate habitat classification with vegetation models that project changes in vegetation through time (Smith, 1986). That is, the habitat requirements of a species are viewed in a hierarchical framework, as follows. Firstly, at the continental scale, bioclimatic models relate the distributions of forest types to relevant climate and physical characteristics. Secondly, within a given forest region, the landscape-scale constraints on habitat may be evaluated by relating such features as forest fragmentation to the distributions of species. Thirdly, at the scale of within-forest patches, habitat may be defined on the basis of structural and compositional features of the vegetation.

Life histories of aquatic species are often very complex, requiring different habitats at differing developmental stages. As for birds, the overall *range* of the species is dependent on large-scale, physical conditions that may be related to province boundaries or other discontinuities. However, the *distribution* of the species (i.e. the apportionment of the species within its range) is constrained by small-scale habitat pattern. It follows that the conditions for suitable habitat at different scales may be described in terms of a hierarchical classification, where the 'potential' for a species' presence is dependent on a set of environmental and habitat requirements defined at those scales. The corollary to this concept is that biodiversity is not only a function of the suitability of the environment as habitat, but also that organisms can influence environmental conditions, and thus the habitats of other species, through feedback mechanisms. For estuarine species, feedbacks may include influences on sedimentation, on inputs of nutrients, or on provision of habitat for other species, especially where organisms, such as oysters, may form biogenic structures.

15.3 Patterns of diversity related to environment

Within our Virginian to Carolinian study area we examine three basic hierarchical levels. Firstly, at the province level, we identify invertebrate and fish assemblages to illustrate how the biota may be associated over the continental shelf and within estuaries. Secondly, at the regional scale, we draw some conclusions about the morphometrics of estuaries that may influence the suitability of these estuaries for various species depending on their different life histories. Thirdly, at the habitat scale, we review the functional significance of oyster reefs and hypothesise that these reefs play a 'keystone' role in estuarine systems.

For consistency we use PCA for the first two levels and to provide insights into the morphometric characteristics of estuaries that may influence biotic distributions. Our analysis of oyster reefs is descriptive and drawn from extensive literature.

15.3.1 Faunal patterns of the Virginian and Carolinian provinces

15.3.1.1 Invertebrates

Fifteen commercial species of molluscs and crustaceans (Table 15.1), whose ranges and life histories were mapped in Ray *et al.* (1980) and described by Odum *et al.* (1984), served as cases for PCA, in order to determine faunal patterns over the broad expanse of the continental shelf. Large-scale faunal patterns become evident when the ranges of species are analysed. These patterns may generally be attributed to physiographical boundary conditions. Benthic species were chosen by virtue of their sessile habit, relative to fishes.

The presence-absence of ranges for each of these species was scored for 341 half-degree latitude by half-degree longitude grid cells, which included all coastal estuaries and the total continental shelf to a depth of 200 m. Three statistically significant components were extracted from this 15 × 341 data matrix, explaining 70% of the total variance. The 341 factor scores for each of the three components were then mapped. Component I (33.2% of the total variance of the sample: Fig. 15.4(*a*)), based on the geography of its factor scores, is shown to be a 'subtropical to temperate inner shelf assemblage'. The strongest positive values are represented in the Carolinian Province, and these values are generally high over the inner shelf and within estuaries. There is a clear faunal break off Cape Cod. The component loadings are large on the following

Table 15.1. *Fifteen species of invertebrates used to define benthic faunal patterns for the USA East Coast continental shelf*

Species	Range within USA	Habitat type
1. Hard clam, *Mercenaria mercenaria*	Maine to Florida. Concentrations in Mid Atlantic Bight	Benthic; enclosed waters gravel to soft mud
2. Ocean quahog, *Arctica islandica*	Maine to Cape Hatteras in waters 20–70 m deep	Benthic, neritic, boreal
3. Soft clam, *Mya arenaria*	Maine to South Carolina but north of North Carolina	Benthic, enclosed waters, rocky gravel to soft mud
4. Surf clam, *Spisula solidissima*	Maine to Cape Hatteras. Concentrated Long Island to Cape Hatteras	Benthic intertidal to 130 m
5. Bay scallop, *Argopecten irradians*	Maine to Florida. Concentrated in Mid Atlantic Bight	Benthic, shallow subtidal, silt/hard sand enclosed waters
6. Callico scallop, *Argopecten gibbus*	Cape Hatteras to Florida. Concentrated at coastal prominences	Benthic, 2–370 m
7. Sea scallop, *Placopecten magellanicus*	Maine to Cape Hatteras. Concentrated in Mid Atlantic Bight	Benthic, sublittoral to 200 m; sand to gravelly rock
8. American oyster (Eastern oyster), *Crassostrea virginica*	Maine to Florida. Concentration in Mid Atlantic Bight historically	Benthic, enclosed waters on shell, rock, hard bottom
9. American lobster, *Homarus americanus*	Maine to Cape Hatteras. Concentrated Maine to Cape Cod	Benthic, neritic, 1–700 m, clay to rocky bottom
10. Spiny lobster, *Panulirus argus*	Cape Hatteras to Florida. Concentrated Cape Kennedy south	Benthic, shallow, neritic rocky/coral bottom
11. Blue crab, *Callinectes sapidus*	Maine to Florida. Concentrated Long Island to Florida	Benthic, enclosed waters, variable bottom
12. Brown shrimp, *Penaeus aztecus*	Massachusetts to Florida. Concentrated Cape Hatteras to Cape Kennedy	Benthic, enclosed waters/neritic, mud bottoms
13. White shrimp, *Penaeus setiferus*	Long Island Sound to Florida. Concentrated Cape Hatteras to Cape Kennedy	Benthic, enclosed waters/neritic, mud bottoms
14. Pink shrimp, *Penaeus duodarum*	Chesapeake Bay to Florida. Concentrated in south Florida	Benthic, enclosed waters/neritic, calcareous/mud bottoms
15. Royal red shrimp, *Pleoticus robustus*	South Carolina to Florida. Concentrated along continental slope	Benthic, neritic deep, 270–460 m

Invertebrate species used for the PCA analysis
Reference: Odum *et al.* (1984).

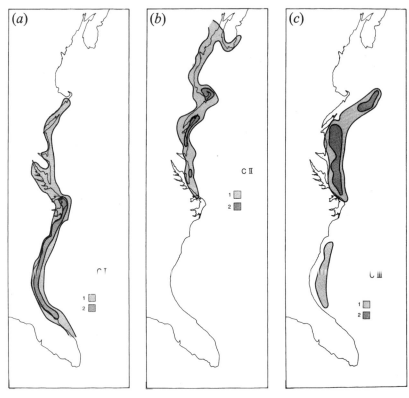

Fig. 15.4. Invertebrate range analysis. See the text for explanation.

subtropical species: penaeid shrimps and blue crab (*Callinectes sapidus*).

Component II (26.1% of variance: Fig. 15.4(*b*)) is shown by its mapped factor scores to be a 'boreal to temperate inner shelf assemblage', and is best represented in the Virginian Province, with a clear faunal break off Cape Hatteras. As with component I the strongest positive values are for the inner shelf and within estuaries. The component loadings are large on the following species, each of which is typical of cool-temperate latitudes: hard clam (*Mercenaria mercenaria*), soft clam (*Mya arenaria*) and American oyster (*Crassostrea virginica*).

Component III (10.8% of variance: Fig. 15.4(c)) is a 'temperate midshelf assemblage', and has the strongest positive values in the Virginian Province, with highest values over the mid- to outer-shelf. The ocean quahog (*Arctica islandica*), surf clam (*Spisula solidissima*) and sea scallop (*Placopecten magellanicus*) are the dominant species in this assemblage; they are generally known to be associated with temperate latitudes and mid-shelf areas. A weak southern portion of this assemblage is also shown on the Figure, being the result of species such as the calico scallop (*Argopecten gibbus*) and royal red shrimp (*Pleoticus robustus*) that occur abundantly on the Carolinian mid-shelf. The disjunct nature of this assemblage is probably an artefact of the few species in the data base.

These three components (assemblages and their geographies) generally illustrate both province-level and cross-shelf biogeographical characteristics. Features worth noting are that the cool-temperate Virginian and warm-temperate Carolinian provinces are transitional from boreal to tropical conditions and that the assemblages cover very large areas, indicative of the sizes of the biogeographical provinces involved and their limiting conditions. Estuarine relationships are not necessarily indicated, nor are they counter-indicated, by the species selected for this analysis or by the biogeographical results.

15.3.1.2 Fishes

An analysis specifically focusing on shelf–estuarine connections is the next step in our delineation of biogeographical pattern and to illustrate some degree of estuarine dependency of the coastal biota. For that task, we chose the best-known estuarine to marine faunal group, namely fishes.

In consultation with C. Richard Robins, then of the University of Miami, we earlier determined that 553 of the 1052 fishes listed in Robins & Ray (1986) were of Virginian and Carolinian distribution (Ray, 1991). Of these, we classified 148 as 'near-shore', 113 as 'near-shore' to 'continental shelf', and 27 as 'near-shore' to 'oceanic-shelf', for a total of 288 species that potentially may enter estuaries. For the present analysis, we have re-evaluated this earlier work to determine that 556 species (52.9%) occur in the Virginian and Carolinian biogeographical provinces. We then estimated that 292 or about half of these occur in nearshore to estuarine environments. Finally, we liberally estimated that 151, or about a fourth of all regional species, are *estuary-dependent*, that is, *obligate* to estuaries at some portion of their life histories (Table 15.2). Put another way, if estuaries were removed these species would be at risk of significant depletion, perhaps to the point of local or regional extirpation. Therefore, this subset of fishes was selected for our analysis so that we could delineate a biogeography indicating a direct connection between shelf and estuarine habitats.

Table 15.2. *Estuarine-dependent fishes of the USA East Coast*

Species	Life-history function on oyster reefs	PCA components I	II	III	IV
1. Blacknose shark, *Carcharinus acronotus*					
2. Bull shark, *Carcharinus leucas*	1				
3. Lemon shark, *Negaprion brevirostris*					
4. Atlantic sharpnose shark, *Rhizoprionodon terraenovae*					
5. Smooth hammerhead, *Sphyrna zygaena*				X	
6. Bonnethead, *Sphyrna tiburo*					
7. Smalltooth sawfish, *Pristis pectinata*					
8. Atlantic guitarfish, *Rhinobatos lentiginosus*					
9. Southern stingray, *Dasyatis americana*	1				
10. Atlantic stingray, *Dasyatis sabina*					
11. Bluntnose stingray, *Dasyatis say*	1				
12. Spiny butterfly ray, *Gymnura micrura*			X	X	
13. Smooth butterfly ray, *Gymnura altavela*					
14. Spotted eagle ray, *Aetobatis narinari*					
15. Bullnose ray, *Myliobatis freminvillei*			X	X	
16. Southern eagle ray, *Myliobatis goodei*				X	
17. Cownose ray, *Rhinoptera bonasus*					
18. Shortnose sturgeon, *Acipenser brevirostrum*	1,3,4		X		X
19. Atlantic sturgeon, *Acipenser oxyrhynchus*	1,3,4	X			
20. Ladyfish, *Elops saurus*					
21. American eel, *Anguilla rostrata*	1,2,4				X
22. Conger eel, *Conger oceanicus*		X			
23. Whip eel, *Bascanichthys scuticaris*					
24. Sooty eel, *Bascanichthys bascanium*					
25. Sailfin eel, *Letharchus velifer*					
26. Shrimp eel, *Ophichthus gomesi*				X	
27. Blueback herring, *Alosa aestivalis*	3,4,5		X		X
28. Alewife, *Alosa pseudoharengus*	3,4,5				
29. Hickory shad, *Alosa mediocris*	3,4,5	X			
30. American shad, *Alosa sapidissima*	5	X			
31. Yellowfin menhaden, *Brevoortia smithi*					
32. Atlantic menhaden, *Brevoortia tyrannus*	5				X
33. Atlantic herring, *Clupea harengus*	3,4,5				X
34. Atlantic thread herring, *Opisthonema oglinum*			X	X	
35. Spanish sardine, *Sardinella aurita*			X	X	
36. Striped anchovy, *Anchoa hepsetus*					
37. Bay anchovy, *Anchoa mitchilli*	3,4,5				X
38. Flat anchovy, *Anchoviella perfasciata*					
39. Gafftopsail catfish, *Bagre marinus*			X	X	
40. Hardhead catfish, *Arius felis*		X	X		
41. Oyster toadfish, *Opsanus tau*	1,2,3,4				

Table 15.2. continued.

Species	Life-history function on oyster reefs	I	II	III	IV
42. Skilletfish, *Gobiesox strumosus*	1,2,3,4,5				
43. Atlantic tomcod, *Microgadus tomcod*					X
44. Ballyhoo, *Hemiramphus brasiliensis*		X	X		
45. Halfbeak, *Hyporamphus unifasciatus*	1,2,3,4				X
46. Atlantic needlefish, *Strongylura marina*	1,2,3,4				X
47. Agujon, *Tylosurus acus*		X	X		
48. Sheepshead minnow, *Cyprinodon variegatus*	1,2,3,4	X	X		
49. Marsh killifish, *Fundulus confluentus*					
50. Mummichog, *Fundulus heteroclitus*	1,2,4		X		X
51. Spotfin killifish, *Fundulus luciae*		X			
52. Striped killifish, *Fundulus majalis*	1,2,3,4		X		X
53. Rainwater killifish, *Lucania parva*	1,2,3,4	X	X		
54. Mosquitofish, *Gambusia affinis*					
55. Sailfin molly, *Poecilia latipinna*					
56. Rough silverside, *Membras martinica*	1,2,3,4				
57. Inland silverside, *Menidia beryllina*	1,2,3,4	X		X	
58. Atlantic silverside, *Menidia menidia*	1,2,3,4		X		X
59. Fourspine stickleback, *Apeltes quadracus*		X			
60. Threespine stickleback, *Gasterosteus aculeatus*		X			
61. Lined seahorse, *Hippocampus erectus*	1,2,3,4				X
62. Northern pipefish, *Sygnathus fuscus*	1,2,3,4		X		X
63. Dusky pipefish, *Sygnathus floridae*	1,2,3,4				
64. Snook, *Centropomus undecimalis*				X	
65. White perch, *Morone americanus*	1,3,4				
66. Striped bass, *Morone saxatilis*	1,4,5		X		X
67. Sand perch, *Diplectrum formosum*					
68. Bluefish, *Pomatomus saltatrix*	1,5				X
69. Atlantic bumper, *Chloroscombrus chrysurus*		X	X		
70. Leatherjacket, *Oligoplites saurus*					X
71. Lookdown, *Selene vomer*					X
72. Atlantic moonfish, *Selene setapinnis*					X
73. Florida pompano, *Trachinotus carolinus*		X	X		
74. Permit, *Trachinotus falcatus*		X	X		
75. Gray snapper, *Lutjanus griseus*		X	X		
76. Spotfin mojarra, *Eucinostomus argenteus*					
77. Silver jenny, *Eucinostomus gula*		X	X		
78. Pigfish, *Orthopristis chrysoptera*	1				
79. Sheepshead, *Archosargus probatocephalus*	1,2,3,4				X
80. Pinfish, *Lagodon rhomboides*		X	X		
81. Spottail pinfish, *Diplodus holbrooki*					
82. Scup, *Stenotomus chrysops*	1,2,3,4				X
83. Silver perch, *Bairdiella chrysoura*					

Table 15.2. continued.

Species	Life-history function on oyster reefs	PCA components I	II	III	IV
84. Southern kingfish, *Menticirrhus americanus*	1				
85. Gulf kingfish, *Menticirrhus littoralis*					
86. Northern kingfish, *Menticirrhus saxatilis*	1	X		X	
87. Atlantic croaker, *Micropogonias undulatus*	1				
88. Black drum, *Pogonias chromis*	1				X
89. Spot, *Leiostomus xanthurus*	1,4,5				
90. Red drum, *Sciaenops ocellatus*	1,3,4	X		X	
91. Spotted seatrout, *Cynoscion nebulosus*	1,4,5				
92. Weakfish, *Cynoscion regalis*	1		X		X
93. Yellow chub, *Kyphosus incisor*		X		X	
94. Bermuda chub, *Kyphosus sectatrix*		X		X	
95. Spadefish, *Chaetodipterus faber*		X		X	
96. Slippery dick, *Halichoeres bivittatus*					
97. Pudding wife, *Halichoeres radiatus*					
98. Tautog, *Tautoga onitis*	1,2,3,4				
99. Cunner, *Tautogolabrus adspersus*					X
100. Emerald parrotfish, *Nicholsina usta*					
101. Striped mullet, *Mugil cephalus*	1,5				X
102. White mullet, *Mugil curema*		X		X	
103. Atlantic threadfin, *Polydactylus octonemus*					
104. Northern stargazer, *Astroscopus guttatus*					
105. Southern stargazer, *Astroscopus y-graecum*					
106. Striped blenny, *Chasmodes bosquianus*	1,2,3,4				
107. Feather blenny, *Hypsoblennius hentz*	1,2,3,4				
108. Rock gunnel, *Pholis gunnellus*					X
109. American sand lance, *Ammodytes hexapterus*		X			
110. Northern sand lance, *Ammodytes dubius*					X
111. Fat sleeper, *Dormitator maculatus*					
112. Darter goby, *Gobionellus boleosoma*					
113. Sharptail goby, *Gobionellus hastatus*					
114. Freshwater goby, *Gobionellus shufeldti*					
115. Emerald goby, *Gobionellus smaragdus*				X	
116. Marked goby, *Gobionellus stigmaticus*				X	
117. Lyre goby, *Evorthodus lyricus*					
118. Naked goby, *Gobiosoma bosci*	1,2,3,4				
119. Seaboard goby, *Gobiosoma ginsburgi*	1,2,3,4				
120. Green goby, *Microgobius thalassinus*	1,2,3,4				
121. Spanish mackerel, *Scomberomorus maculatus*					
122. Smoothhead scorpionfish, *Scorpaena calcarata*					
123. Spotted scorpionfish, *Scorpaena plumieri*					
124. Northern sea robin, *Prionotus carolinus*	1				X

Table 15.2. continued.

Species	Life-history function on oyster reefs	PCA components I	II	III	IV
125. Striped sea robin, *Prionotus evolans*			X		X
126. Leopard sea robin, *Prionotus scitulus*					
127. Bighead sea robin, *Prionotus tribulus*					
128. Inquiline snailfish, *Liparis inquilinus*		X			
129. Eyed flounder, *Bothus ocellatus*					
130. Spottail flounder, *Bothus robinsi*					
131. Three-eye flounder, *Ancylopsetta dilecta*					
132. Gulf flounder, *Paralichthys albigutta*					
133. Summer flounder, *Paralichthys dentatus*	1,5		X		X
134. Southern flounder, *Paralichthys lethostigma*					
135. Broad flounder, *Paralichthys squamilentus*					
136. Fourspot flounder, *Paralichthys oblongus*					X
137. Bay whiff, *Citharichthys spilopterus*					
138. Fringed flounder, *Etropus crossotus*					
139. Winter flounder, *Pseudopleuronectes americanus*	1,2,3,4				
140. Hogchoker, *Trinectes maculatus*			X	X	
141. Blackcheek tonguefish, *Symphurus plagiusa*					
142. Orange filefish, *Aluterus schoepfi*	1,2,3,4				X
143. Scrawled filefish, *Aluterus scriptus*					X
144. Fringed filefish, *Monacanthus ciliatus*					X
145. Planehead filefish, *Monacanthus hispidus*	1,2,3,4				X
146. Gray triggerfish, *Balistes capriscus*					X
147. Trunkfish, *Lactophrys trigonus*			X	X	
148. Bandtail puffer, *Sphoeroides spengleri*					
149. Checkered puffer, *Sphoeroides testudineus*					
150. Northern puffer, *Sphoeroides maculatus*	1,3,4		X		X
151. Striped burrfish, *Chilomycterus schoepfi*	1,2,3,4				

Names of fishes are from Robins & Ray (1986). The Chesapeake Bay estuarine-dependent fishes functionally associated with hard substrate were extracted from Myatt & Myatt (1990). Numbers following the names indicate functions on oyster reefs as follows: 1, feeding; 2, shelter; 3, spawning; 4, nursery; 5, behaviour (e.g. phototaxis, geotaxis, chemotaxis, thigmotaxis, rheotaxis, etc). An 'X' under each of the four components, I–IV, indicates positive vector scores of 0.8 or greater. See the text for further explanation.

To obtain an indication of the extent to which these estuary-dependent fishes occur over the shelf, we scored their presence or absence for each of the 20 1° latitude by 1° longitude grid cells shown in Fig. 15.5. We were careful to omit scoring occasional occurrences, or occurrences that

352   *Land–seascape diversity in USA East Coast estuaries*

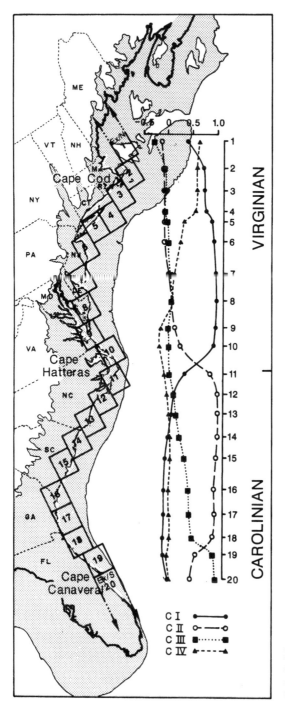

Fig. 15.5. Fish range analysis. Grid cells for analysis are indicated. See the text for explanation.

are not near-shore: e.g. of cold-adapted species that occur in shallow water northward, but only in deeper off-shore waters in the southern portions of their ranges. This matrix of 20 geographical grid cells by 151 estuarine-dependent species was then subjected to a PCA.

In Fig. 15.5 we have also plotted the component loadings for each of the first four components that were found to pass tests of statistical significance at the 95% level. Component I (41.5% of variance) is a 'Virginian assemblage' as shown by the graphed component loadings. An analysis of the species that belong to this assemblage reveals a strong boreal representation with a strong point of inflection at about Cape Hatteras. Component II (37.7% of variance) is a 'Carolinian assemblage' and the species with large factor scores are largely of temperate to subtropical distribution. The northern point of inflection is again near Cape Hatteras. Both components I and II confirm a regionalisation that has long been recognised. Component III (11.5% of variance) clearly indicates a tropical biogeography. Most species have more southerly ranges than is the case for component II; only a few species in this assemblage occur north of the Carolinas. Analogously, component IV (9.3% of variance) somewhat resembles component I, but is a boreal grouping with high component loadings in the Acadian Province.

Table 15.2 indicates the species that belong to each of these four assemblages. Several features are worth noting. Firstly, some species belong to more than one assemblage, as is common for PCA. Secondly, some families of fishes have species that are clearly associated; for example, all catfishes, pompanos, chubs, and filefishes are members of the same components. Thirdly, the analysis, while generally following the invertebrate analysis, implies that these biogeographies may apply as well to estuaries. That is, by contrast with the invertebrate analysis, we specifically selected species that are estuary-dependent so that the biogeography is a special one.

15.3.2 Estuarine morphometrics

The next step is to identify attributes of estuaries that might help define the habitat factors that reinforce the regional biogeographies that we have developed. Estuaries are bodies of water bordered by and partly cut off from the ocean by land-masses that were originally shaped by non-marine agencies (Emery & Stevenson, 1957). Those on our East Coast are generally perpendicular to the coastline and occupy the drowned mouths of submerged steam valleys that resulted from sea level

Table 15.3. *PCA component loadings for USA East Coast estuarine characteristics*

Estuarine attributes	C1	C2	C3	C4	C5
Estuary drainage area	0.66			0.40	
Area of tidal-fresh zone					0.40
Area of mixing zone	0.60				0.54
Area of seawater zone	0.41	0.55	0.41		0.42
Total estuary area	0.78	0.44			
Estuary length	0.53			0.62	
Average width	0.71			−0.46	
Minimum width					
Maximum width	0.68			−0.43	
Average depth			0.53		
High flow stratification		−0.68	0.42		
Low flow stratification		−0.65	0.40		
Freshwater discharge	0.67	−0.64			
Low freshwater flow	0.50	0.67			
50-year flood height	0.60	0.64			
100-year flood height	0.68	0.62			
Estuarine tidal prism	0.42		0.55	0.42	
Tidal range			0.77		

Component loadings with absolute magnitudes equal to or greater than 0.4 are displayed. See the text for further explanation.

rise over the Holocene period. With time, sedimentation, marine erosion and human activities have changed their original shapes and their influences on the shelf. It may safely be assumed that biotic patterns reflect this history.

The previous analyses have illustrated biogeographical patterns over the shelf and have inferred estuarine influences for the shelf biota. We now inquire how Virginian and Carolinian estuaries may differ and how individual estuaries compare with one another. To answer these questions, we relied on the NOAA (1985) report, which presented data on attributes for 95 major USA estuaries, including the 33 Virginian and Carolinian estuaries shown in Fig. 15.2. Table 15.3 lists 18 attributes from NOAA (1985) that may influence biotic pattern; Fig. 15.6 graphically illustrates these attributes and gives the average conditions for all 95 USA estuaries. However, examining each attribute separately would soon overwhelm our understnading of the natural history of the biota. What is required is a means to simplify the variables in a manner more meaningful to fishes. For that purpose we again turn to PCA.

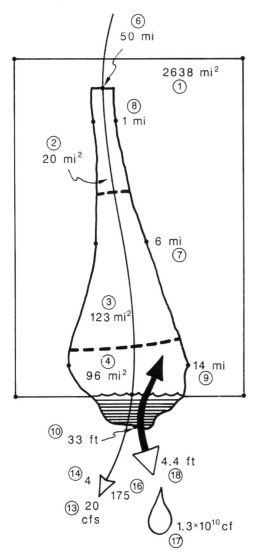

Fig. 15.6. Graphical representation of some major physical attributes of estuaries used in our analysis of estuarine morphometrics. The numbers are averages for major USA estuaries. The circled numbers refer to the following attributes: 1, estuarine drainage-system area; 2, tidal-freshwater area; 3, mixing-zone area; 4, seawater-zone area; (5 not on diagram), is the sum of 2, 3 and 4); 6, estuary length; 7, average estuary width; 8, minimum estuary width; 9, maximum estuary width; 10, average estuary depth; (11 not on diagram), 3-month high flow); (12 not on diagram), 3-month low flow); 13, long-term average daily rate of flow; 14, 7-day, 10-year low flow; (15 not on diagram), 50-year flood); 16, 100-year flood; 17, tidal prism; 18, phase range of tide. Definitions and data are from NOAA (1985). cf, cubic foot; cfs, cubic foot per second; mi, mile.

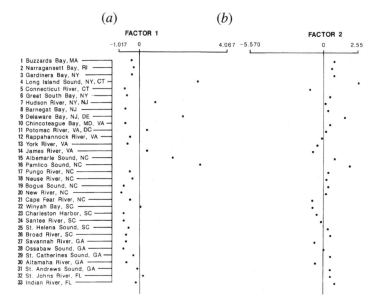
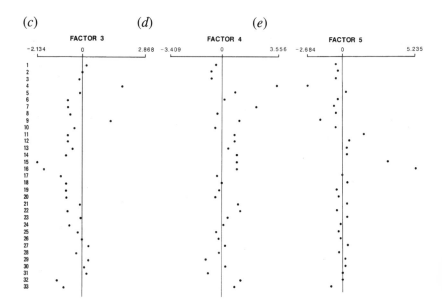

Fig. 15.7. Principal Component Analysis of the morphometrics of Virginian and Carolinian estuaries. The estuaries used in this analysis are identified on the left of the diagram. The numbers at either extreme of each graph are statistics from the PCA; a zero-line has been added for reference.

Therefore, our analyses of estuaries examines how the Virginian and Carolinian estuaries depart from the average conditions shown in Fig. 15.6. The complete data set for the analysis is the 18 morphometric attributes listed in Table 15.3 by the 95 estuarine cases for the USA as a whole. We conducted a varimax PCA analysis on this matrix. Figs 15.7(a)–(e) present the results of this analysis only for the 33 Virginian and Carolinian estuaries. The component loadings with absolute magnitudes equal to or greater than 0.4 are listed in Table 15.3. These component loadings are effectively the correlation coefficients among each of the estuarine variables studied and the component itself.

The first five components explained 27.4%, 19.5%, 11.9%, 9.3% and 6.0% of the total variance, giving a total of 74.1%. These components (= factors) characterise: (a) estuarine dimensions, (b) dominance of marine processes, (c) co-dominance of marine and freshwater processes, (d) fjord-like attributes, and (e) surface area, respectively. The first four components passed a 95% confidence level. The fifth component was of marginal significance.

The characteristics of each estuarine morphometric component and its geographical extent can be detailed, as follows. Component 1 estuaries for which high positive values are shown (Fig. 15.7(a)) have large total dimensions. They are long and wide with extensive catchment areas. Discharges into these estuaries are great while very low discharges are rare. Substantial floods are typical. Because of the large sizes of these estuaries, they also tend to have large tidal prisms. With few exceptions, these estuaries are limited to north of Cape Hatteras, i.e. the Virginian biogeographical province.

High positive values for component 2 (Fig. 15.7(b)) identifies large, embayed estuaries with extensive seawater zones. Fresh water and mixing zones are not extensive. These are poorly stratified estuaries, due to low freshwater discharges into the estuary from terrestrial regions. Low-discharge or drought conditions are typical, as are great floods. These estuaries, in their greatest development, are found in the North Atlantic Bight and are characterised by extensive flooded, river-valley systems. These conditions occur most extensively in the Virginian biogeographical province; Cape Cod Bay, Long Island Sound, Delaware Bay and Pamlico Sound are the prime examples. Had we not taken only individual subestuaries of the Chesapeake Bay system into our analysis (Potomac River to James River, as given in NOAA, 1985), rather than the bay in its full extent, the Chesapeake would also be of this type.

Component 3 estuaries (Fig. 15.7(c)) are marine-dominated, have large seawater zones, and are deep and well stratified. Tidal prisms and tidal ranges are both great. Most of the estuaries with high positive scores for this component are located north of Cape Cod where regional hardrock geology plays a key role in estuarine morphology. From a biogeographical perspective, estuaries that are examples of the component 3 type are characteristic of the boreal biogeographical province.

Component 4 estuaries (Fig. 15.7(d)) are long and narrow, fjord-like systems with large tidal prisms. Most component 4 estuaries occur in New England and to a lesser extent in the middle Atlantic states. The Hudson River and Long Island Sound are the type examples.

Component 5 estuaries (Fig. 15.7(e)) have large surface areas. These estuaries tend to have average values for the other estuarine variables selected for this study. Albemarle and Pamlico sounds are the best Atlantic-Coast examples of this type of estuary.

In this analysis we must keep in mind that the resultant components are not mutually exclusive; i.e. some estuaries may be intermediate or a combination of types. Nevertheless, we found that there is a pattern in estuarine morphometrics. We might make the case that geographically distinct estuarine morphometric types reflect a long-term history of geomorphic processes – coastal plain flooding, glacial outwash and fjord-like coasts – that are linked to long-standing climatic processes and marine processes. Thus, biogeographical and geomorphological connections are not unreasonable where a long history of associations is present. However, unlike our species biogeographical analyses, sharp regionalisations are not possible. To establish such relations, analysis at the continental rather than the regional scale would be more useful. In addition, the estuaries south of Cape Hatteras tend to fall near the grand mean of the 18 variables we selected to characterise the estuaries of the continental USA. Thus, the estuaries of the Carolinian biogeographical province are 'average', whereas Virginian estuaries are more varied.

15.3.3 A critical habitat – the oyster reef

Habitats of temperate estuaries include sea-grasses, marshes, soft bottoms, and hard bottoms. These habitats support discrete communities or assemblages of species. Two ecologically different animal assemblages associated with the marine benthos have been distinguished by Thorson (1957): the epifauna, which are associated with hard bottoms (such as rocks, stones, shells, vegetation, etc) and the infauna, which are

Table 15.4. *Oyster reef species diversity in Chesapeake Bay*

Phylum/common name	Scientific name	Source
Porifera: sponges		
Red beard sponge	*Microciona prolifera*	Hedeen (1986)
Burrowing sponges	*Cliona* spp.	Hedeen (1986)
Cnidaria		
Snail fur	*Hydractinia echinata*	Hedeen (1986)
Ghost anemone (bay)	*Diadumene leucolena*	Hedeen (1986)
Striped anemone	*Haliplanella luciae*	Hedeen (1986)
Sea nettle	*Chrysaora quinquecirrah*	Hedeen (1986)
Common jellyfish	*Aurelia aurita*	Hedeen (1986)
Comb jelly	*Mnemiopsis leidyi*	Hedeen (1986)
Sea gooseberry	*Pleurobrachia* sp.	Hedeen (1986)
Jug stopper (comb jelly)	*Beroe* sp.	Hedeen (1986)
Platyhelminthes: flatworms		
Trematode	*Bucephalus cuculus*	Hedeen (1986)
Oyster flat worm	*Stylochus ellipticus*	Newell & Breitburg (1992)
Bryozoa: moss animals		
Sea hair bryozoan	*Anguinella palmata*	Hedeen (1986)
Cushion moss bryozoan	*Victorella pavida*	Hedeen (1986)
Lacy crust bryozoan	*Conopeum tenuissimum*	Hedeen (1986)
Lacy crust bryozoan	*Electra crustulenta*	Newell & Breitburg (1992)
Red crusts	*Schizoporella unicornis*	MacKenzie (1983)
Coffin box	*Membranipora tenuis*	Newell & Breitburg (1992)
Annelida: segmented worms		
Clamworm	*Nereis succinea*	Hedeen (1986)
Bloodworm	*Glycera* spp.	Hedeen (1986)
Mud worm	*Polydora websteri*	
Oyster mud worm	*Polydora ligni* and spp.	Hedeen (1986)
		Newell & Breitburg (1992)
Fan worm	*Sabella microphthalma*	Hedeen (1986)
Limy tube worm	*Hydroides dianthus*	Hedeen (1986)
Mollusca		
American oyster	*Crassostrea virginica*	Hedeen (1986)
Pyramid shells (Gastropoda)	*Odostomia* spp.	Hedeen (1986)
Common oyster drill	*Urosalpinx cinerea*	Hedeen (1986)
		MacKenzie (1983)
Slipper shells	*Crepidula fornicata*	Gosner (1979); MacKenzie (1983); Hedeen (1986)
	Crepidula plana	MacKenzie (1983)
Hooked mussel	*Ishadium recurvum*	Hedeen (1986)
Bent mussel	*Brachidontes recurvum*	Hedeen (1986)
Giant whelk	*Busycon carica*	Hedeen (1986)
Oyster snail	*Odostomia impressa*	Hedeen (1986)

Table 15.4. continued.

Phylum/common name	Scientific name	Source
Soft shelled clam	*Mya arenaria*	Gosner (1979)
Arthropoda: Joint-legged		
Ivory barnacle	*Balanus improvisus*	Hedeen (1986)
Ivory barnacle	*Balanus eburneus*	MacKenzie (1983); Hedeen (1986)
Little grey barnacle	*Chthamalus fragilis*	MacKenzie (1983)
Amphipods (scuds)	*Corophium lacustre*	Newell & Breitburg (1992)
Grass or shore shrimp	*Palaemonetes purgio*	Gosner (1979); Newell & Breitburg (1992)
Blue crab	*Callinectes sapidus*	Newell & Breitburg (1992)
Rock crab	*Cancer irroratus*	MacKenzie (1983); Hedeen (1986)
Green crab	*Carcinus maenas*	Hedeen (1986)
Black-fingered mud crab	*Eurypanopeus depressus*	Newell & Breitburg (1992)
Black-fingered mud crab	*Panopeus herbstii*	Newell & Breitburg (1992)
Pea, Commensal, or oyster crab	*Pinnotheres ostreum*	Hedeen (1986)
White-fingered mud crab	*Rhithropanopeus harrisii*	Gosner (1979)
Echinodermata: spiny-skin		
Starfish	*Asterias forbesi*	MacKenzie (1983)
Chordata		
Sea squirt (sea grapes)	*Mogula manhattensis*	Hedeen (1986)
Oyster toad fish	*Opsanus tau*	Hedeen (1986)
Cow nose ray	*Rhinoptera bonasus*	Hedeen (1986)
Spot	*Leiostomus xanthurus*	Newell & Breitburg (1992)
Black drum	*Pogonias cromis*	Newell & Breitburg (1992)
Croaker (hard head)	*Micropogonias undulatus*	Hedeen (1986); Newell & Breitburg (1992)
Naked goby	*Gobiosoma bosci*	Newell & Breitburg (1992)
Striped bass	*Morone saxatilis*	Newell & Breitburg (1992)
Great blue heron	*Ardea herodias*	Hedeen (1986)
Oystercatcher	*Haematoptus palliatus*	Hedeen (1986)

The table lists predators, prey and associates. This list is illustrative and is not meant to be comprehensive. Many more species could be listed that use oyster reefs as habitat during portions of their life cycles.

associated with soft bottoms (such as sand or mud-surface layers). Of the two, the epifaunal assemblage reaches a worldwide maximum in shallow water, particularly in the intertidal zone. Here a multitude of microenvironments is created and the inhabitants are exposed to large salinity, temperature, and seasonal changes, factors that play important roles in the number and kinds of associated species that are present.

Karl Möbius observed in the mid 1800s that oysters form complex reef associations (Möbius, 1883). The Eastern oyster, *Crassostrea virginica*, is characteristic of the epifauna of the east coast of the USA. This species ranges from the Gulf of St Lawrence to the Caribbean Sea (Galtsoff, 1964; Abbe, 1992) and colonises hard mud and firm sandy bottoms in intertidal and subtidal areas of the mesohaline portions of estuaries. Over the greater extent of the Virginian and Carolinian provinces, oysters provide the only naturally occurring, hard, epifaunal substrate in an otherwise mobile benthic environment of sand and mud. These structures result from the interactions of biology, geology, and hydrology over varying scales of times, from days, months and years to decades and centuries. These structures also vary over the range of the oyster, with markedly different community associations produced by salinity gradients and seasonal changes (Wells, 1961). Oyster reefs are also important habitat and, in general, oyster reefs support increasing diversities of species from north to south. Tables 15.2 and 15.4 list estuarine-dependent Virginian and Carolinian species that occur at some life-history stage on oyster reefs. This is not to imply obligate dependency, as other substrates (artificial reefs, shells, debris or vegetation) may substitute for attachment. Further, the oyster reef's importance as habitat extends beyond its living period. Dead shells that the reef community produces accumulate over the benthos to stabilise the substrate, restrict infaunal habitat space and contribute to epizoan species richness in taphonomic feedbacks (Kidwell & Jablonski, 1983).

The location of oyster reefs in the marsh–estuarine ecosystem is more than accidental (Bahr & Lanier, 1981). Oysters show highest levels of reproductive productivity two years after major storm events and freshwater flooding (Wilber, 1992). Their association with estuaries has resulted in interdependencies that can influence estuarine development as much as estuarine development influences oyster reef-forming capacity, extent, and resiliency. The geometry of the estuarine basin is critical (Barnes, 1969; Stenzel, 1971; Keck *et al.*, 1973; Dame *et al.*, 1992) and dense populations are associated with large tidal stream meanders (Winslow, 1882; Stenzel, 1971; Keck *et al.*, 1973). In the Virginian Province, oyster reefs are mostly subtidal in 1–6 m depths, while in the Carolinian Province they are principally intertidal.

The survival of the oyster's prodigious production of larvae is dependent on current flows and the vagaries of the estuarine system; this provides the mechanism for ensuring their colonisation in the variable and competitive estuarine environment. An adult may release as many as 40 million eggs, of which 50% are fertilised and become larvae, 20–50 μ in length, which risk 98% mortality during their 3-week pelagic stage. The survivors settle as spat, but mortality may be as high as 99.3% during the 12-week, post-settlement period, at which time they become seed oysters. These seed oysters risk 90% mortality before becoming adults (R.I.E. Newell, personal communication). Another factor is that the larvae are thigmotactic during their settlement stage (Bayne, 1969), and settlement is best on oyster shell (Kenny *et al.*, 1990; Mann *et al.*, 1990). This allows successful oysters over generations of time to build a lattice-like edifice on the benthos for the attachment of spat and other substrate-dependent species.

This quintessential estuarine organism plays many functional roles in the estuarine system (Table 15.5). The edifice that the oyster builds and the community it attracts affects estuarine circulation patterns, tidal flow dynamics, the movements, dispersion, and retention of particles (e.g. sediment, plankton, nekton, and nutrients), nutrient exchange, as well as the settlement, shelter, and feeding of estuarine species. Oyster reefs act as nutrient traps (Dame & Libes, 1993) and exert a fundamental control on the benthic boundary layer (Denny, 1988; Wright *et al.*, 1992). This bottom roughness creates turbulence that extends into the water column to affect phytoplankton encounters important to benthic suspension feeders, as well as larval transport and settlement of organisms. Oysters also play a major role in maintaining water quality by means of their feeding behaviour (Newell, 1988), and as a by-product of their feeding they deposit organic material to the benthos to support the surrounding benthic community and to create sediment (Haven & Morales-Alamo, 1972). They release organic matter in the form of larvae that can feed pelagic, filter-feeding organisms that live in or migrate through the estuary. The reef community also releases nutrients in the form of ammonium ions, which are delivered in tidal channels to marshes (Childers *et al.*, 1993). The oysters themselves actively cleanse the water and themselves of silt through the rapid closure of their valves under conditions of high turbidity, depositing the silt to the benthos by this self-sifting mechanism (Lund, 1957). Finally, oyster reefs support a diverse number of taxa (Table 15.4), and help create and maintain the patterns and processes of estuarine and lagoon systems (Wells, 1961; Galtsoff, 1964; Newell, 1988; Newell & Breitburg, 1992; Childers *et al.*, 1993).

Table 15.5. *Functional roles of oysters in USA East Coast estuaries*

Oysters	Estuaries
Filtration capacity	High turnover rate potential of estuarine water
Nutrient links to other habitats	Releases POC and NH_4^+ and takes up N from marshes
Increases biodiversity	Provides increased niche space for ecological complexity that radiates upward through the system; supports stenohaline species along a salinity gradient; sustains epizoan diversity
Affects water flow patterns	Benthic boundary layer and water column hydrodynamics; particle movements (enhances feeding opportunities, sedimentation, estuarine flushing and particle dispersions)
Influences shoreline processes	Builds and erodes marshland in a continuum of change; buffers against moderate storms and wave action
Increases benthic productivity	Adds nutrients/sediment to benthos to feed demersal feeders
Contributes to estuarine land/seascape patterns	Affects marshland development and benthic infauna community
Seasonal pumping of carbon in form of eggs and larvae	Feeds filter-feeding organisms
Converts plants to useful organic and inorganic forms	Feeds on phytoplankton to build somatic tissue, reproductive tissues, shell, and faeces
A metabolic hot spot	High community metabolism
Dynamic interaction with physical environment	Builds and degenerates in a continuum of change
Contributes to estuarine resiliency	Forms meta-populations and communities as sources to restock disturbed areas; responds to storm events; contributes sediment to build benthic and shoreline habitat; dead shell stabilises benthos
Keeps the benthic water clear around oysters	Active shutting of valves at sustained high rate removes particles in water around oyster
Provides feeding stations	Seasonal migrators can find food, rest, or shelter in and out of estuary; visitors by day or by night to feed

See the text for references and explanation.

Much evidence suggests that oyster reefs are functionally important in the dynamics, energetics, and resiliency of near-shore, estuarine–marsh ecosystems. The reef's rôle in the evolutionary development of estuaries may also be critical. Bahr & Lanier (1981) and Dame et al. (1992) proposed that oyster reefs affect the physiography and hydrological regimes of salt marshes by modifying current velocities and changing sedimentation patterns. Bahr & Lanier also estimated reef-community, metabolic-energy consumption of 27 000 kcal m^{-2} yr^{-1} in a Georgian estuary, suggesting that the oyster reef community is a 'heterotrophic hot spot' in the marsh–estuarine ecosystem. It is also to be emphasised that oyster reefs are not static over time, but that their metapopulations (Hanski & Gilpin, 1991) may respond rapidly to change, existing in a continuum of adaptive development as storms and physical perturbations alter estuarine conditions.

In sum, the Eastern oyster appears to be a classic example of a 'keystone' species at the level of the ecosystem. Structurally and functionally the oyster (individually and the reef it builds) strongly influences species diversity and productivity at the local scale. As a structure, reefs provide habitats that sustain an abundant and diverse range of species. As metabolic 'hot spots', oyster reefs form centres of production and absorption of nutrients. Oyster reefs also contribute to estuarine resiliency and robustness, and serve as metacommunity habitats for species re-establishment after major physical disturbances. Alteration of these functional roles might have widespread consequences for migrating and estuarine-dependent species, as well as for the ecology of the coastal zone.

15.4 Discussion

In this chapter, we have presented examples of patterns of biotic, coastal-zone assemblages, the characteristics of USA east coast estuaries, and a description of the 'keystone' role of the American oyster, *Crassostrea virginica*. These descriptions are at different scales and use different sets of data. However, although the functional relationships among them, as expressed by our model (Fig. 15.3), seem logical, at present it is not possible to develop a quantitative or predictive model for the ecological function of biodiversity in the coastal zone. For example, all the ecological relationships of oysters have changed markedly from historical conditions, due to overfishing, disease, and other factors, suggesting a fundamental change in the Chesapeake Bay ecosystem (Newell, 1988).

Also, food web connectance in fluctuating ecosystems is presumed to be loose (Briand, 1983). Therefore, how the oyster's dramatic decline has affected overall biodiversity is unclear. Despite this caveat, it seems obvious that oyster reefs play a role in the life histories of a significant portion of estuarine species (Tables 15.2 and 15.4). At the very least, the oyster reef's recent demise may be expected to have had a marked effect on estuarine function (Table 15.5.). Without a hard substrate, epifaunal invertebrate species lack secure and dependable areas for attachment. Without reefs, some species of fishes will have reduced habitat for breeding, feeding, and shelter.

With respect to the estuary itself, without reefs the flows of currents will be swifter and less turbulent, which will have an effect on benthic stability, migration routes, sedimentation, turbidity, and shoreline erosion and repair. It is difficult to say just how and to what extent the loss of oyster reefs may have affected, or will affect, estuarine morphometrics. Unfortunately, many of the attributes of estuaries important as habitat of the biota are omitted by NOAA (1985), including small-scale habitat and estuarine physical properties, productivity, and community interactions. NOAA's data indicate large-scale properties that are only partially sufficient to predict the occurrence of species – size of the catchment, extent of the marine-water zone, extent of the tidal prism, fjord-like configuration, and size of the surface area. These attributes are useful for developing an estuarine classification, but it remains to be seen how well they relate to biogeography of the biota, as expressed by our species analyses. Nevertheless, the strong dependency of many coastal species on estuaries and the functional differences of the estuaries themselves lead to the conclusion that estuarine function is a major influence on species distributions, not only within the estuaries themselves, but also over the adjacent shelf.

Monaco *et al.* (1992) have also made an attempt to associate coastal species with estuarine attributes. They have identified groups of West Coast estuaries in the USA with purportedly similar fish assemblages, and have determined that the estuary mouth and area of the seawater zone are determining factors. Monaco & Lowery (1993) are in the process of a similar attempt for the east coast of the USA. The contention that 'estuaries with similar habitats and environmental regimes often support similar species assemblages' (Monaco *et al.*, 1992) seems reasonable, if tautological. However, their estuarine data (NOAA, 1985) are the same as we have used, with the same general problems that we have noted above. More problematically, the invertebrates and fishes these authors have used include many species that are not truly estuarine, but which merely occur in estuaries; these include, for the east coast, such shelf species as skates (*Raja*), pollock (*Pollachius*), and black sea bass (*Centropristis*).

Functional diversity has been defined by Steele (1991) as: 'the variety of different responses to environmental change, especially the diverse space and time scales with which organisms react to each other and to the environment'. In the case of fisheries, this form of biological diversity may be expressed by the large, apparently stochastic fluctuations in the relative abundances of commercial shelf fishes. Our concept for predicting patterns of biodiversity under changing environmental conditions attempts to respond to Steele's observation. We suggest feedbacks among organisms and the environment at several scales. We may conceive the distributions of species in shelf environments as metapopulation aggregates of their estuarine populations. It is possible, therefore, that alterations occurring in estuaries may have major influences on shelf biodiversity that, in turn, could have major consequences for community and ecosystem dynamics. The changes in biodiversity also affect ecological function. In this respect our coastal-zone concept resembles the Bormann & Likens (1979) view of forest ecosystems, which includes inanimate as well as animate processes and systems of controls, and which conceives a forest as a patchwork in which one patch's outputs become another's inputs. Such feedbacks are apparently common within the coastal zone. For example, the phenomenon of 'bootstrapping' allows for accumulation of nutrients, promotes build-up of biomass, and acts as 'a life-support system in harsh environments' that 'can be integral to the dynamics of complex communities' (Stone & Weisburd, 1992). An example of this may be the potentially significant annual inputs (155 kg ha^{-1}, wet weight) of marine-derived animal detritus that are deposited in freshwater systems by spawning clupeid fishes (*Alosa*) that range over the shelf (Garman, 1992). Removal of such large inputs may have a significant effect on nutrient availability. Additionally, the demise of habitats such as oyster reefs may lead directly to cascades of effects up through the hierarchy of subsystems. An example of this may be the suggestion of Dame *et al.* (1992) whereby oyster reefs are capable 'of acting synergistically with external forces to advance the ecological maturity of individual tidal creek systems along the geohydrological continuum'.

15.5 Conclusion

Clearly, there is no satisfactory explanation for functional effects of landscape-level, coastal-zone biodiversity. Nor can there be any one technique for determining estuarine–shelf interactions and species assemblages. However, there are several basic essentials. As Bartholomew (1986) has observed, an understanding of natural history is the fundamental link between levels of organisation. This requires that specific attention must be given to the specific life-history characteristics that 'connect' species range to habitat-scale distribution. With regard to the coastal zone, the species to be used for analysis should, to some demonstrable degree, be obligate on small-scale, estuarine-habitat attributes in order that meaningful relationships can be determined. Simple presence–absence is not a sufficient condition. Furthermore, integrated, multi-scale examinations are necessary to demonstrate to what extent species are influenced by, and may also influence, estuarine conditions.

Recent emphasis on 'ecosystem management' has been made urgent by the all-too-obvious failures of single- or even multiple-species models and practices. The US Marine Mammal Protection Act of 1972 had this intent in calling for 'optimum sustainable population', an ecosystem concept that has been as widely imitated as criticised. This basically requires establishing relationships among different scales of ecological interaction. For the coastal zone, where most commercial fishes and invertebrates occur, the minimal additional data sets and models that are necessary to advance this work include, firstly, descriptions of the dominant species' life histories in terms of their environmental relationships, focusing on particular life-history requirements relevant to estuaries and other enclosed, near-shore environments. Secondly, it will be necessary to determine the controls on boundary conditions among levels in the hierarchy, as expressed both by species assemblages and physical characteristics. Thirdly, the functional relationships must be established among species' distributions and functions of estuaries, including quantitative, landscape-level descriptions of estuarine habitat. For all data sets scale properties must be determined and positive feedbacks must be incorporated into the models. Until and unless these conditions are met, correlations among estuaries and coastal-zone biological diversity will not be meaningful and better understanding of integrated estuarine–shelf associations will not emerge. Most importantly, sustainable ecosystem management of coastal-zone species will continue to be based on too-uncertain ground.

Acknowledgements

The W.L. Lyons Brown Foundation of Louisville, KY, and the Munson and Henry Foundations of Washington, DC, provided the major support for this work. B.P. Hayden and M.G. McCormick-Ray were supported, in part, by the Division of Environmental Biology, National Science Foundation. Dr C.R. Robins, formerly Maytag Professor of Ichthyology, University of Miami, and presently Emeritus Curator, of the Natural History Museum, University of Kansas, provided essential advice on fishes. Robert Dolan, University of Virginia, kindly supplied Fig. 15.1. Tom Tartaglino of Charlottesville provided all graphics, with the exception of Fig. 15.3, which was the work of Robert L. Smith.

References

Abbe, G.R., 1992. Population structure of the eastern oyster, *Crassostrea virginica* (Gmelin, 1971), on two oyster bars in central Chesapeake Bay: Further changes associated with shell planting, recruitment and disease. *Journal of Shellfish Research*, **11**, 421–30.

Bahr, L.M. & Lanier, W.P., 1981. The ecology of intertidal oyster reefs of the South Atlantic coast: A community profile. Washington, DC: USA. Fish and Wildlife Service, Office of Biological Services, FWS/OBS-81/15.

Barnes, H., 1969. Some aspects of littoral ecology: The parameters of the environment, their measurement; competition, interactions, and productivity. *American Zoologist*, **9**, 271–7.

Bartholomew, G.A., 1986. The role of natural history in contemporary biology. *BioScience*, **36**, 324–9.

Bayne, B.L., 1969. The gregarious behaviour of the larvae of *Ostrea edulis* L. at settlement. *Journal of the Marine Biological Association of the United Kingdom*, **49**, 327–56.

Bormann, F.H. & Likens, G.E., 1979. *Pattern and Process in a Forested Ecosystem*. New York: Springer-Verlag.

Briand, F., 1983. Environmental control of food web structure. *Ecology*, **64**, 253–63.

Childers, D.L., McKellar, H.N., Dame, R.F., Sklar, F.H. & Blood, E.R., 1993. A dynamic nutrient budget of subsystem interactions in a salt marsh estuary. *Estuarine Coastal and Shelf Science*, **36**, 105–31.

Dame, R. & Libes, S., 1993. Oyster reefs and nutrient retention in tidal creeks. *Journal of Experimental Marine Biology and Ecology*, **171**, 251–8.

Dame, R., Childers, D. & Keopfler, E., 1992. A geohydrologic continuum theory for the spatial and temporal evolution of marsh-estuarine systems. *Netherlands Journal for Seas Research*, **30**, 63–72.

Denny, M.W., 1988. *Biology and the Mechanics of the Wave-Swept Environment*. Princeton, NJ: Princeton University Press.

Emery, K.O., 1967. Estuaries and lagoons in relation to continental shelves. In *Estuaries*, ed. G.H. Lauff, pp. 9–11. American Association for the Advancement of Science, publication no. 83. Washington, DC: AAAS.

Emery, K.O. & Stevenson R.E., 1957. Estuaries and lagoons. I. Physical and chemical characteristics. In *Treatise on Marine Ecology and Paleoecology*, vol. I, *Ecology*, ed. J.W. Hedgpeth, pp. 673–750. The Geological Society of America, Memoir no. 67, Washington, DC: The Geological Society of America.

Galtsoff, P.S., 1964. The American oyster *Crassostrea virginica* Gmelin. *Fisheries Bulletin of the U.S. Fish and Wildlife Service*, **64**, 1–480.

Garman, G.C., 1992. Fate and potential significance of post-spawning anadromous fish carcasses in an Atlantic coastal river. *Transactions of the American Fisheries Society*, **121**, 390–4.

Gosner, K.L., 1979. *Field Guide to the Atlantic Seashore*. The Perterson Field Guide Series. Boston, MA: Houghton Mifflin Co.

Hanski, I. & Gilpin, M., 1991. Metapopulation dynamics: Brief history and conceptual domain. *Biological Journal of the Linnean Society*, **42**, 3–16.

Haven, D.S. & Morales-Alamo, R., 1972. Biodeposition as a factor in sedimentation of fine suspended solids in estuaries. In *Environmental Framework of Coastal Plain Estuaries*, ed. B.W. Nelson, pp. 121–30. The Geological Society of America, Inc., Memoir no. 133. Washington, DC: The Geological Society of America.

Hayden, P.B., Ray, G.C. & Dolan, R., 1984. Classification of coastal and marine environments. *Environmental Conservation*, **11**, 199–207.

Hedeen, R.A., 1986. *The Oyster. The Life And Lore Of The Celebrated Bivalve*. Centreville, MD: Tidewater Publishers.

Holligan, P.M. & de Boois, H., 1993. *Land-Ocean Interactions in the Coastal Zone (LOICZ): Science Plan*. Global Change Report no. 25. Stockholm: International Geosphere-Biosphere Programme (IGBP).

Keck, R., Mauerer, D. & Watling, L., 1973. Tidal stream development and its effect on the distribution of the American oyster. *Hydrobiologia*, **42**, 369–79.

Kenny, P.D., Michener, W.K. & Allen, D.M., 1990. Spatial and temporal patterns of oyster settlement in a high salinity estuary. *Journal of Shellfish Research*, **9**, 329–39.

Kidwell, S.M. & Jablonski, D., 1983. Taphonomic feedback: Ecological consequences of shell accumulation. In *Biotic Interactions in Recent and Fossil Benthic Communities*, ed. M.J.S. Tueresz & P.L. McCall, pp. 195–248. New York and London: Plenum Press.

Lund, E.J., 1957. Self-silting by the oyster and its significance for sedimentation geology. *Publications of the Institute of Marine Science*, **4**, 320–7.

MacKenzie, C.L. Jr, 1983. To increase oyster production in the Northeastern United States. *Marine Fisheries Review*, **45**, 1–22.

Mann, R., Barber, B.J., Whitcomb, J.P. & Walker, K.S., 1990. Settlement of oysters *Crassostrea virginica* (Gmelin, 1791) on oyster shell, expanded shell and tire chips in the James River, Virginia. *Journal of Shellfish Research*, **9**, 173–5.

Möbius, K., 1883. The oyster. Report of the Commissioner, part VIII, Appendix H. U.S. Commission of Fish and Fisheries, Spencer F. Baird, Commissioner. Washington, DC: Government Printing Office.

Monaco, M.E. & Lowery, T.A., 1993. Comparative analysis of U.S. East Coast estuaries based on fish and invertebrate distributions and estuarine physical and hydrological characteristics. *Estuarine Research Federation Abstracts*, 12th International Estuarine Research Federation Conference: p. 85. (Abstract).

Monaco, M.E., Lowery, T.A. & Emmett, R.L., 1992. Assemblages of U.S. West Coast estuaries on the distribution of fishes. *Journal of Biogeography*, **19**, 251–67.
Myatt, E.N. & Myatt, D.O., 1990. A Study to Determine the Feasibility of Building Artificial Reefs in Maryland's Chesapeake Bay. Prepared for the Maryland Dept. Natural Resources Tidewater Administrations, Fisheries Division. Annapolis, MD. January.
Newell, R.I.E., 1988. Ecological changes in Chesapeake Bay: Are they the result of overharvesting the American oyster, *Crassostrea virginica*? In *Understanding the Estuary: Advances in Chesapeake Bay Research. Proceedings of a Conference, 29–31 March, 1988*, pp. 536–46. Baltimore, MD: Chesapeake Research Consortium Publication no. 129. CBP/TRS 24/88.
Newell, R.I.E. & Breitburg, D., 1992. Oyster Reefs, pp. 61–4, Chesapeake Bay Strategy for the Restoration and Protection of Ecologically Valuable Species by the Ecologically Valuable Species Work Group of the Living Resources Subcommittee of the Implementation Committee of the Chesapeake Bay Program, Annapolis, MD.
NOAA (National Oceanic and Atmospheric Administration), 1985. *National Estuarine Inventory: Data Atlas vol.* **1**: *Physical and Hydrologic Characteristics*. Washington, DC: NOAA/National Ocean Services (NOS) Strategic Assessment Branch.
Odum, W.E., Ticco, P. & McCormick-Ray, M.G., 1984. *East Coast Strategic Assessment Project: Living Marine Resources Data Compendium*. Washington, DC: NOAA/Strategic Assessment Branch.
Patrick, R., 1983. Introduction. In *Diversity*, ed. R. Patrick, pp. 1–5. Benchmark papers in Ecology vol. **13**. Stroudsburg, PA: Hutchinson Ross Publishing Co.
Ray, G.C., 1991. Coastal-zone biodiversity patterns. *BioScience*, **41**, 490–8.
Ray, G.C. & Hayden, B.P., 1992. Coastal zone ecotones. In *Landscape Boundaries*, ed. A.J. Hansen & F. DiCastri, pp. 403–20. New York: Springer-Verlag.
Ray, G.C., McCormick-Ray, M.G., Dobbin, J.A., Ehler, C.N. & Basta, D.J., 1980. *Eastern United States Coastal and Ocean Zones Data Atlas*. Washington, DC: NOAA.
Robins, C.R. & Ray, G.C., 1986. *A Field Guide to Atlantic Coast Fishes of North America*. Boston, MA: Houghton Mifflin Co.
Robbins, C.S., Sauer, J.R., Greenberg, R. & Droege, S., 1989. Population declines in North American birds that migrate to the neotropics. *Proceedings of the National Academy of Sciences of the USA*, **86**, 7658–62.
Smith, T.M., 1986. Habitat simulation models: Integrating habitat classification and forest simulation models. In *Modelling Habitat Relationships of Terrestrial Vertebrates*, ed. J. Verner, M.L. Morrison & C.J. Ralph, pp. 389–94. Madison, WI: University of Wisconsin.
Steele, J.H., 1991. Marine functional diversity. *BioScience*, **41**, 470–4.
Stenzel, H.B., 1971. Bivalvia. Mollusca. In *Treatise on Invertebrate Paleontology*, vol. **3**, Part N, ed. R.C. Moore, pp. N954–N1082. Lawrence, KS: The University of Kansas and the Geological Society of America.
Stone, L. & Weisburd, R.S.J., 1992. Positive feedback in aquatic ecosystems. *Trends in Ecology and Evolution*, **7**, 263–7.
Terborgh, J., 1989. *Where Have All the Birds Gone? Essays on the biology and conservation of birds that migrate to the American tropics*. Princeton, NJ: Princeton University Press.
Thorson, G., 1957. Bottom communities (sublittoral or shallow shelf). In *Treatise on Marine Ecology and Paleoecology*, ed. J.W. Hedgpeth, pp. 461–534. The

Geological Society of America Memoir no. 67. Washington, DC: The Geological Society of America.

Wells, H.W., 1961. The fauna of oyster beds, with special reference to the salinity factor. *Ecological Monographs*, **31**, 239–66.

Wilber, D.H., 1992. Associations between freshwater inflows and oyster productivity in Apalachicola Bay, Florida. *Estuarine, Coastal and Shelf Science*, **35**, 179–90.

Wright, L.D., Boon, J.D., Xu, J.P. & Kim, S.C., 1992. The bottom boundary layer of the bay stem plains environment of lower Chesapeake Bay. *Estuarine Coastal and Shelf Science*, **35**, 17–36.

Winslow, F., 1882. Methods and Results. Report of the Oyster Beds of the James River, Va and of Tangier and Pocomoke Sounds, Maryland and Virginia. U.S. Coast and Geodetic Survey. Washington, DC: Government Printing Office.

Zitko, V., 1994. Principal component analysis in the evaluation of environmental data. *Marine Pollution Bulletin*, **28**, 718–22.

Chapter 16
The development of mariculture and its implications for biodiversity

M.C.M. BEVERIDGE, L.G. ROSS and J.A. STEWART
Institute of Aquaculture, University of Stirling, Stirling, FK9 4LA, UK

Abstract

Mariculture has become an important source of seaweed, shellfish and fish, especially for human food, and production is likely to continue to expand well into the next century.

Mariculture has both direct and indirect impacts on biodiversity through the consumption of natural resources and the production of wastes. Natural resources such as land, water, seed and feed are required, consumption varying with intensity of production. Wastes, comprising uneaten food, faecal and urinary products, chemicals, pathogens and feral animals, are released into the environment, quantities also being largely dependent upon production methods.

Particular attention is drawn to threats to marine biodiversity posed by wetland destruction, use of chemotherapeutants and translocation of exotic plants and animals. Awareness of the impact of mariculture on biodiversity, however, is growing. Legislative, economic and market forces are beginning to persuade producers to reduce the impact through the adoption of a range of measures, including better site selection, husbandry, waste collection and treatment methods. Mariculture technology also offers the possibilities of reducing the impact on marine biodiversity caused by fishing and by entrapment of anthropogenic wastes from coastal conurbations.

16.1 Introduction

World population continues to increase faster than global food supply. While 99% of food comes from terrestrial agriculture (Pimentel & Giampietro, 1994), this disguises the fact that in many, especially developing, countries the bulk of animal protein comes from fish and other aquatic products. Aquatic foods have until recently been derived almost exclusively from capture fisheries sources. In recent years, however, aquaculture has been playing an increasingly important role. Mariculture can be defined as the farming of the marine environment. It is an economic activity that transforms natural resources through inputs of capital and labour into products valued by society. In so doing, wastes are inevitably produced. The impact of aquaculture on the environment and on biodiversity thus arises from these three processes: the consumption of resources, the aquaculture process itself and the production of wastes (Beveridge et al., 1994).

This chapter reviews trends in fisheries production and mariculture. In examining the impact of resource use, the aquaculture process and waste production on biodiversity, it compares mariculture with inland water aquaculture. Ranching and culture-based fisheries are excluded from all but the discussion. Little is said about biodiversity in marine environments per se, as this has been well-covered elsewhere (see Norse, 1993).

16.2 Mariculture and aquatic food production

Between the immediate post-world war period and the early 1970s, world fisheries (fish, shellfish, seaweed; captureculture) production grew at a rate of around 6% per annum. As yields of conventional fish stocks declined, principally through over-fishing, and as the number of new stocks that could be exploited dwindled, growth in production during the subsequent two decades slowed to around 2.5% per annum. In the 1990s production has more or less remained the same at just under 90 million tonnes (FAO, 1994). There is now a consensus that capture fisheries production has reached, or surpassed, sustainable exploitation levels. In its most recent analysis of the situation the Food and Agriculture Organisation (FAO) of the United Nations has concluded that most traditional marine fish stocks have reached full exploitation and that the use of any new method that increases catches will exacerbate over-fishing leading to a further decline in stocks (Christie, 1993; FAO, 1993). Indeed, 4 of the FAO's 17 major marine fishing areas are already classified as over-fished.

Statistics produced by the FAO show that world aquaculture production is currently around 25 million tonnes (FAO, 1996), equivalent to 20% of world fisheries (capture + culture) production by weight and around twice this by value. Production from the marine environment accounts for around 51% of aquaculture production by weight (53% by value) and is growing by some 5% per annum. While only 4% of farmed fish production comes from the sea, all farmed macroalgae, almost all farmed molluscs and more than 90% of farmed crustaceans are produced in the marine environment. The fastest growing sectors of mariculture are in high market value products such as shrimp and fish, production of the former having doubled over the past 5 years. By contrast, farmed production of aquatic plants and molluscs has increased only slowly.

While yields from capture fisheries are likely to remain at the same level or even decline, production from aquaculture is likely to become increasingly important. A conservative estimate is that by the year 2010 some 40 million tonnes, equivalent to 35% of aquatic food production, will come from aquaculture and that up to half of this may be from mariculture, an increase of 50% over present production.

16.3 Mariculture technology

A large and increasing range of plants and animals is being farmed in the sea. Unlike inland water aquaculture, mariculture involves the culture of plants, invertebrates and vertebrates not only for food but also for decoration (shells and pearls) and chemicals (alginates). Systems and methods for the most commonly grown tropical and temperate species are summarised in Table 16.1. The terms 'intensive', 'semi-intensive' and 'extensive' are used here with regard to inputs of food. In extensive aquaculture, the farmed organism is reliant on the environment for food or nutrients while in semi-intensive farming natural food is supplemented with additions of fertiliser and/or food, the latter usually being derived from agricultural by-products such as animal manures and rice bran. In intensive mariculture, all, or almost all, of the nutritional requirements are supplied by the farmer and diets are largely fishmeal based. There is also a correlation between intensity of production, as defined here, and energy consumption.

Unlike inland water aquaculture, where herbivorous/omnivorous species are largely grown using semi-intensive methods, mariculture is increasingly focused on the production of high-value carnivorous fish and crustacean species by intensive methods.

Almost all mariculture production is in sheltered areas of the coastal zone. In the tropics, macroalgae and molluscs are largely grown using longlines or rafts. Shrimp are reared in ponds close to the open sea or adjacent to estuaries and lagoons, while fish are produced both in ponds

Table 16.1. Summary of the principal rearing systems and methods employed in tropical and temperate mariculture

Group	Species	System	Method
Tropical			
Macroalgae	*Laminaria japonica* *Undaria pinnatifida* *Porphyra tenida* *Eucheuma* spp.	Beds, Stake-and-line, Rafts	Extensive
Molluscs	*Crassostrea* spp. *Mytilus* spp. *Pecten yessoensis* *Venerupis japonica* *Solen* spp.	Suspended (rafts, longlines)	Extensive
Crustaceans	*Scylla serata* *Penaeus* spp.	Ponds	Semi-intensive Intensive
Finfish	*Chanos chanos* *Mugil* spp. *Epinephelus* spp. Serranidae *Pagrus major* *Seriola quinqueradiata*	Land-based (ponds) and Water-based (cages)	Intensive
Temperate			
Macroalgae	*Gracilaria*	Beds, rafts	Extensive
Molluscs	*Ostrea edulis* *Crassostrea* spp. *Mytilus* spp. *Tapes* spp.	Bottom (tressels, trays) and Suspended (rafts, longlines)	Extensive
Finfish	*Salmo salar* *Oncorhynchus* spp. *Dicentrarchus labrax*	Water-based (cages) and Land-based (tanks)	Intensive

and, increasingly, in cages. In temperate countries, where mollusc and finfish culture predominates, most rearing systems (cages, rafts and longlines) are located in the sea.

16.4 Impacts of mariculture

There have been many reviews of the environmental impacts of aquaculture (Beveridge, 1984; Gowen & Bradbury, 1987; Beveridge et al., 1991, 1994; Pullin et al., 1993).

16.4.1 Resources

Aquaculture requires land (or areas of the sea-bed), construction materials, water, seed (eggs, fry, etc) and feed or nutrients (Beveridge et al., 1994). Demands differ quantitatively and qualitatively between coastal and inland aquaculture.

16.4.1.1 Land

Inland aquaculture is usually located in areas of agricultural land where the development of ponds can increase habitat diversity (Beveridge et al., 1993). By contrast, tropical marine shrimp and fish culture typically uses previously unexploited areas of land for pond construction. In a study of shrimp farms in Sri Lanka, for example, Jayasinghe & De Silva (1989, cited in Phillips et al., 1993) found that 19% of the area occupied had been agricultural land, while 29% had been bare land and 35% had been wetlands (marshlands and mangroves). Demands for land for shrimp pond construction in Latin America (Ecuador and Colombia) and South-East Asia (China, Thailand, Indonesia and the Philippines) have caused the irreversible destruction of thousands of hectares of mangrove (Fig. 16.1). In the Philippines alone, the area of coastal ponds increased from 61 000 ha in 1941 to 210 000 ha in 1987 (Primavera, 1991, 1993). Sixty per cent of the total reduction in mangrove area in the country is attributable to aquaculture (Primavera, 1991).

It is not just the coastal land that has been converted into ponds that must be considered, but also areas destroyed for roads, dykes and supply canals. Areas of land adjacent to ponds are also often subject to salination, adversely affecting agricultural activities. In ecosystem support terms, it has been calculated that salmon farming requires an area of the sea surface some 40–50 000 times larger than that occupied by the cages in order to supply the feed required (Folke & Kautsky, 1992) while Larsson et al. (1994) have estimated that the area required to support semi-intensive shrimp farming, in terms of food and post-larvae, is around 35–190 times that of the pond area.

Fig. 16.1. Shrimp farm, Sumatra. Mangrove area has been fenced off and cleared for pond and infrastructure construction.

16.4.1.2 Water

Aquaculture uses very large quantities of water, use being dependent upon species and intensity of the operation (Phillips & Beveridge, 1991; Beveridge et al., 1993). Estimates of water used in semi-intensive and intensive shrimp culture, for example, are around 11 000–21 000 m^3 t^{-1} and 29 000–43 000 m^3 t^{-1}, respectively (Phillips et al., 1993). Requirements for salmonids are typically even higher. In pond-based systems surface water is not just borrowed to be returned in a more degraded form but is actually consumed, being lost through seepage and evaporation. In the tropics, such losses can amount to several hundred cubic metres of water per hectare of pond area per day. While losses of seawater on this scale may be unimportant, excessive pumping of groundwater has led to the salination of soils and groundwater and land subsidence (Barg, 1992; Phillips et al., 1993).

16.4.1.3 Seed and broodstock

Although commercially important species such as yellowtail (*Seriola quinqueradiata*) and milkfish (*Chanos chanos* Forskal) can now be produced in hatcheries, with the exception of salmonids, marine fish culture is still largely dependent upon the supply of seed (eggs, larvae, fry and juveniles) from the wild, largely because the economics of production do not justify the development of commercial hatcheries. Shrimp culture, too, remains reliant upon the wild for supplies of post-larvae or gravid broodstock. Molluscs are farmed using both wild-caught spat and hatchery-reared animals, depending on species and circumstances.

The dependence on wild stock in mariculture contrasts sharply with farming in freshwater where most species are reared throughout their life-cycle in captivity. This is desirable from the farmer's point of view since it facilitates selection for economically important traits and is a prerequisite in the manipulation of sex and in the development of year-round supplies of seed. There have been few studies of the impact of demands for seed on wild populations. In West Bengal, it has been estimated that some 64–99% of the fry caught to supply the shrimp fry market, equivalent to some 3×10^9 fish and crustacean fry belonging to 60 species, is by-catch (i.e. non-target animals, often discarded: Banerjee, 1993). Similar observations have been made in the Philippines with regard to the collection of milkfish fry. No studies have been carried out to discover what impact such activities have on biodiversity. While fisheries yields in West Bengal prior to the trade in shrimp post-larvae were higher than today, it is impossible to relate this directly to the development of the industry given other changes in fishing pressure and gear size and environmental degradation.

The impact of mariculture on wild stocks is not necessarily negative. Mollusc farming is likely to result in net local increases in spat availability, although the impact on genetic composition and fitness remain unknown.

16.4.1.4 Feed

As stated above, most marine culture of fish and shrimp is intensive and is dependent upon fishmeal-based diets. Dietary fishmeal and lipid levels used in commercial, compound marine fish feeds (50% protein and 12% lipid) are generally higher than for intensively managed freshwater fish species (Chamberlain, 1993). Tacon (1994) has estimated that in 1990, between 816 and 873 000 tonnes of fishmeal and between 190 and 205 000 tonnes of fish oil were used in aquaculture feeds, equivalent to 13–15% of the total world supply of fishmeal and fish oil. Using data supplied to Tacon (1994) by the International Fishmeal and Oil Manufacturers Association, mariculture appears to have accounted for almost 70% of this value. Moreover, Chamberlain (1993) anticipates that fishmeal use in aquaculture will increase by 50% by the year 2000 so that, assuming fishmeal supplies remain constant, it will become the second biggest consumer after the poultry industry.

Industrial fishing to supply the market for fishmeal is part of the problem associated with fisheries in general and has been specifically implicated in the decline in sand eel (Ammodytidae) stocks and the consequent disruption to bird and mammal populations in the North Sea (Monaghan, 1992).

Extensive aquaculture of molluscs and seaweed, if carried out on a sufficiently large scale, can also affect food-web structure and function. Studies in China, for example, have shown that nutrient levels in the vicinity of seaweed production areas are lower than expected (Phillips, 1990). Mussels and other molluscs remove nitrogen from the water column (e.g. Kaspar *et al.*, 1985). In areas of intense farming activity, such as the Ria de Arosa in northern Spain, filtering activity is so intense that mussels have reduced phytoplankton diversity and replaced copepods as the main pelagic grazing organisms (Tenore *et al.*, 1985, cited in GESAMP, 1991; Barg, 1992).

16.4.2 The aquaculture process

Irrespective of production method, the establishment of a mariculture operation where there had not been one before can cause impacts not only on scenic value, but also on biodiversity. Cage or mussel farming structures can act as fish attractant devices (FADs) or substrates for other species, while the stocks of fish, shellfish and feed held attract predatory or scavenging fish, birds and mammals (Beveridge, 1984, 1988, 1996; Carss, 1989; NCC, 1989; Beveridge *et al.*, 1994). In some instances, immigrants may displace resident populations, as has been claimed with regard to bird populations in some areas of Scotland (NCC, 1989). Losses of farmed stock to predators can be economically significant (Pemberton, 1989; Ruggeberg & Booth, 1989; Pemberton & Shaughnessy, 1993), and the measures used as deterrents often lead to death, either deliberate (shooting) or accidental (entanglement) (Beveridge, 1988; Ross, 1988). It has also been suggested that increased levels of activity around farms in isolated areas may disrupt species sensitive to the presence of humans and noise (NCC, 1989). Quantitative assessments of impacts on flora and fauna, however, have not been made.

16.4.3 Wastes

What is stocked into an aquaculture system or given to farmed animals in the way of food and not removed when they are harvested must be regarded as waste. Thus, not only uneaten food, faecal and urinary wastes, but also chemicals used principally to control disease and microorganisms and animals that escape from the farm must be considered as wastes.

16.4.3.1 Uneaten food, faecal and urinary wastes

Losses vary with species and system, food conversion ratios being a good indicator of losses. Direct measurements, using video recording techniques, suggest that in intensive marine salmon cage farms, some 5–10% of food is typically not eaten (Juell, 1991), although in poorly run farms the figures may be much higher (Beveridge et al., 1991). Losses from marine cages, where rapid dispersal of food by currents outside the reach of the fish can be a problem, are generally higher than from ponds or tanks or from freshwater cage systems (Beveridge et al., 1991). Faecal production from fish and crustaceans is around 230–400 g kg^{-1} food ingested (Beveridge et al., 1991). Much of the solid material remains trapped within pond systems until harvesting when water levels are lowered. As the solid uneaten food and faecal wastes from cages fall towards the bottom, they break up, increasing the area of the sea-bed over which they disperse and the rate of solubilisation of materials into the water column. The major source of dissolved nitrogenous waste is excretory products. Depending on species, dietary protein content and environmental conditions, urinary nitrogen can account for up to 60% of the nitrogen released into the environment by intensive fish production (Barg, 1992).

There have been many studies of fish and shrimp farm effluents and their impacts (Gowen & Bradbury, 1987; Foy & Rossell, 1991a,b; Weston, 1991; Ziemann et al., 1992; Phillips et al., 1993; Briggs & Funge-Smith, 1994). Around cage farms, hypernutrification of the water column is sometimes apparent, although given the degree of water movement at most coastal sites, these are generally only temporary and are apparent principally during periods of slack tide (Aure & Stigebrandt, 1990). Effects on the sea-bed, especially at cage sites, are more readily observable. Loss of habitat heterogeneity is apparent and, as in point source organic pollution discharges, the area under or immediately downstream of cages is most affected. Sediments are largely devoid of oxygen, are highly reducing and may be devoid of macrobenthos in some instances. The anaerobe *Beggiotoa* is often apparent on sediment surfaces. Some 10–20 m distant from the cages, pollution tolerant macrobenthic species such as the annelid *Capitella capitata* occur in high numbers. Within 100–150 m, there is usually a return to undisturbed community structure and function.

Mussel farms increase rates of biodeposition in the immediate vicinity of ropes and rafts, reducing biodiversity among the benthic macroinvertebrate community (Mattsson & Lindén, 1984; Kaspar *et al.*, 1985).

16.4.3.2 Chemicals and drugs

A wide range of chemicals is used by the mariculture industry, including compounds employed in and applied to construction materials (stabilisers, pigments, plasticisers, u.v. absorbents and anti-foulants), pigments incorporated into feeds, and chemotherapeutants. There is particular concern among both scientists and the public about chemotherapeutants because of the amounts used, the ease with which they enter the environment and the lack of knowledge of their effects (Brown, 1989; Michel & Alderman, 1992; Weston, 1996). Even in countries with strict licensing procedures, until recently the emphasis has been on the efficacy of the drug and safety of the consumer rather than on the environment.

The range of anti-microbial compounds, fungicides and parasiticides used is illustrated in Table 16.2. Weston (1996) has estimated that around 30 anti-bacterial compounds are used in aquaculture, use being determined by the species being farmed, pathogens and location. Quantities vary with the intensity of production, the bulk being used in marine finfish culture in Europe, Japan and North America and in crustacean culture in Asia and Latin America. The quantities used can be enormous. Norway is the only country that has kept records of anti-microbial use by fish farmers (Grave *et al.*, 1990; Bangen *et al.*, 1994). In 1987, the year in which anti-bacterial use per unit weight of fish production peaked, some 49 tonnes were introduced into the coastal marine environment by the industry (see Figure 16.2).

Some of the drug-impregnated food is ingested by scavengers; most, however, either diffuses into the water column or becomes incorporated into sediments. Water-borne anti-microbials are rapidly diluted and some are also highly susceptible to photodegradation (Weston, 1996). The sediments, however, can act as a long-term reservoir of drugs and their

Table 16.2. *Chemotherapeutants used in mariculture*

Anti-bacterial agents	Fungicides	Parasiticides
(a) natural antibiotics: Tetracyclines Macrolide antibiotics β-lactams Aminoglycosides Phenicols (b) synthetic antibiotics: Sulphonamides Potentiated sulphonamides Quinolones Nitrofurans	Malachite green Formalin	Hydrogen peroxide Dichlorvos Avermectins Pyrethroids

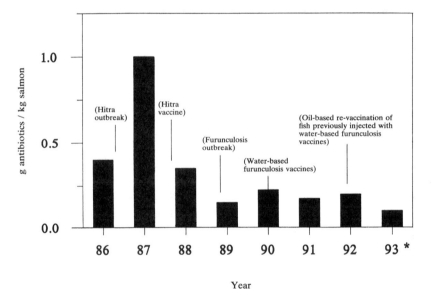

Fig. 16.2. Changes in anti-bacterial use in relation to various disease outbreaks and vaccine developments in Norway. 93*, estimated values. Data from Norsk Medisinaldepot.

residues, compounds such as oxytetracycline and the quinolones oxolinic acid and flumequine persisting for many months after chemotherapy although their bioavailability remains a matter of debate (Weston, 1996).

The impact of anti-microbial compounds can be summarised as the effect on non-target organisms, the effect on sediment chemistry and processes, and the development of resistance. Many studies have looked at non-target organisms that take up anti-microbials through foraging for food around cages, filter-feeding on current-borne feed and faecal particles or direct absorption from the water. High levels of drugs such as oxolinic acid have been detected in wild fish, crabs and mussels several hundred metres from salmon farms during and for up to 2 weeks following treatment (Lunestad, 1992; Samuelsen et al., 1992) and there is concern that such levels pose risks (toxic, allergic and development of resistance) to humans catching and eating these foods (Yndestad, 1992).

While laboratory trials involving exposure to high concentrations of compounds have demonstrated anti-microbial inhibition of sulphate reduction and nitrification (Hansen et al., 1992; Klaver & Mathews, 1994) there is little evidence for quantitative and qualitative changes in sediment flora or in the rates of organic matter degradation under fish cages (Weston, 1996). Many studies have reported increases in resistance, and even multiple resistance, in pathogens as a result of the widespread use of anti-microbials by the industry (Hastings & McKay, 1987; Aoki, 1992; Richards et al., 1992; Kerry et al., 1994). The significance of these findings remains difficult to assess as little is known about the persistence of resistance.

16.4.3.3 Microorganisms and feral animals

The bulk of mariculture production is from a few dozen species, many of which have been translocated. The Pacific cupped oyster *Crassostrea gigas*, for example, is farmed throughout much of the world, far outside its natural range. Despite this, movements of fish and other aquatic organisms seem to have been less in the marine than in freshwater systems, and mariculture is a comparatively minor factor in relation to other causes, both deliberate (improvement of fisheries) and accidental (construction of canals, ballast water transport, etc: Baltz, 1991).

The transfer of exotic marine organisms poses two threats to biodiversity, from the organism itself and from associated pathogens. According to Sindermann (1993), shrimp and salmon viruses and protozoan parasites of molluscs are among the pathogens that have been introduced through stock transfers for aquaculture purposes, resulting in several instances in severe disease outbreaks. Farmed organisms will always escape, cages being particularly vulnerable (Beveridge, 1996). Escapes occur on a small but almost continuous scale during day-to-day management of stock (grading, transfer from one tank/cage/pond to another). Periodic mass releases, involving tens of thousands of animals, can also happen as a result of storm damage or vandalism (e.g. Gausen & Moen, 1991). The fear is that feral organisms will become established and reduce biodiversity through habitat modification, competition and predation and by interbreeding with native stocks. By contrast with freshwater habitats where many feral species have become established and where impacts are well-documented (Welcomme, 1988) there is little evidence that exotic feral marine organisms have become established (Baltz, 1991) or that serious impacts on the environment or on biodiversity have occurred as a result. However, there has also been little study of the fate of feral marine animals. In Chile, for example, where it is likely that hundreds of thousands of farmed salmonids escape each year, there is no information on their impact on marine ecosystems.

The farming of native species reduces the risks of habitat modification and interspecific interactions, but may aggravate the risks to indigenous species through the introduction of non-adaptive genes with consequent reductions in fitness (Ross & Beveridge, 1995). Greatest fears have been expressed with regard to the Atlantic salmon. However, farming of local strains is no answer either. As soon as native animals are brought into the hatchery, they pass through a series of selection bottlenecks, either accidental (disease) or deliberate (selection for growth, shape, colour) and within a generation or two they differ from their wild ancestor.

16.5 Discussion

Mariculture has expanded a great deal in the past 20 years and judging from trends in data published by the FAO there is every indication that intensive fish and shrimp production will continue to expand for the foreseeable future. The argument that intensive aquaculture does not make sense in energy consumption terms (Odum & Arding, 1991; Folke & Kautsky, 1992; Larsson et al., 1994) remains largely irrelevant at present. Moreover, given the lack of political will to make decisions that will result in effective management of wild fish stocks, the advent of culture-based marine fisheries, dependent upon mass stocking of juveniles, seems inevitable. Mollusc and seaweed farming are also likely to continue to expand, albeit at a somewhat slower rate than fish or shrimp farming.

There has been no study of the impact of aquaculture on biodiversity per se, and there is still much to learn. However, a reasonable hypothesis would be that impacts are largely related to resource consumption and waste production (i.e. intensity of production methods), site location and scale (Beveridge et al., 1994). By these criteria, mariculture poses more of a threat to biodiversity than freshwater aquaculture. Inland water production, which is primarily of fish, is largely carried out by semi-intensive methods in ponds constructed in agricultural areas. Demands on natural resources and energy and reliance on drugs and chemicals is less than in marine fish and crustacean culture. Freshwater ponds tend to act as nutrient traps, pollution only posing a risk if ponds are drained without due care prior to harvesting (Beveridge et al., 1993). Extending the argument, intensive marine shrimp and fish culture are of much greater concern than extensive mollusc or macroalgal culture.

Much of the assessment of the environmental impact of mariculture has focused on the effects of uneaten food, faecal and excretory products on the sea-bed. The discharge of organic matter into the coastal zone remains a cause for concern in particular localities. However, effects on the sea-bed are limited in area and appear to be reversible, at least as far as biodiversity at the species level is concerned. While aquaculture may not be a major factor in coastal eutrophication in regional terms (Enell & Ackefors, 1992; see also below), little is known about the effects of vitamins, pigments and trace nutrients on phytoplankton composition or metabolism. It is argued here that more attention should be given to:

Conversion of coastal wetland to shrimp ponds;
The discharge of drugs and chemicals;
Release of exotic genetic material into the environment;
The spread of disease through ungoverned movements of animals
  around the globe.

All four processes pose significant threats to marine biodiversity and the changes caused are poorly understood and/or irreversible.

What can be done to address these problems? First, it should be said that there is a growing awareness, not only among the scientific community but also among resource managers, decision-makers and farmers. Some research is currently being carried out, particularly on the impact of antimicrobial compounds on benthic ecosystems, although there is much less work being done on other issues. The roles played by mangroves in coastal protection, nutrient production/entrapment and as nursery areas for fish and shellfish stocks and the linkages between coastal wetlands and economic well-being need to be better understood. Improved and user-friendly waste dispersion models are required so that developments can be better scaled with respect to sites. It is also timely to consider the impact of stocking on biodiversity before marine culture-based fisheries, currently the subject of much discussion, become a reality.

Already, some positive developments are apparent. There are guidelines concerning the introduction of novel species for aquatic farming (Turner, 1988). The adoption of environmental impact assessment (EIA) procedures for new mariculture developments, in which sensitive sites can be avoided and in which monitoring is a condition of the licence, is apparent in many countries, both north and south. While this helps avoid the worst blunders, only an integrated approach to coastal zone management, in which the linkages between the land and the sea and in which social and cultural values are given due consideration, can properly decide the role for mariculture. Farmers in the northern hemisphere have begun to realise that present production methods are often not sustainable and have begun to modify practices adopting fallowing schedules to allow benthic cage sites to recover, for example. Market forces, too, are beginning to play their part in reducing food waste and in forcing farmers in some areas to consider more ecologically-sound production methods (e.g. pathogen-free shrimp farming in the United States of America; see Chanratchakool *et al.*, 1994). In other areas, however, powerful interests continue to take advantage of poor enforcement of environmental legislation to make a fast profit, over-exploiting areas for several years and abandoning sites once serious environmental and disease problems become apparent (Phillips *et al.*, 1993).

Reductions in organic and chemical wastes have and will continue to occur, despite the fact that the aquaculture industry is ever-intensifying. In salmon farming, for example, nitrogen and phosphorus wastes per unit production have been steadily falling (Fig. 16.3) thanks to improvements in feeding practices and feed formulation. Food conversion ratios (weight of food fed : biomass of fish produced) have dropped from 2.4 : 1 to 1.3 : 1 over 25 years, while dietary nitrogen and phosphorus levels have fallen from around 7.8% to 6.8% and from 1.7% to 0.8%, respectively, over the same period (see Enell & Ackefors, 1992). Scope for further reductions is limited, however; phosphorus inclusion levels in many commercial feeds already approximate dietary requirements and nitrogen levels in diets are, if anything, increasing. Moreover, the reductions in wastes have largely been achieved through the production of more digestible feeds that include greater amounts of fishmeal than their predecessors (Tacon, 1994). Tacon (1994) admits that fishmeal

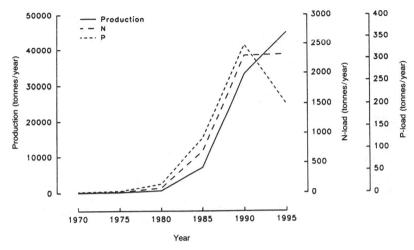

Fig. 16.3. Changes in quantities of wastes (N and P) discharged into Scottish coastal waters by salmon farms in relation to product trends. Data on production from the Scottish Office Agriculture, Environment and Fisheries Department. (See the text for assumptions). ——, production; – –, nitrogen (N); ---, phosphorus (P).

will remain the principal source of dietary protein, although bacterial and fungal single cell proteins and plant protein concentrates (e.g. soya) could replace up to 50% of fishmeal. Market prices for a product with limited supply can only increase, forcing the pace of such changes.

Use of drugs and chemicals in intensive marine salmonid culture has been declining. Evidence from Norway shows that drug use peaked in the late 1980s but has since declined by more than 90% on a per unit weight production basis as a result of the introduction of effective vaccines for Hitra (*Vibrio salmonicida*) and other diseases (see Fig. 16.2). Better husbandry (lower stocking densities, fallowing, stock control measures), however, has also been an important contributory factor. Treatment of mariculture wastes is also a possibility, but is difficult, expensive and in the present climate, in which discharges of wastes into the sea still makes economic sense, it is unlikely to happen. Moving aquaculture away from the sensitive coastal zones is also being widely mooted in many quarters (Fig. 16.4).

In judging the impacts of mariculture, a holistic view must be taken, in which food production from agriculture and from freshwaters is considered. Extensive culture of molluscs and macroalgae, provided it is carried out in a planned way, is a highly resource-efficient means of producing food (Table 16.3). Some suggest a more radical approach to

Fig. 16.4. FarmOcean semi-submersible cage at an exposed site off the west coast of Scotland. The cage has a variable flotation system allowing it to be raised and lowered as environmental and operational conditions dictate.

mariculture development is necessary in which extensive and intensive aquaculture is combined, the wastes from intensive production stimulating the culture of macroalgae or molluscs. This is already being done on a small scale in Thailand and elsewhere where seaweeds and mussels are being grown using wastes from intensive fish and crustacean culture. In the development of aquaculture in the past two decades, much has also been learned about biomanipulation of aquatic environments and it is now apparent that the technology can be used to protect the coastal zone from anthropogenic pollution. In Calcutta, for example, some 3000 ha of fish ponds are currently being used to treat 550 000 m^3 of untreated wastewater per day, in so doing producing 13 000 tonnes of fish for the urban poor and substantially reducing nutrient, biochemical oxygen demand (BOD) and bacterial loadings to the coastal zone (Mara et al., 1993). While such schemes could be extended to other coastal conurbations in the tropics, socio-economic factors are as crucial as technical feasibility (Inui et al., 1991; Muir et al., 1994). In the long term mariculture may also help alleviate pressures on over-fished wild stocks although this remains only a possibility as long as marine fish and shrimp culture remain primarily dependent on fishmeal.

Table 16.3. *Estimates of inputs of fossil fuel energy for the production of various foods*

Food type	kcal fossil energy input $kcal^{-1}$ protein output
Agriculture	
Vegetables	2–4
Sheep	10
Rangeland beef	10
Broiler chickens	22
Pigs	35
Feedlot beef	20–78
Marine fisheries	
Cod	20
Salmon	18–52
Lobster	192
Shrimp	3–198
Mariculture	
Seaweed	1
Salmon ranching	7–12
Salmon cage culture	50

Modified from Folke & Kautsky (1992).

Acknowledgements

We thank colleagues Valerie Inglis, Antonio Noguiera and Kim Jauncey for help and for useful suggestions made during the preparation of this manuscript.

References

Aoki, T., 1992. Chemotherapy and drug resistance in fish farms in Japan. In *Diseases in Asian Aquaculture*, ed. M. Shariff, R.P. Subasinghe & J.R. Arthur, pp. 519–29. Manila: Asian Fisheries Society.

Aure, J. & Stigebrandt, A., 1990. Quantitative estimates of the eutrophication effects of fish farming on fjords. *Aquaculture*, 90, 135–56.

Baltz, D.M., 1991. Introduced fishes in marine systems and inland seas. *Biological Conservation*, 56, 151–77.

Banerjee, B.K., 1993. *The Shrimp Fry By-catch in West Bengal*. Madras, India: Bay of Bengal Project.

Bangen, M., Grave, K., Nordmo, R. & Soli, N.E., 1994. Description and evaluation

of a new surveillance programme for drug use in fish farming in Norway. *Aquaculture*, **119**, 109–18.
Barg, U.C., 1992. *Guidelines for the Promotion of Environmental Management of Coastal Aquaculture Development.* FAO Fisheries Technical Paper, no. 328. Rome: FAO.
Beveridge, M.C.M., 1984. *Cage and Pen Fish Farming. Carrying Capacity Models and Environmental Impacts.* FAO Fisheries Technical Paper, no. 255. Rome: FAO.
Beveridge, M.C.M., 1988. Problems caused by birds at inland waters and freshwater fish farms. In *Prevention and Control of Bird Predation in Aquaculture and Fisheries*, ed. R. Welcomme, pp. 34–73. EIFAC Technical Paper, no. 51. Rome: FAO.
Beveridge, M.C.M., 1996. *Cage Aquaculture*, 2nd edn. Oxford: Fishing News Books.
Beveridge, M.C.M., Phillips, M.J. & Clarke, R.M., 1991. A quantitative and qualitative assessment of wastes from aquatic animal production. In *Advances in World Aquaculture*, vol. 3, ed. D.E. Brune & J.R. Tomasso, pp. 506–33. Baton Rouge, LA: World Aquaculture Society.
Beveridge, M.C.M., Phillips, M.J. & Clarke, R., 1993. Environmental impact of tropical inland aquaculture. In *Environment and Aquaculture in Developing Countries*, ed. R.S.V. Pullin, H. Rosenthal & J.L. Maclean, pp. 213–36. ICLARM Conference Proceedings no. 31. Manila: ICLARM.
Beveridge, M.C.M., Ross, L.G. & Kelly, L.A., 1994. Aquaculture and biodiversity. *Ambio*, **23**, 497–502.
Briggs, M.R.P. & Funge-Smith, S., 1994. A nutrient budget of some intensive marine shrimp ponds in Thailand. *Aquaculture and Fisheries Management*, **25**, 789–812.
Brown, J.H., 1989. Antibiotics: Their use and abuse in aquaculture. *World Aquaculture*, **20**, 34–43.
Carss, D., 1989. Sawbill ducks at fish farms in Argyll, Western Scotland. *Scottish Birds*, **15**, 145–50.
Chamberlain, G.W., 1993. Aquaculture trends and feed projections. *World Aquaculture*, **24**, 19–29.
Chanratchakool, P., Turnbull, J.F. & Limsuwan, C., 1994. *Health Management in Shrimp Ponds.* Bangkok: Aquatic Animal Health Research Institute, Department of Fisheries.
Christie, W.J., 1993. Developing the concept of sustainable fisheries. *Journal of Aquatic Ecosystem Health*, **2**, 99–109.
Enell, M. & Ackefors, H., 1992. Development of Nordic salmonid production in aquaculture and nutrient discharges into adjacent sea areas. *Aquaculture Europe*, **16**, 6–11.
FAO, 1993. *Agriculture: Towards 2010.* FAO Conference, 27th Session, Rome 6–25 November, 1993. Report no. FAO-GIC-93/24. Rome: FAO.
FAO, 1994. *Review of the State of World Fishery Resources, Part 1. The Marine Resources.* FAO Fisheries Circular no. 710, rev. 9, part 1. Rome: FAO.
FAO, 1996. *Aquaculture Production 1986–1994.* FAO Fisheries Circular no. 815, rev. 7. Rome: FAO.
Folke, C. & Kautsky, N., 1992. Aquaculture with its environment: Prospects for sustainability. *Ocean and Coastal Management*, **17**, 5–24.
Foy, R.H. & Rossell, R., 1991a. Loadings of nitrogen and phosphorus from a Northern Ireland fish farm. *Aquaculture*, **96**, 17–30.

Foy, R.H. & Rossell, R., 1991b. Fractionation of phosphorus and nitrogen loadings from a Northern Ireland fish farm. *Aquaculture*, **96**, 31-42.
Gausen, D. & Mocn, V., 1991. Large-scale escapes of Atlantic salmon (*Salmo salar*) into Norwegian rivers threaten natural populations. *Canadian Journal of Fisheries and Aquatic Sciences*, **48**, 426-8.
GESAMP, 1991. *Reducing Environmental Impacts of Coastal Aquaculture*. Report no. 47. Rome: FAO.
Gowen, R.J. & Bradbury, N.B., 1987. The ecological impact of salmonid farming in coastal waters: A review. *Oceanography and Marine Biology Annual Review*, **25**, 563-75.
Grave, K., Engelstad, M., Soli, N.S. & Hastein, T., 190. Utilization of antibacterial drugs in salmonid farming in Norway during 1980-1988. *Aquaculture*, **86**, 347-58.
Hansen, P.K., Lunestad, B.T. & Samuelsen, O.B., 1992. Effects of oxytetracycline, oxolinic acid and flumequine on bacteria in an artificial marine fish farm sediment. *Canadian Journal of Microbiology*, **38**, 1307-12.
Hastings, T.S. & McKay, A., 1987. Resistance of *Aeromonas salmonicida* to oxolinic acid. *Aquaculture*, **60**, 133-41.
Inui, M., Itsubo, M. & Iso, S., 1991. Creation of a new non-feeding aquaculture system in enclosed coastal seas. *Marine Pollution Bulletin*, **23**, 321-5.
Juell, J.-E., 1991. Hydroacoustic detection of food waste - a method to estimate maximum food intake of fish populations in sea cages. *Aquacultural Engineering*, **10**, 207-17.
Kaspar, H.F., Gillespie, P.A., Boyer, I.C. & MacKenzie, A.L., 1985. Effects of mussel aquaculture on the nitrogen cycle and benthic communities in Kenepuru Sound, Marlborough Sounds, New Zealand. *Marine Biology*, **85**, 127-36.
Kerry, J., Hiney, M., Coyne, R., Cazabon, D., NicGabhainn, S. & Smith, P., 1994. Frequency and distribution of resistance to oxytetracycline in micro-organisms isolated from marine fish farm sediments following therapeutic use of oxytetracycline. *Aquaculture*, **123**, 43-54.
Klaver, A.L. & Mathews, R.A., 1994. Effects of oxytetracycline on nitrification in a model aquatic system. *Aquaculture*, **123**, 237-47.
Larsson, J., Folke, C. & Kautsky, N., 1994. Ecological limitations and appropriation of ecosystem support by shrimp farming in Colombia. *Environmental Management*, **18**, 663-76.
Lunestad, B.T., 1992. Fate and effects of antibacterial agents in aquatic environments. In *Problems of Chemotherapy in Aquaculture: From Theory to Reality*, ed. C.M. Michel & D.J. Alderman, pp. 152-61. Paris: Office International de Epizooties.
Mara, D.D., Edwards, P., Clark, D. & Mills, S.W., 1993. A rational approach to the design of wastewater-fed fishponds. *Water Research*, **27**, 1797-9.
Mattsson, J. & Lindén, O., 1984. Impact of cultures of *Mytilus edulis* on the benthic ecosystem in a narrow sound on the Swedish west coast. *Vatten*, **40**, 151-63.
Michel, C. & Alderman, D.J. (eds), 1992. *Chemotherapy in Aquaculture: From Theory to Reality*. Paris: Office International des Epizooties.
Monaghan, P., 1992. Seabirds and sandeels: The conflict between exploitation and conservation in the northern North Sea. *Biodiversity and Conservation*, **1**, 98-111.
Muir, J.F., Walker, D. & Goodwin, D., 1994. *The Productive Re-Use of Wastewater: Potential and Application in India*. Report of the ODA Review

Mission, Calcutta, December 1994. Stirling: University of Stirling, Institute of Aquaculture.
NCC (Nature Conservatory Council), 1989. *Fish Farming and the Safeguard of the Natural Marine Environment of Scotland.* Edinburgh: Nature Conservancy Council.
Norse, E.A. (ed.), 1993. *Global Marine Biological Diversity: A Strategy for Building Conservation into Decision Making.* Washington, DC: Island Press.
Odum, H.T. & Arding, J.E., 1991. *Energy Analysis of Shrimp Mariculture in Ecuador.* Kingston, RI: University of Rhode Island, Coastal Resources Unit.
Pemberton, D., 1989. *The Interaction Between Seals and Fish Farms in Tasmania.* Tasmania: Australian Department of Lands, Parks and Wildlife.
Pemberton, D. & Shaughnessy, P.D., 1993. Interaction between seals and marine fish farms in Tasmania, and management of the problem. *Aquatic Conservation*, **3**, 149–58.
Phillips, M.J., 1990. Environmental aspects of seaweed culture. In *Proceedings of the Regional Workshop on the Culture and Utilization of Seaweeds*, pp. 51–62. Technical Resource Papers no. 2. Bangkok: NACA.
Phillips, M.J. & Beveridge, M.C.M., 1991. Impacts of aquaculture on water resources. In *Advances in World Aquaculture*, vol. 3, ed. D.E. Brune & J.R. Tomasso, pp. 568–91. Baton Rouge, LA: World Aquaculture Society.
Phillips, M.J., Kwei Lin, C. & Beveridge, M.C.M., 1993. Shrimp culture and the environment: Lessons from the world's most rapidly expanding warmwater aquaculture sector. In *Environment and Aquaculture in Developing Countries*, ed. R.S.V. Pullin, H. Rosenthal & J.L. Maclean, pp. 171–97. ICLARM Conference Proceedings no. 31. Manila: ICLARM.
Pimentel, D. & Giampietro, M., 1994. Global population, food and the environment. *Trends in Ecology and Evolution*, **9**, 239.
Primavera, T.H., 1991. Intensive prawn farming in the Philippines; Ecological, social and economic implications. *Ambio*, **20**, 28–33.
Primavera, T.H., 1993. A critical review of shrimp pond culture in the Philippines. *Reviews in Fisheries Science*, **1**, 151–201.
Pullin, R.S.V., Rosenthal, H. & Maclean, J.L. (eds), 1993. *Environment and Aquaculture in Developing Countries.* ICLARM Conference Proceedings no. 31. Manila: ICLARM.
Richards, R.H., Inglis, V., Frerichs, G.N. & Millar, S.D., 1992. Variation in antibiotic resistance patterns of *Aeromonas salmonicida* isolated from Atlantic salmon *Salmo salar* L. in Scotland. In *Problems of Chemotherapy in Aquaculture: From Theory to Reality*, ed. C.M. Michel & D.J. Alderman, pp. 276–87. Paris: Office International de Epizooties.
Ross, A., 1988. *Controlling Nature's Predators on Fish Farms.* Ross-on-Wye, UK: Marine Conservation Society.
Ross, L.G. & Beveridge, M.C.M., 1995. Is a strategy necessary for development of native species for aquaculture? A Mexican case study. *Aquaculture Research*, **26**, 539–47.
Ruggeberg, H. & Booth, J., 1989. *Interactions between Wildlife and Salmon Farms in B.C.: Results of a Survey.* Technical Report Series, no. 67. British Columbia, Canada: Canadian Wildlife Service.
Samuelson, O.B., 1992. The fate of antibiotics/chemotherapeutics in marine aquaculture sediments. In *Problems of Chemotherapy in Aquaculture: From Theory to Reality*, ed. C.M. Michel & D.J. Alderman, pp. 162–73. Paris: Office International de Epizooties.

Sindermann, C.J., 1993. Disease risks associated with importation of nonindigenous marine animals. *Marine Fisheries Review*, **54**, 1-10.
Tacon, A.G.J., 1994. Dependence of intensive aquaculture systems on fishmeal and other fishery resources. *FAO Aquaculture Newsletter*, **6**, 10-16.
Turner, G.E., 1988. *Codes of Practice and Manual of Procedures for Consideration of Introductions and Transfers of Marine and Freshwater Organisms*. EIFAC Occasional Paper no. 23. Rome: FAO.
Welcomme, R.L., 1988. *International Introductions of Inland Aquatic Species*. FAO Fisheries Technical Paper, no. 294. Rome: FAO.
Weston, D.P., 1991. The effects of aquaculture on indigenous biota. In *Advances in World Aquaculture*, vol. 3, ed. D.E. Brune & J.R. Tomasso, pp. 534-67. Baton Rouge, LA: World Aquaculture Society.
Weston, D.P., 1996. Environmental considerations in the use of antibacterial drugs in aquaculture. In *Aquaculture and Water Resources Management*, ed. D.J. Baird, M.C.M. Beveridge, L.A. Kelly & J.F. Muir. pp. 140-65. Oxford: Blackwell Scientific Publications.
Yndestad, M., 1992. Public health aspects of residues in animal products: Fundamental considerations. In *Problems of Chemotherapy in Aquaculture: From Theory to Reality*, ed. C.M. Michel & D.J. Alderman, pp. 494-511. Paris: Office International de Epizooties.
Zieman, D.A., Walsh, W.A., Saphore, E.G. & Fulton-Bennet, K., 1992. A survey of water quality characteristics of effluent from Hawaiian aquaculture facilities. *Journal of the World Mariculture Society*, **23**, 180-91.

Chapter 17
Protecting marine biodiversity and integrated coastal zone management

J.S.H. PULLEN
WWF UK (World Wide Fund For Nature), Panda House, Weyside Park, Catteshall Lane, Godalming, Surrey, GU7 1XR, UK

Abstract

Human pressure on the marine environment has never been so intense. Many human-induced physical, chemical and biological changes are adversely affecting biological diversity, and a range of activities are currently resulting in widespread degradation or even complete destruction of different marine habitats. In particular impacts due to development, and conflicts of interest over resource use, are most acute in the coastal zone. Yet it is here that productivity and biodiversity are generally greatest. Integrated Coastal Zone Management (ICZM) is increasingly recognised as essential for securing the sustainable development of resources and the effective management of coastal and marine areas under national jurisdiction. This chapter describes the typical elements of ICZM and discusses key policy considerations. However the full potential of ICZM for maintaining and enhancing marine biodiversity has yet to be realised.

17.1 Introduction

The protection and management of biological diversity is now firmly established on the international agenda. While ecologists and environmentalists have long recognised the importance of documenting, studying and maintaining biodiversity, it was only in 1992, following years of discussion, that the Convention on Biological Diversity was agreed, and even more recently ratified (UNCED, 1992a). The objectives of the Convention are to conserve biodiversity, to achieve the sustainable use of its components, and to secure the fair and equitable sharing of the genetic resources which that biodiversity represents. The scope of the Convention is stated to include '... marine and other aquatic ecosystems, and the ecological complexes of which they are in part; ...'

The need to protect marine biodiversity was appreciated only some time after concern for the conservation of terrestrial habitats and species first became widespread. The effect of human activity on the seas and oceans is not as noticeable as that on land. For perhaps centuries coastal waters were receiving increasing amounts of human waste, and their fish stocks were being increasingly heavily exploited, without obvious ill effect. Furthermore, with the exception of some small island areas (Johannes, 1982), there was no long-standing tradition of planning and management of marine areas. But now, with human pressure on the marine environment continuing to increase, we can see that the biodiversity of coastal and marine environments is jeopardised. Moreover pressure on the marine environment, and in particular the coastal zone, is unlikely to be relieved in the foreseeable future. It is estimated that the human population of coastal regions will double by 2020 (IUCN/UNEP/WWF, 1991), placing further demands on the coastal/marine environment as a source of renewable resources, as a sink for waste disposal, and as space for urban, industrial and agricultural development.

17.2 Impacts on the marine environment

Impacts on the marine environment resulting in the loss of habitats or species now take a wide variety of form (for a recent review see Norse, 1993). The environment, perhaps especially the marine environment, has a capacity to absorb and recycle a considerable amount of natural waste products (those produced by marine animals themselves, for example) via well-developed detritivore communities. The simplest form of marine pollution arises from local or more widespread overloading of this capacity, through disposal into coastal waters or at sea of excessive amounts of inorganic nutrients (from sewage or agriculture) or organic material (from sewage or industry). Excessive nutrient inputs alter the balance of food chains, eliminate sensitive species, and change the composition of benthic communities, in severe cases completely altering their structure (GESAMP, 1990). In particular, as is well-known, dumping of organic wastes at sea alters the composition of benthic communities in the vicinity. Opportunistic polychaetes and other species characteristic of organically enriched sediments come to dominate (Clark, 1992), while species diversity is greatly reduced. Similar effects occur around sewage outfalls or beneath mariculture cages (Kaspar *et al.*, 1985; Brown *et al.*, 1987; Gowen & Bradbury, 1987).

Addition of inorganic nutrients (supplying nitrogen or phosphorus) by contrast promote dominance of algae. In partially enclosed seas or waters nutrient input triggers blooms of one or a few sometimes toxic species of algae. In some cases toxic phytoplankton have caused fish kills directly (e.g. Halverson & Martin, 1980; White, 1980). It has also been speculated that changes in phytoplankton communities could favour jellyfish food chains at the expense of commercial fish species. In tropical seas reef corals are adapted to clear nutrient-poor (oligotrophic) waters; addition of inorganic nutrients favours increasing growth and dominance by turf and fleshy algae (Laws & Redalje, 1982), leading in severe cases, especially when combined with mortality or overfishing of herbivorous fish and invertebrates, to almost complete loss of coral cover, as for example at Discovery Bay, Jamaica (Hughes, 1994).

Other chemicals, foreign to the marine environment, are frequently toxic, whether they are naturally occurring (as with crude oil and hydrocarbons) or man-made (as with many pesticides). Locally lethal concentrations may directly impact communities by eliminating many or most species. Frequently toxic compounds become concentrated in ascending a food chain, accumulating in and selectively depleting top predators; not only are these conspicuous species (such as otters or seals) often of special public or conservation interest, but their removal may have knock-on consequences for the whole structure of a particular ecological community. Only recently has it been widely appreciated that relatively low concentrations of many toxic compounds may impact species through subtle mechanisms leading to, for example, immune system impairment, reduced reproductive success, or developmental aberration. Oil pollution, for instance, can interfere with detection by lobsters of the sex pheromone that triggers mating (Norse, 1993), and with normal feeding behaviour by sea-anemones (Ormond & Caldwell, 1982). In a recent study of the North American Great Lakes undertaken by WWF and the Canadian Institute for Research on Public Policy (Colborn *et al.*, 1990), reproductive and endocrine problems were found to be affecting populations of 16 species, including beluga whales, bald eagles, turtles and various fish. In each case elevated concentrations of pesticides and industrial chemicals were found in the species; in fact in beluga whale the levels of these chemicals were so high that the animals qualified as 'toxic waste'!

Even inert waste may impact marine life. Certainly it is now responsible for significant deterioration in the aesthetic quality of marine and coastal environments. It is estimated that there are now 3.6 million objects large enough to be sighted by eye afloat in the Mediterranean alone (McCoy, 1988). And large amounts of debris from shipping and other sources are washing ashore and disfiguring the coastline, even in many remote areas (e.g. Benton, 1995). Plastic has become a particular problem since it is so durable, and such large amounts are being made. Plastic debris in particular also maims and kills (Shomura & Yoshida, 1985; Shomura & Godfrey, 1990). Marine mammals, seabirds, turtles, fishes and even invertebrates such as crabs and lobsters all become entangled, especially in lost or discarded fishing gear and plastic rings from drinks packs. For example, entanglement kills thousands of northern fur seals (*Callorhinus ursinus*) in the Pribilof Islands, Alaska, every year (Fowler, 1987). Marine animals also ingest a wide variety of materials. Juvenile marine turtles, which live and feed in areas where both floating vegetation and debris concentrate, may be particularly prone to digest debris such as cigarette filters and plastic (Carr, 1987).

Even when marine pollution was not a significant problem, human activities have degraded and destroyed marine and coastal environments. In western countries coastal development has altered the character both of the open coast, which has been favoured for residential and tourist use, and of estuaries which have been reconstructed as ports or 'reclaimed' for industrial projects. This has resulted in extensive loss of coastal and near-shore habitat. In most developed countries a high proportion of salt-marsh has been lost to agriculture or development (Long & Mason, 1983), a process that is still continuing. In Britain it has been calculated that loss of salt-marshes and of mud- and sand-flats has occurred at mean rates of 0.2% and 0.7% per year for at least the last 200 years (English Nature, 1992).

In tropical countries sedimentation or direct infilling linked to coastal construction work for housing, roads or industry has commonly caused degradation or destruction of both coral reefs and sea-grass beds. Most seriously, considering their high productivity and economic value, there has been dramatic loss of valuable and productive mangrove habitat during the last 40–50 years, estimated for example at over 50% within the Indian Ocean Region (IUCN/UNEP, 1985). While clear-felling for commercial timber exploitation has been the greatest single cause of this destruction, conversion for commercial, industrial, residential or agricultural use has also been common. In particular, during the last decade, there has been wide-spread clearing of mangroves to construct ponds for shrimp and fish aquaculture (see Chapter 16). The Philippines alone is estimated to have lost nearly 70% of its mangroves since the 1920s (Primavera, 1991), over half of it for conversion to mariculture ponds, though a few mangrove restoration projects have also recently been established. Such coastal development has not been shown conclusively to lead to the extinction of any species, but a number of rare marine invertebrates and plants, dependent on such habitats as coastal lagoons and shingle banks, have been threatened or endangered, and some may be extinct. This has been as a result of engineering works, flood defences, water abstraction and pollution, often occurring in combination making the identification of specific factors difficult if not impossible.

The large-scale removal of biomass has also been highly destructive of marine life. Fishing practices have become so efficient that stocks can easily be over-exploited, even when intended to be managed sustainably, as with North Sea herring, or West Atlantic cod. Uncontrolled exploitation can drive populations so low that even if not exterminated they may take many years to recover. The structure and dynamics of associated food webs may also be altered. Where one species is exploited more than another their relative abundance will be changed, as in the North Atlantic (Sherman, 1990). Conversely the removal of predator species may lead to increased abundance of some prey or competitor species, as when removal of large sharks off South Africa, in order to protect swimmers, led to increases in the numbers of smaller sharks, and as a consequence a decline in the abundance of some of their prey fish species (van der Elst, 1979). Removal of predator species may also lead to reduced species diversity (Paine, 1980; Pimm, 1980), and perhaps affect the resilience of the community to natural or anthropogenic stresses. Removal of keystone species may even result in complete shifts in community (Beddington, 1984), as when loss of sea-otters (Simenstad *et al.*, 1978) or invertebrate feeding fish (McClanahan & Muthiga, 1988; Watson & Ormond, 1994) has been associated with population outbreaks of herbivorous sea-urchins, creating sea-urchin barrens among kelp forests, sea-grass beds or coral reefs (Hay, 1984; Hutchings & Wu, 1987).

Over-exploitation has not been restricted to species of fish. The world's great whale populations were of course decimated by commercial whaling, and at least one species, the Atlantic grey whale, exterminated (Mead & Mitchell, 1984). Seal and bird species, such as the great auk and the Caribbean monk seal, have also been hunted to extinction, while others such as the Mediterranean monk seal, the Dalmatian penguin, the Christmas frigate bird, the Madagascar fish eagle and all the species of marine turtle are critically endangered. Invertebrates are also exploited and overexploited, for food and for ornament. As well as long established fisheries for pearl oyster (*Pinctada* spp.) and trochus (*Trochus niloticus*), exploited for their mother-of-pearl, there is a huge trade in shells to be sold as souvenirs to tourists (Wells et al., 1983). These are exported from such countries as the Philippines, Indonesia, Haiti and Mexico, not only to Western countries, but to other coral reef areas. It is thought that as many as 5000 species may be involved, many of which are probably over-exploited locally, if not throughout their range.

Fishing activity may also be directly damaging of marine habitats. Trawling disturbs or damages sea-bed habitat, especially where seagrasses or algae are dominant, and in major trawling grounds each piece of sea-bed may be trawled over several times per year (McAllister, 1991; Riemann & Hoffmann, 1991). Trawling may also unintentionally damage other scarcer habitats that it is not intended to exploit. In the United Kingdom problems have been experienced with such damage to mature horse-mussel beds in Strangford Lough (Brown, 1988), and to fragile long-lived species such as sea-fans and Ross corals off the Pembrokeshire coast and in Lyme Bay, Dorset (Fowler, 1989; Edwards, 1993). In the tropics extensive coral mortality is caused by dynamiting for fish, which has become endemic in many African and Asian countries. Also highly destructive of the reef habitat is Muro-Ami fishing, a technique that originated in Japan but has become common in some other parts of South-East Asia; it involves large numbers of swimmers thrashing the water with sticks and smashing corals to drive fish into a large surround net (Gomez et al., 1987).

The introduction of non-native species, for example via ships' ballast water, fish farming, research centres or home aquariums, is recognised as a major threat to marine and coastal biological diversity by the Subsidiary Body on Scientific, Technical and Technological Advice (SBSTTA) under the Biodiversity Convention (SBSTTA, 1995). Considerable evidence has been documented of cases where catastrophic or overwhelming changes in natural ecosystems have occurred in terrestrial and freshwater systems (Drake et al., 1989) and increasing information is available on similar changes in the marine environment. Better known cases include the introduction of an Asian clam (*Potomocorbula amurensis*) into San Francisco Bay where it is considered to now be a dominant benthic organism (Carlton et al., 1990) and the north-west Atlantic comb jelly (*Mnemiopsis leidyi*) in the Black Sea.

A final set of threats that deserve not to be ignored are those that arise from widespread or even global changes that human action is bringing about. Thinning of the ozone layer brought about as a consequence of the use and release into the atmosphere of chlorofluorocarbons (CFCs) and some other compounds could cause damage to phytoplankton and some zooplankton including the larval stages of fishes through an increase in the amount of biologically damaging ultra-violet-B radiation reaching the surface of the Earth.

Even more dramatic are likely to be the effects of global warming, now generally accepted as fact, occurring as a result of emissions and accumulation in the atmosphere of greenhouse gases, especially carbon dioxide. General warming itself, expected to be by an average of 1–3°C within the next century, is likely to have a major impact on some marine organisms. During the last decade widespread bleaching of corals (loss of symbiotic microalgae, often followed by mortality), has been observed in corals in parts of both the Pacific and Caribbean Oceans, associated with peak ocean temperatures 1–2°C above the normal maximum for those areas (Glynn et al., 1988; Jokiel & Coles, 1990). Global warming is also causing rising sea levels. Although there are uncertainties over the precise rate, because of lack of knowledge about the behaviour of some glaciers and ice-shelves, it is predicted that by 2030 mean sea level will be 8–30 cm, and by 2100, 31–110 cm higher than at present. This change, as well as having serious consequences for human populations, will inevitably constrict and threaten a high proportion of intertidal habitats, including especially salt-marsh and mangrove. Even more widespread changes of marine environment will occur if, as has been suggested (Broecker, 1987), continued warming were to result in a shut-off or reorganisation of deep ocean circulation; this would involve failure of the Gulf Stream in the Atlantic, and a permanent El Niño type condition in the Pacific. There is evidence that such changes occurred perhaps repeatedly associated with interglacial warmings during the Pleistocene (Lehman & Keigwin, 1992). Inevitably such changes would not only render large areas of ocean unsuitable for both pelagic and benthic species adapted to present conditions, but would disrupt many reproductive and life cycles, attuned as they are to existing seasonal changes and current patterns.

17.3 Effects on marine biodiversity

Assessing the effect of these and a wide variety of other impacts on marine biodiversity presents a considerable problem. First there is the difficulty of identifying the many types of marine animal and plant. The systematics are frequently difficult or incomplete, and the scientific literature sometimes obscure and often scattered. Moreover the phyletic variety is higher than on land. Of 34 animal phyla, 29 occur in the oceans, 14 occurring only there; this compares with 11 present on land, of which only 1 is endemic (Angel, 1992; Funch & Kristenen, 1995). Even in the best worked marine areas, not all the phyla have been well explored. For example Picton *et al.* (1994) in the Marine Species Directory list over 6500 species recorded from European coastal waters, but state that even for this region the protists, platyhelminthes and entoprocts have not been properly studied. In deep-sea and high diversity tropical habitats many invertebrate groups remain relatively unknown, and, as discussed in earlier Chapters, there is even uncertainty regarding the order of magnitude of the total diversity of species likely to be found in the deep ocean. The possibility of even further phyla new to science cannot be ruled out, as witnessed by the recent description of a new species, *Symbion pandora*, found on the mouthparts of the Norway lobster, and assigned to a completely new phylum, the Cycliophora (Funch & Kristensen, 1995).

Partially linked to these taxonomic difficulties, confirmed extinctions of marine species are few, and almost exclusively of marine mammals and birds. The four mammals are the Atlantic grey whale, *Eschrichtius robustus* and the Caribbean monk seal, *Monachus tropicalis*, both mentioned above, and also the sea mink, *Mustela macroda*, and, most famously, Steller's sea cow, *Hydrodamalis gigas*. Six birds whose loss seems certain are the great auk, *Pinguinus impennis*, Steller's spectacled cormorant, *Phalacrocorax perspicillatus*, the Labrador duck, *Camptorhynchus labradorius*, the Auckland Island merganser, *Mergus australis*, the Guadeloupe storm petrel, *Oceanodroma macrodactyla*, and the Bonin night heron, *Nycticorax caledonicus crassirostris*. Only one marine invertebrate is known to have become extinct during recent times, the eelgrass limpet, *Lottia alveus alveus*, which was an obligate associate of the eelgrass, *Zostera marina*, and disappeared in the 1930s after the extensive loss of eelgrass beds due to wasting disease (Carlton *et al.*, 1991). Another invertebrate extinction may be that of the cerithid, *Cerithidea fuscata*, last collected in San Diego Bay, California, in 1935 (Carlton *et al.*, 1991). A locally endemic fire coral, *Millepora boschmai*, was thought to have become extinct in 1982–3 following an exceptionally strong El Niño event (Glynn & de Weerdt, 1991), but a further population has subsequently been discovered (Glynn & Feingold, 1992).

A greater range of species are known to be threatened, or to have very specialist habitat requirements or a very limited range that is itself threatened. An extreme case quoted by Norse (1993) are the whale lice (*Cyamus*), each species of which lives on the skin of a particular species of whale. A very large number of invertebrates have been described from only one or a few locations. Such is the limited expertise in some invertebrate groups that in perhaps a majority of such cases we have little idea whether these species are genuinely rare, or whether it is the taxonomists with the knowledge and time to study them who are in short supply! A real concern must be that while at present populations of many marine fish and invertebrates may still be viable because of the dispersion of planktonic larvae to other suitable localities, once a proportion of critical breeding areas has been lost, populations may collapse, even though some suitable habitat remains. Extinctions of marine species might then snowball, their rate becoming comparable to those observed on land.

While extinction rates are a popular and easily comprehensible measure of the loss of biodiversity, local declines in species number and loss of habitats characterised by high biodiversity and high productivity are much more significant in functional terms, and are becoming widespread in coastal waters, as on land. This implies a loss in renewable resources, and of the service capacity of the environment (e.g. to recycle waste), as well as a decline in aesthetic and amenity value that can only result in a negative effect on the quality of life of human populations. Focused and co-ordinated research programmes, involving a wide range of specialist expertise from large numbers of institutions and organisations, will be essential if global and local loss of marine biodiversity is to be tracked more accurately, or its effects on the human environment and condition understood more fully. Our lack of understanding of the causes and consequences of marine diversity, and of the impact upon it of human activities, should not however be considered an excuse for lack of the action required to husband that diversity. That is a point increasingly appreciated via the growing application of the 'precautionary principle', and specifically made in the Biodiversity Convention, which both recognises these difficulties, and stresses that where there is a suspected threat to biological diversity, lack of full scientific knowledge should not be used as a reason for postponing appropriate protective or preventative measures.

17.4 Protecting marine biodiversity

The coming into force of the Biodiversity Convention is the most recent of a series of developments, both formal and informal, which have occurred in response to the accelerating environmental crisis. Most of these have been concerned to conserve threatened species and resources both on land and in the sea. As mentioned in the introduction to this volume, the first Global Environment Conference in Stockholm in 1972 led to the formation of the United Nations Environment Programme. The Report of the World Commission on Environment and Development (1987) gave currency to the concept of sustainability that has since been adopted as a critical tenet by many national governments, and given international recognition at the Earth Summit in Rio de Janiero. Some international moves have been directed specifically at the marine environment, notably those conventions directed at controlling marine pollution and the recently agreed Global Programme of Action on the Protection of the Marine Environment from Land-based Activities. The Biodiversity Convention is self-evidently the first international agreement to be directed specifically at protecting biological diversity; and in November 1995 a new Decision on marine and coastal biodiversity was negotiated, known as the Jakarta Mandate. While the wording of this Decision is relatively weak it does recognise integrated marine and coastal area management as being fundamental for the protection of marine and coastal biodiversity and it sets in motion the establishment of an international roster of experts on marine and coastal biodiversity.

The Biodiversity Convention advocates the conservation of biodiversity, and the sustainable use of its components. It identifies a number of measures that need to be taken by states, within national strategies or programmes, in order to achieve this. Included are:

Establishing a system of protected areas or areas where special
  measures need to be taken.
Regulating or managing biological resources important for
  biodiversity conservation, whether inside or outside protected areas,
  in order to ensure their conservation.
Rehabilitating degraded ecosystems.
Legislating for the protection of threatened species and promoting
  their recovery.
Preventing the introduction of exotic or non-native species.
Regulating and managing relevant processes and activities known to
  have an adverse effect on biological diversity.
Encouraging relevant practices of indigenous and local communities.

All these measures are important if not critical to the conservation of marine habitats and species. It must be emphasised that a detailed understanding of mechanisms causing loss of biodiversity is not essential for their implementation, although increased scientific knowledge will certainly assist.

Of the above measures, the establishment of a system of marine protected areas (MPAs) must be regarded as a priority for conserving marine biodiversity and productivity. The number of MPAs has increased dramatically over the last few decades. Many of the earliest marine parks and reserves were established in the 1960s. By 1974 Bjorklund was able to list over 130 mostly marine protected areas; today there are over 1300 (Kelleher *et al.*, 1995), but there remain regions and countries where either the establishment of MPAs or the proper protection and management of existing ones is very inadequate. Even in Britain, in the 13 years since legal recognition of MPAs was made possible, only three small marine reserves have been formally recognised, although new developments as a result of EU legislation could result in as many as 35 new designations.

Selection of sites for MPAs is critical. Attention should be focused both on regions or areas of high diversity (diversity 'hotspots'), and on species, communities, habitats and ecosystems that are either endemic or at risk. Critical habitats, areas that are important as feeding or breeding areas, or as spawning or nursery grounds, or that have high productivity or are the location of processes that drive that productivity (Ray, 1976), should also be protected. However, the designation of MPAs cannot be undertaken in isolation. Because of the movement of organisms, of air, and especially of water and the materials it transports, numerous activities carried out far away may nevertheless impact a protected area. In addition many marine species move considerable distances during their life-cycle and require protection at all stages.

Regulating the exploitation of biological resources outside MPAs is also essential, not only to secure the sustainable use of those resources, but to maintain the ecological integrity of marine environments. Given that many, if not most, fish stocks are currently exploited at levels that are unsustainable, fishing effort needs to be urgently reduced for the benefit of those human populations that are physically or economically dependent on those stocks. Further, if, as described above, fishing can also have damaging direct or indirect effects on marine environments, then fishing methods or effort may need to be controlled to sustain the ecological function of the relevant habitats.

Recovery of marine habitats is in general poorly understood, although some progress has been made with coastal habitat recovery programmes. Mangroves can often be replanted fairly easily, provided their supply of brackish or saltwater is intact and not polluted; indeed planting of seedlings following felling in areas that have not naturally recolonised is a standard forestry practice in Malaysia and some other countries (Ong, 1982; Field & Dartnall, 1987). Sea-grass beds have been restored by transplanting propagules, although since sediments often become more mobile when the stabilising effect of the sea-grass roots is lost, reintroduced plants may need to be anchored down (Phillips, 1980). Artificial reefs have been constructed in many areas, sometimes, as in several parts of South-East Asia, on a relatively large scale (White et al., 1990). However, the extent to which such reefs function as true coral reefs, or can develop a comparable diversity, is often very questionable. In particular, artificial reefs will not become well-colonised by corals or function as productive habitats in areas where the natural reefs have been damaged by pollution or ecological imbalance. Anthropogenic impacts need first to be controlled or removed, following which recovery may well occur naturally, depending on the proximity of healthy reefs from which fish and invertebrates may recolonise. With deeper habitats, encouragement of natural recovery may be the only viable approach.

Species-oriented protection programmes are not easy to promote in the marine environment. Even when a community or population is clearly in decline, determining the cause with any certainty is frequently difficult, if not impossible, after the event. A range of measures is generally required and these need to be given statutory backing. Enforcement is often a critical consideration (witness the problems of fisheries enforcement!). Special measures are necessary for highly migratory species, as well as for exploited ones.

The introduction of non-native species presents a particular problem in the marine environment where once a species has been introduced to another ocean or region it may be easily spread by currents, if not under its own propulsion. Introductions cause most harm where exotic species are able to out-compete native ones. The spread of pathogens or parasites is a similar concern. Perhaps first to receive attention were the species introduced into the Mediterranean from the Red Sea following the construction of the Suez Canal. Over 250 have now been recorded, among which some have partially displaced native eastern Mediterranean species, and at least one has become the basis of a significant commercial fishery (Spanier & Galil, 1991). Elsewhere introduced species have altered habitat distribution or caused widespread changes in community composition, for example the European periwinkle, *Littorina littorea*, on introduction to North America in the early 19th century (Carlton,

1989). More recently ballast water carried by ships has been the principal mode by which organisms have been transported across or between oceans. Dozens of invertebrates have colonised new regions in this way, sometimes becoming super-abundant; for example the comb-jelly, *Mnemiopsis leidyi*, from the North-West Atlantic has been introduced into the Black Sea where there are now reported to be millions of tons. Other introductions have been intentional or linked to intentional introductions; thus two oyster species have been introduced to western Europe, along with their predators, parasites and pathogens. Strong measures will be necessary to control the rates of both unintentional and intentional introductions.

The regulation and management of activities, for example pollution or sedimentation, known to have an adverse effect on the marine environment, is fundamental to sustaining the health of the oceans; establishment of MPAs will have little benefit in the face of upstream pollution. Given that in practice the pressure to exploit marine and coastal resources, albeit sustainably, is irresistible, and that there is increasing pressure on the marine environment from rapidly expanding human populations in the coastal zone, then it probably has to be accepted that some detrimental activities will continue for the foreseeable future. These activities should however be regulated and managed in such a manner that the impact on the environment is minimised, for example by developing alternative processes for dealing with waste materials. Activities resulting in irreversible changes, thus jeopardising biodiversity, should be eliminated.

Last, but by no means least, the role of local and indigenous people in the management of the marine environment, as of terrestrial resources, must be recognised. It is these local communities who are dependent on the marine environment for food, income and livelihood. Whether legally recognised or not, in one sense it is their environment, and they are the principal stakeholders. More than that, it is the everyday activities and decisions of these people that normally, through their number if not their extent, represent the principal management interventions. Further such stakeholders, as among some Pacific Islands, often have a traditional wisdom of seeking to husband their resources through appropriate management practices (Johannes, 1982). Where such concepts are dormant they must be re-awakened, where they are new they must be established. This involves not only education and consultation, but involvement of the local community in decision-taking and management (Wells & White, 1995). This is no easy task, but must be considered a priority.

17.5 Integrated coastal zone management

Each of these measures may on their own go some way towards achieving the objectives of protecting biodiversity and securing sustainable use. In the marine and coastal environment, however, it is especially important that these separate measures be integrated both with each other and with other planning and administrative activities into an integrated management framework. This is so firstly because of the connectivity of the marine environment: impacts, such as pollution or coastal engineering, arising in one location, are inevitably transported to another, and similarly organisms present within a protected area are likely, to an even greater extent than on land, to have been dependent, for part of their life-cycle, on other areas that are not protected. Secondly, all the anthropogenic demands placed on the ocean and on coastal seas originate ultimately on land: most pollution and comparable impacts, including eutrophication and sedimentation, have their source on land, often at considerable distance, but can be transported from within a catchment to have an impact on the coast or at sea; and the distribution of the physical and biological effects of fishing, shipping and tourism depends on the location of accommodation, ports and other facilities. The wider management of all these activities and potential impacts is required, and must form the basis of any national or international policy for the protection of biodiversity.

An approach that provides a framework for this wider integration, and one advocated by the Biodiversity Convention (UNCED, 1992a), is known as Integrated Coastal Zone Management (ICZM), a term agreed on by a 1989 workshop of the Coastal Area Management and Planning Network, and now widely accepted (Coastal Area Management and Planning Network, 1989). The need for ICZM has been endorsed, among others, by the coastal zone management subgroup of the International Panel on Climate Change (IPCC CZMS, 1992), by the World Coast Conference (1993), and in particular by the United Nations Conference on Environment and Development (UNCED, 1992b) within Agenda 21: the Programme of Action for Sustainable Development. Chapter 17 of Agenda 21 calls upon coastal states to develop capabilities for ICZM and to implement national ICZM programmes.

A clear, generally accepted explanation of ICZM is difficult to find. The World Coast Conference (1993) defined it as involving 'the planning and management of coastal systems and resources, taking into account traditional, cultural and historical perspectives, and conflicting interests and uses; it is a continuous and evolutionary process for achieving sustainable development'. In reality this definition is little more helpful than most of its predecessors. In fact the key element of ICZM as it is generally used is the formulation of an integrated plan for the use and development of a section of coast, together with its hinterland and adjacent coastal seas; the plan seeks to protect the environment and resources of that coastal zone but must, above all, take into account the fact that the activities and actions undertaken in one location for one purpose are likely to have an impact upon the environment and resources that ought to be enjoyed or utilised in both the same and other locations for other purposes. Consequently the development of a successful ICZM plan requires a full appreciation of the ecological and other values of each part of the area in question, and a thorough understanding of the many ways in which different actions and activities can impact or degrade the natural value and resources, both of the area where they are undertaken, and of other even relatively distant areas. The same principles are applicable to fully terrestrial regions, when the more general term Integrated Area Management (IAM) is sometimes used.

At the World Coast Conference in 1993 it was concluded that the quest for a unique recipe for ICZM was misguided, and that efforts to define one should be re-orientated. For, as the report of the meeting describes, 'whenever it was decided that a top-down central government driven approach was most effective, national experiences to the contrary were held up. And as soon as certain technologies were proposed as the most useful, experiences favouring other technologies were cited'. Nevertheless the report does summarise in very general terms those elements typically involved both in developing an ICZM plan, and in implementing it. In general ICZM involves reviewing the resources present, the activities in demand, the habitats and wildlife requiring development, and the opportunities for sustainable development. Any activity or development should be undertaken in an environmentally sensitive manner.

More specifically, effective ICZM normally involves most or all of the following elements (see Sorensen & McCreary, 1990): the establishment of a system for regulating land use, the development of a map-based area or regional land-use plan, the protection of areas of greatest ecological, landscape and amenity value, and the undertaking of thorough environmental impact assessments (EIA) on both individual development proposals and on the provisions of the broader plan. EIAs should lead to recommendations for ameliorating proposed developments, and should allow for the possibility of recommending that development should not take place. Other elements that are often considered desirable are: integration with the national economic plan, integration with local sectoral plans (e.g. for transport, commerce and industry), the imposition of a shoreline setback or exclusion (i.e. no development is permitted within one or more set distances of the high tide line, often with the shore being made available for public access), and the establishment (and if possible enforcement) of policies and guidelines in order to minimise or prevent environmental impact. Within the plan a strategic view needs to be taken, both at the level of the plan, and at a national level.

Considerable variation exists in the institutional arrangements used to pursue ICZM (Gubbay, 1989; Sorenson & McCreary, 1990). Any of national government, regional government, local government, a semi-independent government agency, or a local agency may to a greater or lesser extent take the lead role in promoting and overseeing the process. In addition the different levels of government, separate agencies and non-government organisations may be statutory consultees, whose approval may either be required or desirable, while other organisations and individuals may be consulted on a voluntary basis, both to obtain advice and to engender support for planning policies and decisions. Almost certainly it is essential that one level of government or one agency is identified as having primary responsibility for the ICZM plan and its implementation, though whether it is best for this to be a central government department or agency, or a local branch or body, will vary considerably, depending on the size, character and culture of the country concerned. Whatever the lead institution, it is also highly desirable that an effective system of consultation both with special interest groups and the wider community be put in place; this is a step endorsed by Agenda 21 as a key part of the 'Local Agenda 21' process. As indicated above, indigenous peoples and coastal communities frequently have special knowledge and skills, and often traditional practices that may prove invaluable in the protection of marine biodiversity.

One of the benefits of ICZM frequently overlooked is the way in which it can, especially when such elements are formally incorporated, help focus public awareness and attention on the coastal zone, the flora, fauna and resources that are present, and the activities that take place (Gubbay, 1992). The principal advantage of ICZM is that it provides a framework for integrating the management of activities, developments, and resources, and of the environment and biodiversity across the land/sea interface. The conservation of marine biodiversity and the sustainable use of its components is possible within a framework of ICZM that minimises environmental degradation, controls pollution, encompasses a network of protected areas, and introduces other measures for the sustainable use of biological resources.

17.6 Conclusion

There is undoubtedly insufficient known about marine biodiversity, in particular about the full extent and distribution of that diversity, about its functional role within different ecosystems, and about the potential for local loss of diversity and resources to accelerate out of control. But better management of our coasts and oceans cannot wait for a full understanding; we already known sufficient of what ought to be done to initiate more effective action.

References

Angel, M.V., 1992. Managing biodiversity in the oceans. In *Diversity of Oceanic Life, an Evaluative Review*, ed. M.N.A. Peterson, pp. 1–110. Washington, DC: The Center for Strategic and International Studies.

Beddington, R.R., 1984. The response of multispecies systems to perturbations. In *Exploitation of Marine Communities*, ed. R.M. May, pp. 209–25. Report of the Dahlem Workshop on Exploitation of Marine Communities. Berlin: Springer-Verlag.

Benton, T.G., 1995. From castaways to throwaways: Marine litter in the Pitcairn Islands. *Biological Journal of the Linnean Society*, **56**, 415–22.

Bjorklund, M., 1974. Achievements in marine conservation. I. Marine parks. *Environmental Conservation*, **1**, 205–23.

Broecker, W.S., 1987. The biggest chill. *Natural History*, **96**, 74–82.

Brown, J.R., Gowen, R.J. & McLusky, D.S., 1987. The effect of salmon farming on the benthos of a Scottish sea loch. *Journal of Experimental Biology and Ecology*, **109**, 39–51.

Brown, R.A., 1988. Bottom trawling in Strangford Lough, problems and policies. Paper presented to the Third North Sea Seminar: Distress Signals, Signals from the Environment in Policy and Decision Making, Rotterdam, 31 May–2 June 1989.

Carlton, J.T., 1989. Man's role in changing the face of the ocean: Biological invasions and implications for conservation of near-shore environments. *Conservation Biology*, **3**, 265–73.

Carlton, J.T., Thompson, J.K., Schemel, L.E. & Nichols, F.E., 1990. Remarkable invasion of San Francisco Bay (California, USA) by the Asian clam *Potamocorbola amurensis*. Introduction and dispersal. *Marine Ecology Progress Series*, **66**, 81–94.

Carlton, J.T., Vermeij, G.J., Lindberg, D.R., Carlton, D.A. & Dudley, E., 1991. The first historical extinction of a marine invertebrate in an ocean basin: The demise of the eelgrass limpet, *Lottia alveus*. *Biological Bulletin*, **180**, 72–80.

Carr, A., 1987. Impact of nondegradable marine debris on the ecology and survival outlook of sea turtles. *Marine Pollution Bulletin*, **18**, 352–6.
Clark, R.B., 1992. *Marine Pollution*, 3rd edn. Oxford: Clarendon Press.
Coastal Area Management and Planning Network, 1989. *The Status of Integrated Coastal Zone Management: A Global Assessment.* Summary report of a workshop, Charleston, South Carolina, 4–9 July. Miami, FL: Rosensteil School of Marine Sciences, University of Miami.
Colborn, T.E., Davidson, A., Green, S.N., Hodge, R.A., Jackson, C.I. & Liroff, R., 1990. *Great Lakes, great legacy?* Washington, DC: The Conservation Foundation.
Drake, J.A., Mooney, H.A., di Castri, F., Groves, R.H., Kruger, F.J., Rejmanek, M. & Williamson, M. (eds), 1989. *Biological Invasions: A Global Perspective.* SCOPE Report no. 37. Chichester: John Wiley and Sons.
Edwards, J., 1993. *Lyme Bay. A Report on the Nature Conservation Importance of the Inshore Reefs and the Effects of Fishing Gear.* Exeter: Devon Wildlife Trust.
English Nature, 1992. *Campaign for a Living Coast.* Peterborough: English Nature.
Field, C.D. & Dartnall, A.J. (eds), 1987. *Mangrove Ecosystems of Asia and the Pacific: Status, Exploitation and Management.* Townsville: Australian Institute of Marine Science.
Fowler, C.W., 1987. A review of seal and sea lion entanglement in marine fishing debris. Paper presented to the North Pacific Rim Fishermen's Conference on Marine Debris, Kailua-Kona, Hawaii, 13–16 October, 1987.
Fowler, S.L., 1989. *Nature Conservation Implications of Damage to the Seabed by Commercial Fishing Operations.* Report to the NCC. Peterborough: Nature Conservation Council.
Funch, P. & Kristensen, R.M., 1995. Cycliophora is a new phylum with affinities to Entoprocta and Ectoprocta. *Nature*, **378**, 711–14.
GESAMP (Joint Group of Experts on the Scientific Aspects of Marine Pollution), 1990. *The State of the Marine Environment.* Oxford: Blackwell Scientific Publications.
Glynn, P.W. & de Weerdt, W.H., 1991. Elimination of two reef-building hydrocorals following the 1982–83 El Niño warming event. *Science*, **253**, 69–71.
Glynn, P.W. & Feingold, J.S., 1992. Hydrocoral species not extinct. *Science*, **257**, 1845.
Glynn, P.W., Cortez, J., Guzman, H.M. & Richmond, R.H., 1988. El Niño 1982/83 associated coral mortality and relationship to sea surface temperature deviations in the tropical Eastern Pacific. In *Proceedings of the 6th International Coral Reef Symposium, Townsville*, vol. 3, pp. 237–43. Townsville, Queensland: Sixth International Coral Reef Symposium Committee.
Gomez, E.D., Alcala, A.C. & Yap, H.T., 1987. Other fishing methods destructive to coral. In *Human Impacts on Coral Reefs: Impacts and Recommendations*, ed. B. Salvat, pp. 67–75. French Polynesia: Atenne Museum EPHE.
Gowen, R.J. & Bradbury, N.B., 1987. The ecological impact of salmonid farming in coastal waters: A review. *Oceanography and Marine Biology Annual Review*, **25**, 563–75.
Gubbay, S., 1989. *Coastal and Sea Use Management: A Review of Approaches and Techniques.* Ross-on-Wye: Marine Conservation Society.
Gubbay, S., 1992. Progress and opportunities for advancing marine nature conservation through coastal zone management in the United Kingdom. *Aquatic Conservation: Marine and Freshwater Ecosystems*, **2**, 357–63.

Halverson, M. & Martin, D.F., 1980. Studies of cytoclysis of *Chattonella subsalsa*. *Florida Science*, **43**, 35.
Hay, M.E., 1984. Patterns of fish and urchin grazing on Caribbean coral reefs: Are previous results typical? *Ecology*, **65**, 446–54.
Hughes, T.P., 1994. Catastrophes, phase shifts, and large scale degradation of a Caribbean coral reef. *Science*, **265**, 1547–51.
Hutchings, P.A. & Wu, B.L., 1987. Coral reefs of Hainan Island. *South China Sea. Marine Pollution Bulletin*, **18**, 25–6.
IPCC CZMS (International Panel on Climate Change. Coastal Zone Management Subgroup), 1992. *Global Climate Change and the Rising Challenge of the Sea*, ed. L. Bijlsma, J. O'Callaghan, R. Hillen, R. Misdorp, B. Mieremet, K. Ries, J.R. Spradley & J. Titus. The Hague: Ministry of Transport and Public Works.
IUCN/UNEP (International Union for the Conservation of Nature/United Nations Environment Programme), 1985. *Management and Conservation of Renewable Marine Resources in the Indian Ocean Region: Overview*. UNEP Regional Seas Reports and Studies, no. 60. Nairobi: UNEP.
IUCN/UNEP/WWF (International Union for the Conservation of Nature/United Nations Environment Programme/World Wide Fund For Nature), 1991. *Caring for the Earth: A Strategy for Sustainable Living*. Gland: IUCN.
Johannes, R.E., 1982. Traditional conservation methods and protected marine areas in Oceania. *Ambio*, **11**, 258–61.
Jokiel, P.L. & Coles, S.L., 1990. Response of Hawaiian and other Indo-Pacific reef corals to elevated temperatures. *Coral Reefs*, **8**, 155–62.
Kaspar, H.F., Gillespie, P.A., Boyer, J.C. & Mackenzie, A.L., 1985. Effects of mussel aquaculture on the nitrogen cycle and benthic communities in Kenepura Sound, Marlborough Sound, New Zealand. *Marine Biology*, **85**, 127–36.
Kelleher, G., Bleakley, C. & Wells, S. (1995). *A Global Representative System of Marine Protected Areas*. Vol. **1**: *Antarctic, Arctic, Mediterranean, Northwest Atlantic, Northeast Atlantic and Baltic*. Washington, DC: Great Barrier Reef Marine Park Authority, The World Bank.
Laws, E.A. & Redalje, D.G., 1982. Sewage diversion effects on the water column of a subtropical estuary. *Marine Environmental Research*, **6**, 265–79.
Lehman, S.J. & Keigwin, L.D., 1992. Sudden changes in North Atlantic circulation during the last deglaciation. *Nature*, **356**, 757–62.
Long, S.P. & Mason, C.F., 1983. *Saltmarsh Ecology*. Glasgow: Blackie.
McAllister, D.E., 1991. Questions about the impact of trawling. *Sea Wind*, **5**, 28–32.
McClanahan, T.R. & Muthiga, N.R., 1988. Changes in Kenyan coral reef community structure and function due to exploitation. *Hydrobiologia*, **166**, 269–76.
McCoy, F.W., 1988. Floating megalitter in the Eastern Mediterranean. *Marine Pollution Bulletin*, **19**, 25–8.
Mead, J.G. & Mitchell, E.D., 1984. Atlantic Gray Whales. In *The Gray Whale, Eschrictius robustus*, ed. M.L. Jones, S.L. Swartz & S. Leatherwood, pp. 33–53. Orlando: Academic Press.
Norse, E.A. (ed.), 1993. *Global Marine Biological Diversity: A Strategy for Building Conservation into Decision Making*. Washington, DC: Island Press.
Ong, J.E., 1982. Mangroves and aquaculture in Malaysia. *Ambio*, **11**, 252–7.
Ormond, R.F.G. & Caldwell, S., 1982. The effect of oil pollution on the reproduction and feeding behaviour of the common sea-anemone *Actinia equina*. *Marine Pollution Bulletin*, **130**, 118–22.

Paine, R.T., 1980. Food webs: Linkages, interactions strength and community infrastructure. *Journal of Animal Ecology*, **49**, 667–85.
Phillips, R.C., 1980. Transplanting methods. In *Handbook of Sea-grass Biology: An Ecosystem Perspective*, ed. R.C. Phillips & C.P. McRoy, pp. 41–56. New York: Garland.
Picton, B.E., Ball, B.J., Bowler, M. & Howson, C.M. (eds), 1994. *The Marine Conservation Society and Ulster Museum Directory of Marine Flora and Fauna*. Ulster: Ulster Museum.
Pimm, S.L., 1980. Food web design and the effects of species deletion. *Oikos*, **35**, 139–49.
Primavera, J.H., 1991. Intensive prawn farming in the Philippines: Ecological, social and economic implications. *Ambio*, **20**, 28–33.
Ray, G.C., 1976. Critical marine habitats. In *Proceedings of the International Conference on Marine Parks and Reserves*, pp. 15–59, IUCN Publications New Series no. 37. Morges: IUCN.
Riemann, B. & Hoffmann, E., 1991. Ecological consequences of dredging and bottom trawling in the Limfjord, Denmark. *Marine Ecology Progress Series*, **69**, 171–8.
SBSTTA (1995). Final Report of the First Meeting of the Subsidiary Body on Scientific, Technical and Technological Advice (SBSTTA) Under the Biodiversity Convention, Paris, 4–8 September, 1995.
Sherman, K., 1990. Productivity, perturbations, and options for biomass yields in large marine ecosystems. In *Large Marine Ecosystems: Patterns, Processes and Yields*, ed. K. Shermann, L.M. Alexander & B.D. Gold, pp. 206–9. Washington, DC: American Association for the Advancement of Science.
Shomura, R.S. & Godfrey, M.L. (eds), 1990. *Proceedings of the Second International Conference on Marine Debris, Honolulu, Hawaii*. Honolulu, HI: U.S. Department of Commerce.
Shomura, R.S. & Yoshida, H.O. (eds), 1985. *Proceedings of the Workshop on the Fate and Impact of Marine Debris, Honolulu, Hawaii*. Honolulu, HI: U.S. Department of Commerce.
Simenstad, C.A., Estes, J.A. & Kenyon, K.W., 1978. Aleuts, sea-otters, and alternate stable-state communities. *Science*, **200**, 403–11.
Sorensen, J.C. & McCreary, S.T., 1990. *Coasts: Institutional Arrangements for Managing Coastal Resources and Environments*. Renewable Resources Information Series. Coastal Management Publication no. 1, 2nd edn. Washington, DC: National Park Service, U.S. Department of the Interior.
Spanier, E. & Galil, B.S., 1991. Lessepsian migration: A continuous geographical process. *Endeavour, New Series*, **15**, 102–6.
UNCED (United Nations Conference on Environment and Development), 1992a. *Convention on Biological Diversity*. New York: UN.
UNCED (United Nations Conference on Environment and Development), 1992b. *Agenda 21*. New York: UN.
van der Elst, R.P., 1979. A proliferation of small sharks in the shore-based Natal sport fishery. *Environmental Biology of Fishes*, **4**, 349–62.
Watson, M. & Ormond, R.F.G., 1994. Effect of an artisanal fishery on the fish and urchin populations of a Kenyan coral reef. *Marine Ecology Progress Series*, **109**, 115–29.
Wells, S.M., Pyle, R.M. & Collins, N.M., 1983. *The IUCN Invertebrate Red Data Book*. Gland: IUCN.
Wells, S.M. & White, A.T., 1995. Involving the community. In *Marine Protected*

Areas. Principles and Techniques for Management, ed. S. Gubbay, pp. 61–84. Conservation Biology Series. London: Chapman & Hall.

White, A.T., Chou, L.M., De Silva, M.W.R.N. & Guarin, F.Y., 1990. *Artificial Reefs for Marine Habitat Enhancement in Southeast Asia*. Manila: International Centre for Living Aquatic Resources.

White, A.W., 1980. Recurrence of kills of Atlantic herring (*Clupea harengus harengus*) caused by dinoflagellate toxins transferred through herbivorous zooplankton. *Canadian Journal of Fisheries and Aquatic Science*, **37**, 2262–5.

World Coast Conference, 1993. *Preparing to Meet the Coastal Challenges of the 21st Century*. Conference Statement, Noordwijk, The Netherlands. The Hague: Ministry of Transport, Public Works & Water Management.

World Commission on Environment and Development (1987). *Our Common Future*. Oxford: Oxford University Press.

Chapter 18
Conserving biodiversity in North-East Atlantic marine ecosystems

KEITH HISCOCK
Joint Nature Conservation Committee, Monkstone House, Peterborough, PE1 1JY, UK

Abstract

Maintaining biodiversity is a central tenet of wildlife conservation. To conserve marine life, information on the 'resource' (the habitats, communities and species) needs to be collected and used in a structured way in order to identify the nature conservation importance of sites and to manage areas to conserve their important features. This includes the organisation of descriptive data through systems of classification and the establishment of criteria for assessing the nature conservation importance of locations. Management requires information on the functioning of marine ecosystems, but especially an understanding of how vulnerable particular habitats, communities and species are to different activities being undertaken in the area. Research needs to tackle some difficult applied questions that will not always suit the requirements of publishing and of short project length.

18.1 Introduction

'Nature conservation' is defined here as: 'the regulation of human use of the global ecosystem to sustain its diversity of content indefinitely' (NCC, 1984). Conservation requires information on the resource and a structured approach to decision-making so that actions are soundly based and defensible. Conserving biodiversity also requires political initiatives and will. Recent such initiatives have included the United Nations Convention on Biodiversity (published in Britain by the Foreign and Commonwealth Office, 1995) and the European Communities Directive on the Conservation of Natural Habitats and of Wild Fauna and Flora (Council of the European Communities, 1992). Several of the recommendations made at the 1994 General Assembly of The World Conservation

Union also directly address marine conservation issues (IUCN, 1994a). Conservation of biodiversity as a central part of nature conservation is achieved through three main approaches:

Management for conservation of representative examples of the range of biotopes present within biogeographically/physiographically similar areas of coast.

Management for conservation of species that are rare, scarce and/or threatened and require special conservation measures.

Application of a general duty of care and principles of sustainable use to maintain general environmental quality and ensure availability of resources in perpetuity.

18.2 Information requirements and their application

Five main areas of information are required for management of the marine environment including site-based conservation:

Physical and biological resource data.
Physical and chemical environmental information.
Information on the structure of marine communities and on the key elements in their functioning.
Information on natural variability.
Information on effects of human activities.

This information will require a structured approach to its application to conservation and management through:

Classification systems for resource data.
Criteria to identify the nature conservation importance of an area including those to identify a comprehensive series of marine protected areas.

In conserving marine ecosystems including habitats, communities and species, we are constrained by poor knowledge of the resource, the functioning of that resource and the importance of human activities to that resource. With such difficulties, it is necessary to take a very practical approach using the information we have to best effect and often to support 'best available judgement' rather than definitive certainty.

18.2.1 Site assessment and selection

Using information that describes the habitats, communities and species present, locations can be assessed for nature conservation importance and some will be selected for management as marine protected areas. Criteria to identify the nature conservation importance of marine areas have been developed (for instance, for the Council of Europe (Mitchell, 1987); for the International Maritime Organisation (IMO, 1991); for the IUCN (Kelleher & Kenchington, 1992) and those included in the European Union Habitats Directive (Council of the European Communities, 1992)). To take full advantage of the information now available in Great Britain through the work of the Marine Nature Conservation Review (MNCR) (Hiscock, 1996) and to include newly developed concepts of rarity and sensitivity, new criteria and protocols are being developed by the Joint Nature Conservation Committee in the United Kingdom and will be published in due course. The main scientific criteria that should be considered for use in identifying nature conservation importance to conserve biodiversity are noted below.

18.2.1.1 Representativeness

Conservation of biodiversity can be based on protecting representative examples from the range of natural or semi-natural habitats and associated communities (biotopes) within a biogeographically distinct area or the boundaries of a national territory. This approach anticipates that, by protecting the full range of biotopes, the full range of species will also be protected. The identification and conservation of representative biotopes is necessarily crude because of the sparse information on the distribution and abundance of all species. The basic tool for identifying representative sites is the classification of biotopes. The development of a classification is being undertaken by the MNCR in Great Britain and is being expanded, through the European Union LIFE-funded project 'BioMar', to be capable of extension to a North-East Atlantic classification (Connor *et al.*, 1995).

18.2.1.2 Diversity

Diversity in terms of richness of different types is most frequently what is meant by this term. Number of species in an area together with the abundance of each organism, can be analysed using various formulae to calculate 'diversity indices' but these are not generally relevant to conservation assessment and the term 'Diversity' as used in nature conservation cannot be equated with these indices. Locations with high species richness are bound to stand out as of higher value than examples of similar habitats with a lower diversity but including the same species. In undertaking assessment of species richness, it is important to compare the same or similar habitats as some are inherently poor in species. Diversity of different habitats or biotopes within an area is also important particularly when ensuring that the full range of biotopes, and preferably the best examples, in a biogeographically or physiographically similar area are represented in a reasonably small number of protected sites. Conserving diversity of size can be important if, for instance, large sexually mature individuals are being decimated by a fishery leaving no reproductive stock. Conserving genetic diversity is also important in maintaining healthy populations.

18.2.1.3 Rarity

'Rarity' has a high perceived importance in public and political terms. The degree of rarity of a species, community or habitat is often clear to an experienced biologist but difficult to quantify as a criterion. It is important to acknowledge that different species will exhibit different sorts of rarity. Species that have a large size would not be expected to be so numerous as those with a small size. Some species will naturally occur in very small numbers although over a wide range of habitats or extensive habitats and therefore a large area. Others will be very habitat specific and, if the habitat is unusual, will be found in only a few locations although possibly in high densities. Others may be rare because they are relicts of a much wider distribution in previous times and conditions. Low numbers and restricted distribution may be a wholly natural characteristic for a particular species and should not infer any likelihood of threatened status. Conclusions of the significance of rarity in a nature conservation context therefore need to take account of the type of distribution and abundance that would be expected in a particular species and any historical information about past numbers. However, wherever a species, or a biotope, is only known from a small number of locations, those locations will be significant from a nature conservation point-of-view.

Inventories of fauna and flora and of biotopes provide the basis, if they are sufficiently comprehensive and at a sufficiently fine mapping scale, to establish which are rare or scarce species or biotopes from a distributional point-of-view. Although such information is generally much less complete for marine than for most terrestrial species, similar criteria to those used for terrestrial species to assess rarity are being developed and are being trialled for conspicuous benthic species around Great Britain. Obtaining records that reflect the 'real' occurrence of species, especially inconspicuous rare species, rather than the distribution of recorders is never going to be easy. If we are to get as close as possible to a scheme that maps distribution in a meaningful way for inconspicuous as well as conspicuous species, the following must be key attributes.

Information must be drawn from the widest range of sources: surveys; the records of taxonomic specialists; museum collections; competent volunteer/amateur recording schemes.

Information must be recorded at the finest scale possible (six figure Ordnance Survey grid reference in Great Britain/two decimal points of minutes of latitude/longitude) but to coarser scales if that is all that is available.

Interpreting IUCN guidelines (most recently revised in IUCN, 1994b) in a Great Britain context, nationally 'rare', 'scarce' or 'uncommon' species are identified on the basis of their percentage occurrence in 10 x 10 km squares of the Ordnance Survey National Grid. Using this approach, a series of British Red Data books have been produced for terrestrial, freshwater and lagoonal species (e.g. Bratton, 1991). If this mapping approach is extended to in-shore areas within the three nautical mile (approximately 5.5 km) limit of territorial seas (which approximates to the zone under the influence of coastal processes), a 'nationally rare' species would occur in eight or fewer squares, and a 'nationally scarce' species in 9–55 squares (Sanderson, 1996). If this mapping approach is adopted for assessing rarity, the use of 10 x 10 km squares should be retained for the marine environment throughout the North-East Atlantic even though different national mapping grids will be applied. Units of latitude and longitude should not be used as the same units of longitude get progressively narrower with increasing distance away from the equator and therefore the area of each unit becomes smaller.

Rarity must be used in association with species richness in assessing the importance of a site. A site with low species richness but several rare species may be more highly rated than one of similar habitats with a large number of common, widely distributed species.

Although no quantitative criteria have been developed for biotopes, some habitats are likely to occur at only a few locations and, particularly if certain species only occur, or only occur in abundance, there, those biotopes are likely to be important. The occurrence of chalk coasts is an example of both rarity in Great Britain and international importance. Chalk coasts comprise less than 0.6% of the coastline of Great Britain and are colonised by assemblages of filamentous green algae that include species known only from these habitats (Fowler & Tittley, 1993).

If a country has the largest or a significant proportion of the population of a species within its national territory (even if the population is a large and thriving one), it will have a special responsibility for the conservation of that species. Such a situation exists for the Atlantic grey seal, *Halichoerus grypus*, in Great Britain, which holds 40–45% of the world population (Hiby *et al.*, 1992). There will also be special responsibility in relation to migratory species. Estuaries in Great Britain regularly support 44 species of migratory waterfowl including up to 100% of the world population of some species and subspecies (N. Davidson, personal communication).

18.2.1.4 Risk of extinction

Rarity criteria have been developed partly to support quantitative measures for 'risk of extinction'. Although global extinction is a very rarely recorded event in marine species (in the North-East Atlantic, Steller's sea cow and the great auk are the only marine species known with certainty to have been made extinct by man in recent centuries), the lack of data might also mean that species are being extinguished without record. A framework for the classification of species according to their risk of extinction is provided in the IUCN Red List Categories (IUCN, 1994b). The IUCN categories 'critically endangered', 'endangered' and 'vulnerable' rely on quantitative information being available on recent population decline, on extent of occurrence in km^2 worldwide or on numbers of mature individuals known to be alive and whether declining. It is likely that such criteria could only be applied to some marine vertebrates living or spending a key part of their life close in-shore (for instance, seabirds: Avery *et al.*, 1994). The quantitative criteria described earlier to identify nationally rare or scarce marine species do not provide an indication of risk of extinction. Almost all marine species would fall into the IUCN Red List 'data deficient' category and so risk of extinction is not currently a practical criterion to pursue with regard to benthic and most pelagic species.

18.2.1.5 Sensitivity

Species or habitats are likely to be sensitive if they:

are fragile (brittle);
are susceptible to pollution;
are long-lived and recruit poorly;
are slow to reach maturity;
have poor recruitment;
have poor larval dispersal or no larval stage;
are unable to move away.

However, our knowledge of which species are sensitive because of longevity or larval biology is very sparse. Studies of the effects of pollution and physical disturbance, including experimental studies, reveal most about sensitive species whilst long-term monitoring is required to understand growth rates, longevity and recruitment rates of species.

Sensitivity is here considered synonymous with 'fragility' – the term used previously in conservation criteria.

Sensitivity is most likely to be used as a strong criterion in site selection by taking account of vulnerability of the sensitive species or habitats to activities and impacts. For instance, the tall seapen, *Funiculina quadrangularis*, is sensitive because it has a brittle rachis and does not retract into the sediment. However, it is vulnerable only when an activity such as bottom fishing with mobile gear is undertaken where it occurs. Future work might include mapping the sensitivity of coastal areas by 'tagging' biotope types with sensitive species and matching those records to the different impacts and activities in or proposed for those areas.

18.2.1.6 Dependency

The dependency of a species, community or ecological process on a particular location (for instance, a feeding, breeding, sheltering area or a migration corridor) or structure (for instance, a kelp forest, a sea-grass bed, a maerl bed) makes that location important particularly if there are no (or very few) alternative locations for a species or community to survive in.

18.2.1.7 Irreplaceability

Some locations will include habitat features, biotopes and species that if destroyed in some way, will not be capable of replacement. This may be because the habitat could not be restored or replaced or because species present have poor or local recruitment, are relict (and therefore have no nearby sources of larvae for recruitment) or, for some other reason, will not re-establish once lost. Where the irreplaceable features are rare or even unique, the impetus for conservation action must be high as loss would be a blow for biodiversity.

18.2.1.8 Naturalness

In general, the more natural a site, the better. Since naturalness is found so widely in the marine environment, compared to terrestrial systems, this criterion will be less relevant than foregoing ones. However, a site may be considered less valuable for conservation because of the presence of non-natural features including artificial substrata, pollution or non-native species that affect community composition. As on land, man-made features in the marine environment sometimes harbour unusual communities and rare species. In this case they may be considered of high nature conservation importance. For instance, one of the few Sites of Special Scientific Interest in Great Britain with marine biology currently cited as important includes Abereiddy Quarry in West Wales. Abereiddy Quarry is a flooded slate quarry in which sheltered-water biotopes and those characteristic of de-oxygenated conditions have developed (Hiscock & Hoare, 1975).

18.2.1.9 Extent

In identifying sites for protection, preference will be given to sites with larger examples of highly rated or rare biotopes. It is also necessary to consider the size of site required to ensure that the unit to be managed is 'viable'. Conservation of 'functional units' in which the ecological processes that support the communities of interest are confined is desirable. However, such units are rare in nature – especially in the sea. The closest a site might come to a functional unit is a lagoon, estuary or other enclosed area.

Identification of sites for conservation also involves use of practical criteria such as 'situation', 'recorded history', 'research and educational potential', 'restoration potential', 'intrinsic appeal', 'vulnerability', 'urgency' and 'feasibility' (see Mitchell, 1987 for an explanation of their application).

18.2.2 Management

Information on environmental conditions and on the structure and functioning of communities is used in preparing management plans and minimising adverse effects of development and pollution. Conservation in marine systems is fundamentally more difficult than on the land because of the much greater degree of interconnectedness and lack of clear boundaries. The following might be considered units for conservation management.

1. Large marine ecosystems (e.g. the North Sea, the western English Channel).
2. Physiographically or (as near as marine systems come) functionally distinct systems (e.g. a stretch of rocky coastline, a large open bay, an estuary, an island).
3. Shore/sea-bed 'types' (e.g. a wave sheltered shore from high water to the off-shore plain).
4. Biotopes (e.g. shallow eulittoral rock pools with coralline crusts and *Corallina officinalis*).
5. Species.

This is almost a hierarchy, but not quite, as many coastal types, although physiographically distinct, are not functionally discrete. Also, species are often 'managed' by specific and universal legislative protection rather than site-specific protection. Management of large marine ecosystems is often achieved through widespread measures limiting inputs or activities universally. Because scientific analysis is often not able to establish cause and effect between (likely) changes in the ecosystem and the activities of man, a precautionary approach often has to be applied. However, some areas are clearly adversely affected by man and restoration is an important aspect of management. In particular, for functionally discrete systems such as an estuary, there may be a clear and quantifiable benefit in terms of biodiversity following clean-up (for instance, recovery following clean-up in the Tyne estuary: Hardy *et al.*, 1993).

At the level of sea-bed types or biotopes, localised disturbance that damages habitats, communities and species needs to be avoided or controlled. Adverse effects on the natural character of the sea-bed, often including reduction in biodiversity, may be brought about by, for instance: physical damage to habitats and species by mobile fishing gear; removal of a habitat by dredging; smothering of habitats by dumping including land claim; disposal of harmful effluents, or educational/scientific collecting activities. Locations with sensitive habitats, communities and species will need a higher level of protection than more robust or less sensitive parts of a protected area.

Monitoring provides essential feedback to management. This includes monitoring of environmental conditions (which may account for observed changes), monitoring of the special features of the area (which indicates if the features for which the area is important are still present, improving or degrading), monitoring human activities that may damage features of importance and monitoring of compliance with regulations.

18.3 Some conflicts and dilemmas

18.3.1 Charisma and conservation

From a practical point-of-view, it is far easier to obtain conservation action for charismatic vertebrate species than for most algae and invertebrates. The fate of charismatic megafauna such as cetaceans, seals and seabirds attracts great public sympathy and therefore political interest. 'Marketing' of nature conservation usually relies on such species. With some vertebrate species, animal welfare issues may also become confused with nature conservation requirements.

Charismatic species provide a focus for marine conservation and their protection, whatever the motive, can incidentally provide protection for the habitats that they occupy. However, the conservation of biodiversity cannot depend solely or mainly on the chance presence of charismatic species. Fortunately, conservation of 'biodiversity' is currently seen as a 'good thing' politically and, from a scientific point-of-view, it is this conservation of the whole range of habitats and species that should be encouraged.

18.3.2 Increasing biodiversity?

There are bound to be many issues that science cannot resolve but come close to philosophical considerations. We know that introducing hard substrata into otherwise sedimentary environments (for instance, jetty piles in estuaries, oil production platforms in the North Sea) increases the biodiversity in that area. But we are not yet deliberately introducing new habitats for nature enhancement. Introducing non-native species also increases diversity but this is tampering with nature and frequently, as in the case of the slipper limpet, *Crepidula fornicata*, diversity is subsequently reduced over extensive areas because of domination by the non-native or, as in the case of the toxic marine alga, *Heterosigma akashiwo*, other organisms may be killed by the newcomer. Strenuous efforts are therefore appropriate to prevent biodiversity being increased by the introduction of non-natives.

18.3.3 Research objectives

Much research is designed to produce results within a time-scale that facilitates rapid publication. The often quick and certain approach means that fairly simple accessible systems that are likely to respond to experimental study are usually selected. Also, results from surveillance (or monitoring) that indicate little change with time are often considered less interesting than reporting systems that change substantially and trying to establish why they change. In the tedious analysis of surveillance results, the temptation must be to concentrate on changeable species or communities. However, the most vulnerable marine biotopes and species are likely to be those that are highly stable with time and probably fragile. We need to establish which those are and what are the likely consequences of the activities of man including pollution – even if the work is going to be unattractive for a journal or will take years to generate unequivocal results.

18.4 Future requirements

International collaboration is required to further develop systems of classification that will provide a structure for comparison of different areas and application of criteria for assessment and site selection. Local effort is required to understand better how different systems function and what happens when they are disturbed including opportunistic studies during or following extreme events. Field and laboratory studies including monitoring are required to establish which species and communities are likely to be robust through to those likely to be highly vulnerable in relation to different inputs, activities and impacts. We also need to understand the role of species in different communities including the importance (or otherwise) of biodiversity in ecosystem function.

Acknowledgements

Bill Sanderson undertook the work to establish rarity criteria and, with David MacDonald, Colin McLeod and Sue Scott, has reviewed earlier drafts of the manuscript. Developing marine site assessment and selection criteria is being undertaken with colleagues in the UK nature conservation agencies and preparation of material for that exercise has helped in writing this chapter.

References

Avery, M., Gibbons, D.W., Porter, R., Tew, T., Tucker, G. & Williams, G., 1994. Revising the British Red Data List for birds: The biological basis of U.K. conservation priorities. *Ibis*, **137**, S232–9.

Bratton, J.H. (ed.), 1991. *British Red Data Books: vol. 3. Invertebrates Other Than Insects.* Peterborough: Joint Nature Conservation Committee.

Connor, D.W., Hiscock, K., Foster-Smith, R.L. & Covey, R., 1995. A classification system for benthic marine biotopes. In *Biology and Ecology of Shallow Coastal Waters* (Proceedings of the 28th European Marine Biology Symposium, Crete, September 1993), ed. A. Eleftheriou, A.D. Ansell & C.J. Smith, pp. 155–65. Fredensborg: Olsen and Olsen.

Council of the European Communities, 1992. Council Directive 92/43/EEC of 21 May 1992 on the Conservation of Natural Habitats and of Wild Fauna and Flora. *Official Journal of the European Communities, Series L*, **206**, 7–50.

Foreign and Commonwealth Office, 1995. *Convention on Biological Diversity. Open for signature at Rio de Janeiro from 5 to 14 June 1992 and thereafter at the headquarters of the United Nations at New York from 15 June 1992 until 4 June 1993.* London: HMSO. Cm 2915.

Fowler, S.L. & Titley, I., 1993. *The Marine Nature Conservation Importance of British Coastal Chalk Cliff Habitats.* English Nature Research Report no. 32. Peterborough: English Nature.

Hardy, F.G., Evans, S.M. & Tremayne, M.A., 1993. Long-term changes in the marine macroalgae of three polluted estuaries in north-east England. *Journal of Experimental Marine Biology and Ecology*, **172**, 81–92.

Hiby, L., Duck, C. & Thompson, D., 1992. Seal stocks in Great Britain: Surveys conducted in 1990 and 1991. *NERC News*, **20**, 30–1.

Hiscock, K. (ed.), 1996. *Marine Nature Conservation Review: Rationale and Methods.* Coasts and Seas of the United Kingdom, MNCR series. Peterborough: Joint Nature Conservation Committee.

Hiscock, K. & Hoare, R., 1975. The ecology of sublittoral communities at Abereiddy Quarry, Pembrokeshire. *Journal of the Marine Biological Association of the United Kingdom*, **55**, 833–64.

IMO (International Maritime Organisation), 1991. *Report of the Marine Environment Protection Committee on its Thirty-first Session. MEPC 31/21, Annex 17. Procedures for the Identification of Particularly Sensitive Sea Areas.* London: International Maritime Organisation.

IUCN (International Union for the Conservation of Nature and Natural Resources), 1994a. *Resolutions and Recommendations. 19th Session of the General*

Assembly of IUCN – The World Conservation Union. Buenos Aires, Argentina. 17–26 January 1994. Gland, Switzerland: IUCN.
IUCN (International Union for the Conservation of Nature and Natural Resources), 1994b. *IUCN Red List Categories*. Gland, Switzerland: IUCN.
Kelleher, G. & Kenchington, R., 1992. *Guidelines for Establishing Marine Protected Areas*. A Marine Conservation and Development Report. Gland, Switzerland: IUCN.
Mitchell, R., 1987. *Conservation of Marine Benthic Biocenoses in the North Sea and the Baltic. A Framework for the Establishment of a European Network of Marine Protected Areas in the North Sea and the Baltic*. Nature and Environment Series, no. 37. Strasbourg: Council of Europe for European Committee for the Conservation of Nature and Natural Resources.
NCC (Nature Conservancy Council), 1984. *Nature Conservation in Great Britain*. Shrewsbury: Nature Conservancy Council.
Sanderson, W., 1996. Rarity of marine benthic species in Great Britain: Development and application of assessment criteria. *Aquatic Conservation: Marine and Freshwater Ecosystems*, **6**, 245–56.

Author index

Page numbers in *italics* indicate authors cited in the reference lists.
For all papers cited, only the first-named author has been indexed.

Abbe, G.R., 361, *368*
Abbott, R.T., 183, *194*, 263, *270*
Abe, N., 193, *194*
Abele, L.G., 22, *32*, 130, 142, *142*
Abrams, P.A., 232, 242, 243, *249*
Aebischer, N.J., 332, *333*
Allen G,R., 219, 227, 231, 236, *249*
Aller, J.Y., 152, 158, 159, 160, *172*
Altena, C.O., 186, 194, *194*, 263, *270*
Andersin, A.-B., 332, *333*
Anderson, G.R.V., 229, 230, *249*
Angel, H.H., 36, *65*
Angel, M.V., 9, *15*, 45, 47, 58, 59, 60, 61, *65*, 70, 71, *92*, 104, *117*, 128, *142*, 401, *410*
Ankar, S., 325, 329, *333*
Anon., 331, *333*
Aoki, T., 383, *389*
Arnaud, P.M., 129, 137, *142*
Arntz, W.E., 128, 129, 130, 132, 135, 141, *142*
Aschan, M., 25, 26, *32*
Auffret, G., 157, *172*
Aure, J., 380, *389*
Avery, M., 420, *426*
Avise, J.C., 277, 284, 289, *290*
Azuma, M., 185, *195*

Bäck, S., 326, *333*
Backeljau, T., 295, 298, 314, *315*
Backus, R.H., 41, *65*
Badcock, J., 59, *66*
Bahr, L.M., 361, 364, *368*
Baker, A. de C., 55, *66*
Baltz, D.M., 383, 384, *389*
Banerjee, B.K. , 378, *389*

Bangen, M., 381, *389*
Barange, M., 54, *66*
Barg, U.C., 377, 380, *390*
Bargelloni, L., 138, *143*, 259, *270*
Barker, P.F, 135, *143*
Barnes, H., 361, *368*
Barnes, R.F.K., 149, *172*
Barthel, D., 132, *143*
Bartholomew, G.A., 367, *368*
Bayne, B.L., 362, *368*
Beaumont, A.R., 295, *315*
Beddington, R.R., 398, *410*
Bell, J.D., 212, *213*, 232, *249*
Bellwood, D.R., 226, *249*
Benton, M.J., 6, *15*
Benton, T.G., 397, *410*
Berger, W.H., 160, *172*
Bernes, C., 325, *333*
Bert, T.M., 283, *290*
Beveridge, M.C.M., 373, 376, 379, 380, 383, 385, *390*
Bieri, R., 70, *92*
Billett, D.S.M., 101, *117*
Birkeland, C., 179, 180, 181, 182, 184, 193, *195*
Bjorklund, M., 404, *410*
Blake, J.A., 97, 98, 99, 116, 117, *117*
Blot, M., 295, *315*
Bock, W.J., 3, *15*
Bogdanov, I.P., 263, *270*
Bohnsack, J.A., 232, 244, *249*
Bonsdorff, E., 329, *333*
Booth, D.A., 154, *172*
Bormann, F.H., 366, *368*
Botton, M.L., 152, *172*
Bouchet, P., 113, *117*

429

Bouchon-Navaro, Y., 231, 237, *249*
Brandt, A. , 109, *117*, 137, *143*
Bratton, J.H., 419, *426*
Brewer, A. , 14, *15*
Brey, T., 7, *15*, 104, 108, 109, *117*
Briand, F., 365, *368*
Briggs, D.E.G., 3, *15*
Briggs, J.C., 4, *15*, 219, *249*, 260, *270*
Briggs, M.R.P., 380, *390*
Brinton, E., 41, *66*
Britton, J.C., 186, *195*
Broecker, W.S., 50, *66*, 400, *410*
Brouard, F., 231, *249*
Brown, J.H., 6, 10, *15*, 381, *390*
Brown, J.R., 395, *410*
Brown, R.A., 398, *410*
Bucklin, A., 70, *92*
Bugge, T., 115, *117*
Bullock, L.H., 247, *249*
Buroker, N.E., 280, 281, 282, *290*
Burton, R.S., 278, 279, 280, *290*

Caley, K.J., 297, 298, 300, 301, 311, 313, *315*
Campbell, J.W., 44, *66*
Carlson, C.A., 58, *66*
Carlton, J.T., 399, 401, *410*
Carney, R.S., 96, *117*
Carr, A., 397, *411*
Carss, D., 379, *390*
Cernohorsky, W.O., 186, 187, 194, *195*, 263, *270*
Chamberlain, G.W., 378, *390*
Chanratchakool, P., 386, *390*
Charlesworth, D., 287, *290*
Chesson, P.L., 242, *250*
Cheung, S.G., 184, *195*
Childers, D.L., 362, *368*
Choat, A.H., 243, *250*
Christie, W.J., 373, *390*
Clark, R.B., 395, *411*
Clarke, A., 20, *32*, 130, 135, 136, 137, 138, 141, *143*, 260, *270*, *270*
Clarke, R.D., 231, *250*
Clayton, M.A., 136, *143*
Coastal Area Management and Planning Network, 407, *411*
Cody, M.L., 19, *32*, 202, 205, 206, *213*
Cohen, J.E., 14, *15*, 48, *66*
Colborn, T.E., 396, *411*
Colwell, R.K., 10, *15*, 126, *143*
Connell, J.H., 152, *172*, 234, 238, 244, *250*
Connor, D.W., 417, *426*
Connor, E.F., 126, *143*
Cook, R.E., 9, *15*
Cornell, H.V., 111, 113, 114, *117*, 130, 143
Cotgreave, P., 9, *15*

Council of European Communities, 415, 417, *426*
Cowen, R.K., 243, *250*
Cowling R., 12, *15*
Coyne, J.A., 294, 310, *315*
Crame, J.A., 136, 137, *143*, 261, 263, 267, 270, *270*, *271*
Crisp, J.D., 279, *290*
Crossland, C.J., 181, *195*
Crossland, S., 300, *315*
Crothers, J., 5, *15*
Crow, J.F., 282, *290*
Csanady, G.T., 154, *172*
Cunningham, C.W., 284, *290*
Currie, D.J., 9, *15*, 104, *117*
Curtis, L.A., 186, *195*
Curtis, M.A., 128, *143*

Dahl, E., 115, *118*, 321, *333*
Dale, G., 243, *250*
Dame, R., 361, 362, 364, 366, *368*
David, P.M., 87, 89, *92*
Dayton, P.K., 22, 32, 128, 129, 130, 131, *143*, 151, 152, *172*, 236, *250*
de Baar, H.J.W., 44, *65*
de Broyer, C., 129, *143*
DeAngelis, D.L., 180, *195*
Dell, R.K., 128, 130, 141, *143*, 263, *271*
DeMaster, D.J., 159, *172*
Denman, K.L., 45, *66*
Denny, M.W., 362, *368*
Desbruyères, D., 152, 162, *172*
Dickson, R.R., 154, 155, 156, 162, 164, 167, *172*
Diel, S., 54, *66*
Dixon, A.F.G., 7, *15*
Dodge, J.D., 104, *118*
Doherty, P.J., 234, 236, 241, 244, *250*
Donaldson, T.J., 223, *250*
Done, T.J., 248, *250*
Drake, J.A, 399, *411*
Dunton, K., 128, 132, 133, 134, 137, *144*

Eastman, J.T., 138, *144*, 259, *271*
Ebeling, A.W., 237, 243, *250*
Eckman, J.A., 157, *173*
Eckman, J.E., 101, 103, *118*
Edwards, C.A., 295, 313, *315*
Edwards, J., 399, *411*
Eggleton, P., 7, *15*
Ehrlich, P.R., 277, *290*, 331, *334*
Eitteim, S.L., 155, *173*
Ekman, S., 35, *66*
Elliott, A.J., 155, *173*
Elmgren, R., 319, 321, 325, 326, 327, 328, 329, 333, *333*, *334*
Embry, T.M., 2, *16*
Emery, K.O., 341, 353, *368*, *369*

Author index

Emmet, A.M., 5, *16*
Endler, J.A, 278, *290*
Enell, M., 385, 386, *390*
English Nature, 397, *411*
Erwin, T.L., 276, *291*
Etter, R.J., 28, *33*, 96, 97, 99, 102, 103, 104, 113, 116, 117, *118*, 151, 158, *173*
Ewens, W.J., 288, *291*

Fagetti, E.G., 79, *92*
Fairbanks, R.G., 53, *66*
FAO, 373, *390*
Farrell, B.D., 259, 268, *271*
Fasham, M.J.R., 47, *66*
Fauquet, C.M., 4, *16*
Fautin, D.G., 221, *250*
Feigenbaum, D., 84, *92*
Felsenstein, J., 294, 300, 306, *315*
Fenchel, T., 149, *173*
Field, C.D., 405, *411*
Fischer, A.G., 104, *118*, 127, 139, *144*
Fisher, R.A., 19, *33*, 282, *291*
Fitzmaurice, M., 332, *334*
Fleminger, A., 52, *66*, 79, 82, *92*
Flessa, K.W., 127, 141, *144*, 273, *273*
Folke, C., 376, 384, 389, *390*
Foote, M., 3, *16*
Forbes, E., 296, *315*
Foreign and Commonwealth Office, 415, *426*
Fowler, A.J., 231, *250*
Fowler, C.W., 397, *411*
Fowler, S.L., 399, *411*, 420, *426*
Foy, R.H., 380, *390*
France, R., 104, *118*
Franklin, I.R., 276, 288, *291*
Frith, D., 186, *195*
Funch, P., 70, *92*, 401, *411*
Furnestin, M.-L., 52, *66*
Futuyuma, D.J., 293, *315*

Gage, J.D., 9, 13, *16*, 28, 29, 31, *33*, 96, 109, *118*, 158, 160, 162, 163, 167, *173*
Galtsoff, P.S., 280, *291*, 361, 362, *369*
Galzin, R., 212, *213*
Gardner, W.D., 154, 155, *173*
Garman, G.C., 366, *369*
Gaston, K.J., 4, *16*, 146, *147*, 245, *250*
Gaudie, A., 32, *33*
Gauld, I.D., 9, *16*
Gausen, D., 384, *391*
Gentry, A.H., 207, *213*
GESAMP, 379, *391*, 395, *411*
Giovannoni, S.J., 4, *16*
Gladfelter, W.B., 232, 234, 237, 238, *250*, *251*
Glynn, P.W., 182, *195*, 400, 401, *411*
Golikov, A.N., 263, *271*
Gomez, E.D., 399, *411*

Gooday, A.J., 158, *173*
Gosliner, T.M., 203, *213*
Gosling, E.M., 293, 295, *315*
Gosner, K.L., 359, 360, *369*
Gosse, P.H., 294, *315*
Gowen, R.J., 376, 380, *391*, 395, *411*
Grahame, J., 297, 299, 310, 314, *315*, *316*
Grant, W.D., 155, 160, *173*
Grassle, J.F., 4, *16*, 22, 26, 27, 29, 31, *33*, 101, 103, 110, 117, *118*, 124, 128, 142, *144*, 148, 150, 152, 153, 162, *173*, 326, *334*
Grave, K., 381, *391*
Gray, J.S., 25, 28, *33*, 107, *118*, 332, *334*
Greig-Smith, P., 36, *66*
Grigg, R.W., 180, *195*
Groombridge, B., 3, 4, *16*
Gross, T.F., 155, *173*
Gubbay, S., 409, *411*
Gyllensten, U., 279, *291*

Hadfield, M.G., 115, *118*
Hain, S., 141, *144*
Hall, S.J., 157, 158, *173*
Hallock, P., 179, 180, 182, *195*
Halverson, M., 396, *412*
Hammer, M., 331, *334*
Hannaford Ellis, C.J., 297, *316*
Hansen, P.K., 383, *391*
Hanski, I., 364, *369*
Hardy, F.G., 423, *426*
Harmelin-Vivien, M.L., 222, 232, 237, 244, 245, *251*
Harper, J.L., 3, *16*
Harrison, S., 130, *144*
Hart, A., 310, *316*
Hastings, T.S., 383, *391*
Haury, L.R., 36, 37, *66*, 85, 86, *92*
Haven, D.S., 362, *369*
Hawksworth, D.L, 2, *16*
Hay, M.E., 398, *412*
Hayami, I., 263, *271*
Hayden, P.B., 338, *369*
Haylor, G.S., 294, *316*
Hecht, A.D., 260, *271*
Hecker, B., 96, *118*
Hedeen, R.A., 359, 360, *369*
Hedgecock, D., 279, 288, *291*
Hedgpeth, J.W., 132, *144*
Hedrick, P., 278, *291*
Heezen, B.C., 154, *174*
Heller, J., 297, *316*
Hernroth, L., 325, *334*
Hessler, R.R., 20, *33*, 109, *118*, 160, 161, 167, *174*
Hiby, L., 420, *426*
Highsmith, R.C., 182, 185, *195*
Hiscock, K., 417, 422, *426*

Hixon, M.A., 241, 245, *251*
Hobson, E.S., 230, *251*
Hochachka, P.W., 96, *118*
Hoffman, M.T., 10, *16*
Holbrook, S.J., 241, *251*
Holligan, P.M., 338, *369*
Hollister, C.D., 101, *118*, 155, 156, 157, *174*
Hopkins, T.L., 55, 56 *66*
Hourigan, T.F., 222, 227, *251*
Howard, R., 3, *16*
Hughes, T.P., 396, *412*
Hugueny, B., 113, *118*
Hulbert, S.H., 29, 31, *33*, 97, 100, 106, 111, *118*, 160, *174*
Hull, S.L., 299, 310, *316*
Huston, M., 10, *16*, 19, 22, 23, 32, *33*, 102, 119, 126, *144*, 204, 205, 211, *213*
Hutchings, P., 219, *251*, 398, *412*
Huthnance, J.M., 154, *174*

ICBP, 246, *251*
IMO, 417, *426*
Inui, M., 388, *391*
IOC CDNI3, 407, 412
Isley, A.E., 157, *174*
IUCN, 416, 419, 420, *426*, 427
IUCN/UNEP, 397, *412*
IUCN/UNEP/WWF, 395, *412*

Jablonski, D., 113, *119*, 267, *271*
Jackson, J.B.C., 185, *196*, 211, *213*, 226, *251*, 331, *334*
James, B.L., 296, *316*
Janson, K., 298, 309, *316*
Jansson, B.-O., 324, *334*
Jansson, K., 327, *334*
Jernelöv, A., 332, *334*
Johannes, R.E., 189, *196*, 395, 406, *412*
Johannesson, K., 297, 299, 310, 311, *316*, 326, *334*
Johnson, G.L., 155, *174*
Johnson, R.G., 152, 155, *174*
Jokiel, P., 219, 223, 224, 225, *251*, 400, *412*
Jones, D.S., 186, *196*
Jones, G.P., 234, 235, 237, 243, 245, *251*
Josefson, A.B., 113, *119*
Juell, J.-E., 380, *391*
Jumars, P.A., 22, *33*, 95, 101, 103, *119*, 151, 152, 153, 157, 158, 159, 160, *174*

Kami, H.T., 236, *252*
Kankaala, P., 325, *334*
Kantor, Yu.I., 263, *271*
Karl, S.A., 281, 282, *291*
Karl, S.L., 299, 313, *316*
Kaspar, H.F., 379, 381, *391*, 395, *412*
Kauffman, E.G., 213, *213*, 267, *271*

Kautsky, H., 325, *334*
Kawaguti, S., 185, *196*
Kay, E.A., 178, 183, 184, 185, 187, *196*
Keck, R., 361, *369*
Kelleher, G., 404, *412*, 417, *427*
Keller, G., 267, *271*
Kendall, M.A., 127, 141, *144*
Kenny, P.D., 362, *369*
Kerr, R.A., 155, *174*
Kerry, J., 383, *391*
Kidwell, S.M., 361, *369*
Kikkawa, J., 226, *252*
Kinsey, D.W., 181, 182, *196*
Klaver, A.L., 383, *391*
Klein, H., 156, *174*
Kloet, G.S., 5, *16*
Knight, A.J., 298, 299, 314, *316*
Knoll, A.H., 50, *67*
Knowlton, N., 13, *16*, 202, *213*, 214, 293, 295, 314, *317*
Knox, G.A., 129, 130, 132, 133, *144*
Kohn, A.J., 178, 183, 187, 188, 189, 190, 194, *196*, 202, 205, 207, 208, 209, 210, 211, 212, *213*, 214, 219, *252*, 263, 264, 265, *271*
Kontar, E.A., 156, *174*
Kukert, H., 101, 103, *119*, 152, *174*
Kulbicki, M., 226, 228, *252*
Kuroda, T., 184, 187, *196*

Lambshead, P.J.D., 150, *175*
Lampitt, R.S., 101, *119*, 158, *175*
Lande, R., 276, 277, 289, *291*
Larson, R.J., 241, *252*
Larsson, J., 376, 384, *391*
Laws, E.A., 396, *412*
Lawton, J.H., 248, *252*, 331, *334*
Lee, S.Y., 183, 184, 185, *196*
Lehman, S.J., 400, *412*
Leppäkoski, E., 327, *334*
Levin, L.A., 101, 103, *119*, 152, *175*
Levin, S.A., 116, *119*
Leviten, P.J., 210, *214*
Levitus, S., 42, *67*
Lewin, R., 125, *144*
Lewis, M.R., 179, *197*
Lewontin, R.C., 282, *291*
Lidgard, S., 139, *144*
Lipps, J.H., 137, *144*
Littler, M.M., 179, 181, *197*
Lochte, K., 157, *175*
Long, S.P., 397, *412*
Lonsdale, P., 155, 162, *175*
Lovén, S., 321, *334*
Loya, Y., 32, *33*,
Luckhurst, B.E., 232, 233, 237, *252*
Lund, E.J., 362, *369*
Lunestad, B.T., 383, *391*

Author index

MacDougall, N., 162, *175*
Maciolek, N., 97, 98, 116, *119*
MacKenzie, C.L. Jr., 359, 360, *369*
MacNae, W., 32, *33*
Magurran, A.E., 130, *144*, 160, *175*
Mallet, J., 294, *317*
Mann, R., 362, *369*
Mara, D.D., 388, *391*
Margulis, L., 2, *16*
Mattsson, J., 381, *391*
May, R.M., 3, 4, 14, *16*, *17*, 32, *33*, 70, *92*, 124, *144*, 149, *175*, 234, 244, *252*, 331, *335*
Mayr, E., 278, *291*, 293, 310, 311, *317*
McAllister, D.E., 217, 218, 219, 221, 225, 246, 247, *252*, 399, *412*
McArthur, R.H., 233, *252*
McCave, I.N., 157, 160, *175*
McClanahan, T.R., 398, *412*
McCoy, E.D., 224, *252*
McCoy, F.W., 396, *412*
McDonald, J.H., 326, *335*
McGowan, J.A., 7, *17*, 41, 46, 67, 84, *92*, 104, *119*, 128, *144*
McManus, J.W., 220, *252*
McRoy, C.P., 32, *33*
Mead, J.G., 398, *412*
Medlin, L.K., 70, *92*
Mee, L.D., 327, *335*
Mellor, C.A., 104, *119*
Memmott, J., 12, *17*
Mercier, H., 156, *175*
Michel, C., 381, *391*
Milicich, M.J., 236, *252*
Mill, P.J., 297, *317*
Miller, C.B., 46, *67*
Miller, M.C., 158, *175*
Minelli, A., 124, *144*
Mitchell, R., 417, 423, *427*
Möbius, K., 361, *369*
Moe, R.L., 136, *144*
Monaco, M.E., 365, *369*
Monaghan, P., 379, *391*
Moore, P.G.M., 32, *33*
Mork, J., 326, *335*
Morse, D.R., 228, *252*
Morton, B., 182, 183, 184, 185, 186, 189, 190, 191, *197*
Mougenot, D., 166, *175*
Mühlenhardt-Siegel, U., 141, *145*
Muir, J.F., 388, *391*
Mulchern, P.J., 156, *175*
Müller, H.-G., 55, *67*
Muscatine, L., 180, *197*
Myatt, E.N., 351, *370*
Myers, R.F., 229, *252*
Myrberg, A.A., 241, *252*

Naeem, S., 332, *335*
NCC, 379, *392*, 415, *427*
Nee, S., 248, *253*
Nei, M., 294, *317*
Neigel, J.E., 284, 289, *291*
Nelson, C.M., 263, *271*
Newell, R.I.E., 359, 360, 362, 364, *370*
Newman, W.A, 186, *197*
Nicol, D., 263, *271*
NOAA, 354, 355, 357, 365, *370*
Norse, E.A., 373, *392*, 396, 401, *412*
North Sea Task Force,, 329, *335*
Nowell, A.R.M., 157, *175*
Noyes, J.S., 7, *17*
Nunney, L., 288, *291*
Nyffeler, F., 157, *175*

O'Brien, E.M., 126, *145*
Odum, H.T., 384, *392*
Odum, W.E., 344, 345, *370*
Oliver, P.G., 184, 185, *197*, 263, *271*
Olivera, B.M., 208, *214*
Olsgard, F., 25, *33*
Omori, M., 71, *92*
Ong, J.E., 405, *412*
Orive, M.E., 288, *291*
Ormond, R.F.G., 221, 228, 232, 233, 238, 248, *253*, 396, *412*
Osman, R.W., 152, *175*

Pagel, M.D., 10, *17*, 126, *145*
Paine, R.T., 398, *413*
Paine, R.Y., 39, *67*
Palmer, M.W., 126, *145*
Palumbi, S.R., 50, *67*, 147, *147*, 293, *317*
Pandolfi, J.M., 224, *253*
Parish, J., 49, *67*
Pascoe, P.L., 296, *317*
Paterson, G.L.J., 99, *119*, 159, *175*
Patrick, R., 341, *370*
Paxton, J.R., 219, *253*
Pearse, J., 48, *67*, 70, 91, *92*
Pearson, D.L., 9, *17*, 233, *253*
Pearson, T.H., 25, *33*, 182, *197*
Pemberton, D., 379, *392*
Peters, R.H., 228, *253*
Petersen, K.S., 327, *335*
Petraitis, P.S., 153, *175*
Phillips, M.J., 376, 377, 380, 386, *392*
Phillips, R.C., 405, *413*
Pianka, E.R., 10, *17*, 126, *145*
Picken, G.B., 131, *145*
Picton, B.E., 401, *413*
Pierrot-Bults, A.C., 70, 71, 77, 79, 82, 84, 87, *92*
Pimentel, D., 372, *392*
Pimm, S.L., 124, *145*, 331, 332, *335*, 398, *413*

Pine, R.H., 13, *17*,
Pineda, J., 103, *119*
Pingree, R.D., 154, *175*, *176*
Pitcher, G.C., 58, *67*
Ponder, W.F., 263, *271*
Poore, G.C.B., 20, 28, 32, *34*, 124, *145*
Potts, D.C., 222, *253*
Powell, A.W.B., 263, *271*
Preston, F.W., 131, *145*
Primavera, J.H., 397, *413*
Primavera, T.H., 376, *392*
Prior D.B., 154, *176*
Pulliam, H.R., 243, *253*
Pullin, R.S.V., 376, *392*

Qi Zhongyan, 184, 185, *197*

Rand, D.M., 260, *271*
Randall, J.E., 227, *253*
Raup, D.M., 267, *272*
Ray, G.C., 54, *67*, 275, *291*, 338, 339, 344, 347, *370*, 404, *413*
Reeb, C.A., 281, 282, *291*
Reeve, M.R., 84, *93*
Rehder, H.A., 263, *272*
Reichelt, R.E., 209, 210, 212, *214*
Reid, D.G., 183, 185, *197*, 296, 297, 298, 300, *317*
Reid, J., 42, *67*, 71, 73, 74, *93*
Reid, W.V., 126, *145*
Reist, J.D., 300, *317*
Remane, A., 321, 323, 325, *335*
Rex, M.A., 7, *17*, 20, 23, 24, *34*, 96, 97, 102, 104, 105, 107, 108, 110, 113, *119*, *120*, 128, 142, *145*
Rhoads, D., 28, *34*
Rice, A.L., 101, *120*, 152, 153, *176*
Richard, G., 183, 184, 185, *197*
Richards, K.J., 157, *176*
Richards, R.H., 383, *392*
Richardson, M.D., 127, 132, 141, *145*
Richardson, M.J., 155, 156, 160, *176*
Ricklefs, R.E., 13, *17*, 95, 111, 112, 114, *120*, 202, *214*, 259, *272*
Riemann, B., 399, *413*
Riemann, F., 157, *176*
Righton, D., 233, *253*
Risk, M.J., 233, *253*
Robb, J.M., 154, *176*
Robbins, C.S., 343, *370*
Roberts, C.M., 219, 220, 228, 230, 232, 233, 246, 247, 248, *253*
Robertson, D.R., 231, 237, 241, *254*
Robins, C.R., 347, 351, *370*
Röckel, D., 205, *214*
Rohde, K., 10, *17*, 110, *120*, 146, 147, *147*, 217, *254*
Rolán-Alvarez, E., 310, 314, *317*

Rosen, B.R., 178, *197*, 222, *254*
Rosenzweig, M.L. , 9, 11, *17*, 125, *145*, 146 *147*
Rosewater, J., 263, *272*
Ross, A., 379, *392*
Ross, L.G., 384, *392*
Roughgarden, J., 113, *120*, 237, *254*
Rowe, G.T., 96, 102, *120*
Roy, K., 104, *120*, 127, *145*, 146, *147*
Ruggeberg, H., 379, *392*
Russ, G.R., 219, 229, 231, *254*
Russell, B.C., 234, *254*
Ryman, N., 326, *335*
Ryther, J.H., 179, 182, *198*

Sale, P.F., 203, 212, *214*, *215*, 230, 232, 233, 234, 235, 236, 237, 238, 242, 244, 245, *254*, *255*
Salvat, B., 183, 184, *198*
Samuelsen, O.B., 383, *392*
Sanders, H.L., 21, 22, 23, *34*, 95, 110, *120*, 128, 142, *145*, 151, *176*
Sanderson, W., 419, *427*
Sars, G.O., 18, *34*
Saunders, P.M., 156, *176*
Savidge, G., 44, *67*, 179, *198*
SBSTTA, 399, *413*
Schaff, T.R., 101, *120*
Schlegel, M., 2, *17*,
Schmitz, W.J., 156, *176*
Schneider, J.A., 185, *198*
Schneider-Broussard, R.M., 283, 284, 285, *291*
Scott, P.J.B., 182, *198*
Sebens, K.P., 203, 204, 211, *215*
Segerstråle, S.G., 321, 322, 323, *335*
Sepkoski, J.J.Jr, 49, *67*, 115, *120*, 264, *272*
Seshappa, G., 296, 297, *317*
Shackleton, N.J., 50, *67*
Shepard, F.P., 154, 165, *176*
Sheppard, C.R.C., 32, *34*, 224, 228, *255*
Sherman, K., 398, *413*
Shomura, R.S., 397, *413*
Shulman, M.J., 241, *255*
Siciński, J., 129, *145*
Sidall, S.E., 184, *198*
Silva, P.C., 20, *34*
Simberloff, D., 276, 277, 289, *292*
Simenstad, C.A., 398, *413*
Sinclair, M., 38, 53, *67*, 82, 87, *93*
Sindermann, C.J., 383, *393*
Skibinski, D.O.F., 295, *317*
Slatkin, M., 277, 282, *292*
Smethie, W.M., 41, *67*
Smith, C.L., 234, 235, *255*
Smith, C.R., 95, 101, *120*, 152, 153, 156, 159, *176*
Smith, G.B., 221, *255*

Author index

Smith, J.T., 263, *272*
Smith, S.L., 54, *67*
Smith, T.M., 343, *370*
Snelgrove, P.V.R., 101, *120*, 157, *176*
Solbrig, O.T., 32, *34*
Solé-Cava, A.M., 294, *317*
Sorensen, J.C., 408, 409, *413*
Sørenson, T.A., 206, *215*
Sorokin, Yu.I., 203, *215*
Soulé, M.E., 276, 288, *292*
Sournia, A., 70, *93*, 184, *198*, 325, *335*
Southard, J., 158, *176*
Southward, A.J., 332, *335*
Spanier, E., 405, *413*
Springsteen, F.J., 187, *198*
Stanley, S.M., 260, 263, 264, 267, 268, *272*
Steele, J.H., 236, *255*, 319, 329, 332, *335*, 366, *370*
Stehli, F.G., 19, 20, *34*, 127, 139, 140, 141, *146*, 260, 261, *272*
Stenzel, H.B., 361, *370*
Stephenson, W., 26, *34*
Stevens, G.C., 10, *17*, 103, 104, *121*, 126, *146*, 243, *255*
Stilwell, J.D., 263, *272*
Stommel, H., 36, *68*
Stone, L., 366, *370*
Stork, N.E., 123, *146*, 149, *176*
Stuart, C.T., 99, 110, 111, 112, 113, 114, *121*
Stuiver, M., 40, *68*
Sundberg, P., 294, 301, *318*
Svavarsson, J., 99, 100, 115, *121*
Sysoev, A.V., 263, *272*

Tacon, A.G.J., 378, 386, *393*
Taft, B.A., 167, *176*
Talbot, F.H., 231, 232, 235, *255*, *256*
Tantanasiriwong, R., 187, *198*
Taylor, J.D., 127, *146*, 179, 182, 183, 185, 186, 187, 189, 190, 191, 192, 193, 194, *198*, *199*
Tedengren, M., 333, *335*
Telford, M.J., 87, *93*
Terborgh, J., 10, 11, *17*, 125, *146*, 343, *370*
Terlau, H., 208, *215*
Thiel, H., 157, 164, *177*
Thistle, D., 103, *121*, 152, 153, 158, 159, 160, 161, 162, 163, 166, 167, *177*
Thorpe, J.P., 298, *318*
Thorpe, S.A., 155, *177*
Thorson, G., 20, *34*, 127, 139, 141, *146*, 149, *177*, 360, *370*
Thresher, R.E., 224, 228, 241, *256*
Thurston, M.H., 131, *146*
Tilman, D., 332, *335*
Titova, L.V., 263, *272*

Tokioka, T., 82, 84, 90, *93*
Tong, L.K.Y., 183, 184, 191, 192, *199*
Tsi, C.Y., 184, *199*
Tsuchiya, M., 183, 184, *199*
Turley, C.M., 102, *121*
Turner, G.E., 386, *393*
Tyler, P.A., 159, *177*
Tyler-Walters, H., 313, *318*

UNCED, 394, 407, *413*
Underwood, A.J., 113, *121*
UNEP/IUCN, 218, *256*
Utter, F.M., 279, *292*

Väinölä, R., 326, *335*
Vale, F.K., 96, 102, *121*
Valentine, J.W., 136, *146*, 270, *272*
van der Elst, R.P., 398, *413*
van der Spoel, S., 41, 42, 50, *68*, 84, 85, 86, 87, 90, *93*
Van Veghel, M.L.J., 295, *318*
Vangriesheim, A., 155, *177*
Varvio, S.L., 295, *318*
Vawter, L., 284, *292*
Vermeij, G.J., 134, 137, *146*, 179, 183, 193, *199*, 259, *272*
Veron, J.E.N., 219, *256*
Victor, B.C., 223, 224, 234, 235, 236, *256*
Voipio, A., 320, 321, 325, 326, *335*
Voβ, J, 134, 135, *146*
Vrba, E.S., 193, 194, *199*

Waldner, R.E., 243, *256*
Walker, B.H., 331, *335*
Wallentinus, I., 321, 322, 325, 326, 329, *335*
Walsh, G.E, 32, *34*
Waples, R.S., 288, *292*
Ward, R.D., 298, *318*
Warner, A.J., 37, *68*
Warwick, R.M., 27, *34*
Warwick, T., 313, *318*
Watson, M., 241, *256*, 398, *413*
Weatherly, G.L., 155, 156, *177*
Webb, P.-N., 136, *146*
Wei, K.-Y., 265, 267, 268, *272*
Welcomme, R.L., 384, *393*
Wells, F.E., 184, 185, 186, 187, *199*
Wells, S.M., 399, 406, *413*
Wells, W.H., 361, 362, *371*
Weston, D.P., 380, 381, 382, 383, *393*
Wheatcroft, R.A., 157, *177*
White, A.T., 405, *414*
White, A.W., 396, *414*
White, B.N., 41, 50, 55, *68*
White, M.G., 130, *146*
Whittaker, R.H., 19, *34*, 130, *146*, 206, *215*
Wiebe, W.J., 181, *199*

Wiegmann, B.M., 12, *17*
Wiens, J.A., 234, 244, *256*
Wilber, D.H., 361, *371*
Wilkins, N.P., 297, 298, *318*
Wilkinson, C.R., 181, *200*, 247, *256*
Willén, T., 325, *336*
Williams, A.B., 283, *292*
Williams, D.McB., 182, *200*, 229, 230, 231, 232, 233, 236, *256*, *257*
Williams, G.C., 286, *292*
Williamson, M., 7, 8, 9, *17*
Wilson, E.O., 3, 4, 6, 10, *17*
Winslow, F., 361, *371*
Winston, J.E, 129, *146*
Witkowski, A., 327, *336*
Wolff, W.J., 321, *336*
Woodland, D.J., 223, *257*
Woodroffe, C.D., 203, *215*
Woodwood, F.I., 125, *146*
World Coast Conference, 407, *414*
World Commission on Environment and Development, 403, *414*
Wright, D.H., 11, 13, *17*
Wright, J.C. , 87, *93*
Wright, L.D., 362, *371*
Wright, S., 277, 278, 282, 288, *292*
Wu R.S.S., 182, *200*
Wu, Wen-Lung, 184, 185, *200*
Wytrki, K.E., 156, *177*

Yingst, J.Y., 159, *177*
Yndestad, M., 383, *393*
Yoder, J.A., 115, *121*

Zachos, J.C., 50, *68*
Zajac, R.N., 152, *177*
Zaslavskaya, N.I., 298, *318*
Zaykin, D.V., 306, *318*
Ziemann, D.A., 380, *393*
Zitko, V., 338, *371*

Species index

Page numbers in *italics* refer to Figures.

Acanthochromis polyacanthus, 241
Acanthurus spp., 241
 A. lineatus, 232
 A. sohal, 232
Acipenser brevirostrum, 348
 A. oxyrhynchus, 348
 A. sturio, 327
Actinia equina, 294
 A. prasina, 294
Adudefduf sordidus, 238
 A. vaigiensis, *238*
Aetobatis narinari, 348
Aforia spp., 262, *263*
Airus felis, 348
Alosa spp., 366
 A. aestivalis, 348
 A. pseudoharengus, 348
 A. mediocris, 348
 A. sapidissima, 348
Aluterus schoepfi, 351
 A. scriptus, 351
Amblyglyphidon leucogaster, *238*
Ammodytes dubius, 350
 A. hexapterus, 350
Anchoa hepsetus, 348
 A. mitchilli, 348
Anchoviella perfasciata, 348
Ancylopsetta dilecta, 351
Anguilla anguilla, 330
 A. rostrata, 289, 348
Anguinella palmata, 359
Anodontia spp., 185
Apeltes quadracus, 349
Arca ventricosa, 184
Archosargus probatocephalus, 349
Arctica islandica, 345, 346
Ardea herodias, 360

Argopecten gibbus, 345, 347
 A. irradians, 345
Arius felis, 289, 348
Astarte spp., 135
Asterias forbesi, 360
 A. rubens, 322
Astroscopus guttatus, 350
 A. y-graecum, 350
Atlantoserolis spp., 138
Aurelia aurita, 359

Babylonia spp., 186, 193, 194, 262, *263*, 264
 B. areolata, 186
 B. lutosa, 186
Bagre marinus, 348
Bairdiella chrysoura, 349
Balanus amphrite, 191, *192*
 B. eburneus, 360
 B. improvisus, 327, 360
Balistes capriscus, 351
Barbatia spp., 183, 191
 B. virescens, *192*
Bascanichthys bascanium, 348
 B. scuticaris, 348
Beggiotoa spp., 381
Benthosema glaciale, 52
Beroe spp., 359
Bothus ocellatus, 351
 B. robinsi, 351
Brachidontes spp., 183, 191
 B. recurvum, 359
 B. variabilis, *192*
Brevoortia smithi, 348
 B. tyrannus, 348
Bucephalus cuculus, 359
Bullia spp., 183

437

Species index

Bursa granularis, 189, *190*
  B. rana, 193
Busycon carica, 359

Calanoides carinatus, 54
Calanus finmarchicus, 54
  C. helgolandicus, 82
Callinectes sapidus, 345, 346, 359
Callorhinus ursinus, 397
Camptorhynchus labradorius, 401
Cancer irroratus, 360
Cantharus spp., 189
  C. undosus, 188
Capitella spp., *188*
  C. capitata, 381
Carcharinus acronotus, 348
  C. leucas, 348
Carcinus maenas, 322, 360
Centropages aucklandicus, 52
Centropomus undecimalis, 349
Centropristis spp., 366
Cephalopholis spp., 241
Ceratoscopelus spp., 60
Cerithidea fuscata, 401
Chaetodipterus faber, 350
Chaetodon paucifasciatus, 245–6
Chama spp., 183
  C. iostoma, 184
Chamelea gallina, 295, 314
  C. striatula, 295
Chanos chanos, 375, 377
Chasmodes bosquianus, 350
Chicoreus brunneus, 191
  Ch. microphyllus, 191
Chilomycterus schoepfi, 351
Chloroscombrus chrysurus, 349
Chromis atripectoralis, *238*
  C. caeruleus, *238*
  C. weberi, *238*
Chrysaora quinquecirrah, 359
Chrysiptera spp., *238*
  Ch. biocellata, *238*
  Ch. unimaculata, *238*
Chthalamus fragilis, 360
Citharichthys spilopterus, 351
Cladophora spp., *324*
Clausocalanus paupulus, 79
Cliona spp., 359
Clupea harengus, 326, 348
Codakia spp., 185
Conchoecia antarctica, 63
  C. australis, 63
  C. borealis, 63
  C. obtusata, 63
Conger oceanicus, 348
Conopeum tenuissimum, 359
Conus spp., 183, 189, 201–2, 205–13
  C. arenatus, 209

C. chaldaeus, *188*, 209
C. coronatus, *188*, 209
C. dorreensis, *190*
C. ebraeus, *188*, 209
C. eburneus, 209
C. flavidus, 209
C. frigidus, *188*, 209
C. lividus, 209
C. marmoreus, 209
C. miles, 209
C. miliaris, *188*, 209
C. musicus, 209
C. rattus, *188*
C. sanguinolentus, 209
C. sponsalis, *188*, *190*, 209
C. textile, 209
Corallina officinalis, 423
Coralliophila costularis, 191
Corculum spp., 185
Coregonus albula, 330
Corophium lacustre, 360
Crassostrea spp., 375
Crassostrea virginica, 280, 345, 346, 350, 361, 364
Crepidula fornicata, 359, 425
  C. plana, 359
Cristaserolis spp., 138
Cronia avellana, 189, *190*
  C. margariticola, 191
Ctena spp., 185
Cyamus spp., 401
Cynoscion nebulosus, 350
  C. regalis, 350
Cypraea spp., 262
Cyprinodon variegatus, 349

Dascyllus trimaculatus, *238*
Dasyatis americana, 348
  D. sabina, 348
  D. say, 348
Dendropoma spp., *188*
Diadumene leucolena, 359
Dicathais orbita, 189, *190*
Dicentrarchus labrax, 375
Diplectrum formosum, 349
Diplodus holbrooki, 349
Distorsio reticulata, 193
Dodecaceria spp., *188*
Dormitator maculatus, 350
Dreissena bugensis, 327
  D. polymorpha, 327
Drupa clathrata, *188*
  D. morum, *188*
  D. ricina, *188*
  D. rubusidaeus, *188*
Drupella rugosa, 191

Electra crustulenta, 359

Species index

Elodea canadensis, 327
Elops saurus, 348
Engina mendicaria, 188
Epinephelus spp., 375
Epinephelus itajara, 247
Equetus spp., 241
Eschrichtius robustus, 401
Etropus crossotus, 351
Eucheuma spp., 375
Eucinostomus argenteus, 349
   E. gula, 349
Eukrohnia bathyantarctica, 77
   Eu. fowleri, 77
   Eu. hamata, 63, 77, 89
Euphausia lucens, 54
Eurypanopeus depressus, 360
Evorthodus lyricus, 350

Fimbria spp., 185
Fragum spp., 185
Fucus serratus, 322
   F. vesiculosus, 322, 325
Fundulus confluentus, 349
   F. heteroclitus, 349
   F. luciae, 349
   F. majalis, 349
Funiculina quadrangularis, 421
Furcellaria lumbricalis, 322
Fusitriton spp., 262, 263

Gadus morhua, 322
Gambusia affinis, 349
Gasterosteus aculeatus, 349
Glabroserolis spp., 138
Glycera spp., 359
Glyptocephalus cynoglossus, 330
Gobiesox strumosus, 349
Gobionellus boleosoma, 350
   G. hastatus, 350
   G. shufeldti, 350
   G. smaragdus, 350
   G. stigmaticus, 350
Gobiosoma bosci, 350, 360
   G. ginsburgi, 350
Goniopora spp., 181
Gracilaria spp., 375
Gutturnium muricinum, 189
Gymnura altavela, 348
   G. micrura, 348

Haematopus palliatus, 360
Haemulon spp., 241
Halichoeres bivittatus, 350
   H. radiatus, 350
Halichoerus grypus, 420
Haliplanella luciae, 359
Harpa spp., 263, 264
Haustellum spp., 262, 263

Hemicardia spp., 185
Hemimysis anomala, 327
Hemiramphus brasilensis, 349
Heterosigma akashiwo, 425
Hetrokrohnia mirabilis, 77
Hippocampus erectus, 349
Homarus americanus, 345
Hydractinia echinata, 359
Hydrodamalis gigas, 401
Hydroides dianthus, 359
Hyporamphus unifasciatus, 349
Hypsoblennius hentz, 350

Idotea balthica, 322
Ilyanassa obsoleta, 186
Ishadium recurvum, 359
Isognomon spp., 183

Krohnitta pacifica, 89
   Kr. subtilis, 77, 89
Kyphosus incisor, 350
   K. sectatrix, 350

Labidocera spp., 52
Lactophrys trigonus, 351
Lagodon rhomboides, 349
Laminaria japonica, 375
   L. saccharina, 322
Leiostomus xanthurus, 350, 360
Leptoserolis spp., 138
Letharchus velifer, 348
Liparis inquilinus, 351
Littorina arcana, 293, 297–314
   L. littorea, 322
   L. neglecta, 293, 297, 299, *307*, 311, 313–14
   L. (Neritrema) patula, 296, 297
   L. (Neritrema) rudis, 296
   L. (Neritrema) saxatilis, 293, 296–314
   L. (Neritrema) tenebrosa, 296, 298
   L. nigrolineata, 293, 296–9
   L. obtusata, 297
   L. rudissima, 297
   L. sitkana, 298
Lottia alveus alveus, 401
Lucania parva, 349
Lunulicardia spp., 185
Lutjanus griseus, 349

Macoma balthica, 322, 326
Mancinella echinata, 191
Marenzelleria viridis, 327
Meganyctiphanes norvegica, 52
Melanogrammus aeglefinus, 330
Membranipora tenuis, 359
Membras martinica, 349
Menidia beryllina, 349
   M. menidia, 349

Species index

Menippe adina, 283–6
  M. mercenaria, 283–6
Menticirrhus americanus, 350
  M. littoralis, 350
  M. saxatilis, 350
Mercenaria mercenaria, 345, 346
Mergus australis, 401
Microciona prolifera, 359
Microgadus tomcod, 349
Microgobius thalassinus, 350
Micropogonias undulatus, 350, 360
Millepora boschmai, 401
Mitra spp., *190*
  M. litterata, 188
Mnemiopsis leidyi, 327, 359, 399
Modiolus modiolus, 135, 295
Mogula manhattensis, 360
Monacanthus ciliatus, 351
  M. hispidus, 351
Monachus tropicalis, 401
Monastraea annularis, 295, 296, 314
Monodonta spp., 191, *192*
Monoporeia affinis, 332
Morone americanus, 349
  M. saxatilis, 349, 360
Morula anaxeres, 188
  M. granulata, 189
  M. marginatra, 188
  M. musiva, 191, *192*
  M. uva, *188*
Mugil spp., 375
  M. cephalus, 350
  M. curema, 350
Murex spp., 262, *263*
Mustela macroda, 401
Mya arenaria, 325, 327, 360
Myliobatis freminvillei, 348
  M. goodei, 348
Mytilus spp., 375
  M. desolationis, 295
  M. edulis, 295, *322*, 325
  M. galloprovincialis, 295
  M. trossulus, 295, 326

Nacella concinna, 295
Nassarius spp., 193
Negaprion brevirostris, 348
Nephrops norvegicus, 330
Neptunea spp., 262, *263*
Nereis succinea, 359
Nicholsina usta, 350
Nucella lapillus, 296
Nycticorax caledonicus crassirostris, 401
Nyctiphanes capensis, 54

Oceanodroma macrodactyla, 401
Odostomia spp., 359
  O. impressa, 359

Oithona hebes, 52
Oligoplites saurus, 349
Oncaea dentipes, 52
Oncorhynchus spp., 375
Ophichthus gomesi, 348
Opisthonema oglinum, 348
Opsanus tau, 348, 360
Orconectes limosus, 327
Orthoprostis chrysoptera, 349
Ostrea edulis, 375

Pagrus major, 375
Palaemonetes purgio, 360
Pandalus borealis, 330
Panopeus herbstii, 360
Panulirus argus, 345
Paralichthys albigutta, 351
  P. dentatus, 351
  P. lethostigma, 351
  P. oblongus, 351
  P. squamilentus, 351
Parapercis hexopthalma, 240
Pareuthria spp, 262, 263
Patella vulgata, 322
Patelloida spp., 191
Pecten yessoensis, 375
Penaeus spp., 375
  P. aztecus, 345
  P. duodarum, 345
  P. setiferus, 345
Perca fluviatilis, 322
Perna spp., 183, 184
  P. perna, 184
  P. viridis, 184
Phalacrocorax perspicillatus, 401
Pholis gunnellus, 350
Pinctada margaritifera, 184
Pinguinus impennis, 401
Pinnotheres ostreum, 360
Placopecten magellanicus, 345, 346
Plectropomis leopardus, 241
Pleoticus robustus, 345, 347
Pleurobrachia spp., 359
Pleuronectes platessa, *322*, 330
Poecilia latipinna, 349
Pogonias chromis, 350, 360
Pollachius spp., 366
Pollachius virens, 330
Polydactylus octonemus, 350
Polydora ligni, 359
  P. redeki, 327
  P. websteri, 359
Pomacentrus spp., 241
  P. albicaudatus, 238
  P. amboinensis, 237
  P. mollucensis, 237
  P. philippinus, *238*
Pomatomus saltatrix, 349

Species index

Pontella spp., 52
Porites spp., 181
Porphyra tenida, 375
Potamopyrgus jenkinsi, 327
Potomocorbula amurensis, 399
Prionotus carolinus, 350
P. evolans, 351
P. scitulus, 351
P. tribulus, 351
Pristis pectinata, 348
Prochloron spp., 180
Pseudopleuronectes americanus, 351
Pterosagitta draco, 77, 77, 78, 84, 89

Raja spp., 366
Rhinobatos lentiginosus, 348
Rhinoptera bonasus, 348, 360
Rhithropanopeus harrisii, 360
Rhizoprionodon terraenovae, 348
Rutilus rutilus, 326

Sabella microphthalma, 359
Saccostrea spp., 183, 191
Saccostrea cucullata, 191, *192*
Saduria spp., *324*
Sagitta batava, 82
S. bedoti, 89
S. bierii, *81*
S. bipunctata, 77, 84, *89*
S. decipiens, 77, *89*
S. enflata, 77, *89*
S. euxina, 82
S. friderici, 82
S. gazellae, 84, *89*
S. hexaptera, 77, *89*
S. lyra, 77, 84, *89*
S. macrocephala, 77, 79, *89*
S. marri, 89
S. maxima, 77, 84, *89*
S. minima, 77, 84, *89*
S. pacifica, 79, *80*
S. planctonis planctonis, 77, 84, *89*
S. planctonis zetesios, 77, *89*
S. pseudoserratodentata, 79, *81*, 89
S. pulchra, 89
S. regularis, 89
S. robusta, 89
S. serratodentata, 77, 79, 84, *89*
S. serratodentata atlantica, *80*
S. serratodentata serratodentata, *80*
S. setosa, 82, *83*
S. sibogae, 77
S. tasmanica, *81*, 90
Salmo salar, 326, 375
Sargassum spp., *180*
Sardinella aurita, 348
Schizoporella unicornis, 359
Sciaenops ocellatus, 350

Scomber scomberus, 330
Scomberomorus maculatus, 350
Scorpaena calcarata, 350
S. plumieri, 350
Scorpaenopsis oxycephala, *240*
Scylla serata, 375
Selene setapinnis, 349
S. vomer, 349
Septifer spp., 183, 189, *190*
S. virgatus, 191
Seriola quinqueradiata, 375, 377
Serolina spp., 138
Siphonaria spp., *188*, 191, *192*
Solea solea, 330
Solen spp., 375
Sphoeroides maculatus, 351
S. spengleri, 351
S. testudineus, 351
Sphyrna tiburo, 348
S. zygaena, 348
Spirobis spp., *188*
Spisula sapidissima, 346
Spisula solidissima, 345
Sprattus sprattus, 326
Stegastes spp., 241
Stenotomus chrysops, 349
Strombus spp., 262, *263*
Strongylocentrotus droebachiensis, *322*
Strongylura marina, 349
Stylochus ellipticus, 359
Sygnathus floridae, 349
S. fuscus, 349
Symbion pandora, 401
Symphurus plagiusa, 351

Tapes spp., 375
Tautoga onitis, 350
Tautogolabrus adspersus, 350
Thais armigera, *188*
T. clavigera, 191, *192*
T. intermedia, *188*
T. luteostoma, 191
Tigriopus californicus, 278, 279
Trachinotus carolinus, 349
T. falcatus, 349
Trichoplax spp., 124
Tridacna spp., 180, 183, 185
Tridacna maxima, 184
Trinectes maculatus, 351
Trophon spp., 262, *263*
Turricula nelliae, 193
Tylosurus acus, 349

Undaria pinnatifida, 375
Undulina spp., 52
Urosalpinx cinerea, 359
Urticina eques, 294
U. felina, 294

Vasum turbinellus, 188
Venerupis japonica, 375
Vibrio salmonicida, 387
Victorella pavida, 327, 359

Xenia spp., *240*
Xyloplax spp., 124

Zostera marina, 325, 401

Subject index

Page numbers in *italics* refer to figures.

abundances, ranked, 47, 160, *170*, *171*
adaptation, 276, 279
adaptive potential, losses, 289
aesthetic value, xvii
Agenda 21, 407, 409
allozyme allele frequency, 278, 280–1, 283
allozyme polymorphisms, 299
analyses
 ANOVA, 113
 canonical variate, 300, *304*
 multiple regression, 114
 principal components, 238, 300, 301, *303*, 337, 338, 344, *356*
Antarctic, amphipods, *131*
 prosobranchs,, *131*
Arctic, species poorness, 129
Atlantic Ocean, *100*, *105*
Atlantic
 North, 97–100, 102, 104, *106*–8, 110–15
 South, 104, *106*–10
atolls, 183, 184, 212
 Cocos-Keeling, 183, 184, 185, 187
 Enewetak, 183, 185, 187

ballast water, 399, 406
Baltic Sea
 characterisation, 320–1
 climate, 321
 endemic species, 326
 energy flows, 327
 fishes, 326
 food webs, *324*
 macrofaunal richness gradients, 325
 phytoplankton richness, 325
 Pleistocene history, 321
 pollution, 332
 recent faunal changes, 327
 salinity gradients, 322
 species' ranges, *323*
 zooplankton communities, 325
barnacles, 180, 185
barriers to
 reproduction, 312
 dispersal 153, 178, 228, 246, 285
 distribution, 70, 79
 reproduction, 294
bathymetric gradients, 20–9
 abiotic, 61
 clines in bioluminescence, 59
 clines in colouration, 59
 maxima in species richness, 23, 231
 profiles of mean body size, 58, 84
 richness profiles, *24*, *25*, 57–63
 zonation, 59
bed forms, 156
benthic boundary layer (BBL), 154, 157
benthic nepheloid layer (BNL), 154, 157
benthic storms, 101, 155
Berger–Parker index of dominance, 161
Biodiversity Action Plan, xix
biodiversity
 alpha, 19, 27, 212, 217
 and productivity, 10, 14
 beta, 19, 212, 217
 biomass effects, 84–87
 causes of variations, 10–13
 comparison of marine with terrestrial, 10
 definition, 35, 274
 ecological, 2–3
 ecosystem function, 338
 gamma, 19, 139–41, 217
 importance of scale, 7
 latitudinal clines, 124–8, 141–2
 morphological, 2–3

443

444 Subject index

biodiversity – cont.
  of Baltic Sea, 321–6
  of coral reefs, 124, 201–13, 219–48
  of fishes, 217
  of gastropod molluscs, 19
  of mangroves, 19
  origins, 48–53, 259
  phylogenetic, 2
  polar–tropical gradients, 6–10, 260
  value, xvi–xviii
biogenic structure, 101, 343
biogeographical zones, 132, *133*
biogeographical patterns, north versus south, 140–1
biogeography
  of euphausiids, *57*
  of mysids, *57*
  oceanic, 36
  pelagic, 39–48
biological species concept, 293
bioturbation, 101, 164–5
bivalve molluscs, 104–8, *106*, 168, 180
blooms, 44, 85
  diatoms, 58
body size, 228, *229*
bootstrapping, 366
Bottom Water formation, 40
boundaries, 337
box-cores, 160, 166–7
burrows, 152

camouflage, *240*
canyons, *see* Setubal Canyon
carbonate skeletons, 141
carcases, 101, 152
carrion, 186, 189
carrying capacity, 152, 234
Cenozoic, 51, 136, 268
Cephalopoda, 90
Chaetognatha, 70, 71–85
Challenger, HMS, 128
chromosomal polymorphisms, 296
Circum-Antarctic Current, 136
climate changes, 32, 136, 269–70, 332
clines, 20, 124–8, 141–2
co-evolution, 193
co-existence, 12
co-occurrences, 208
coastal geomorphology, 54, 70
coastal upwelling, 43
Coastal Zone Colour Scanner (CZCS), 179
coastal zone
  management, 386, 407–10
  classification, 337
  definition, 338
colonisation, 362
  of arctic, 134–5
community
  structure, 233–4
  concept, 53
competition, 19, 102, 230, 234, 241–2, 244
  recruitment interaction, 235
competitive displacement, 102
conservation
  conflicts, 424–5
  of biodiversity, 275
  of reef fishes, 245–8
continental margins, 179
convective overturn, 43–4
Convention on Biodiversity, xviii, 407, 415
  measures, 403
  objectives, 394–5
coral bleaching, 400
coralline algae, 181
corals, 180, *see* Reefs, coral
Cretaceous, 262, *263*
Cretaceous–Tertiary (K-T) boundary, 50, 64, 267
cross shelf patterns, *56*, 230
cryptic species, 70, 79, 282
currents
  near bed, 151, 156, 162, 163, 6
  western boundary, 155–6
cyanobacteria, 181

debris from shipping, 396
degradation, of reefs, 247
deposit feeders, 158
detritus, 187, 189
development time, 279
diapause, 46
diel vertical migration, 47, 59–61
disjunct distributions, 347
disparity, 3, 48, 64
dispersal, 113, 159, 211, 230
  ability, 224
  larval, 223, 285
dissolved organic matter (DOM), 58
distributional boundaries, Cape Hatteras, 353, 359
distributional ranges, 89, 205, 211, 269, *352*
disturbance, 152, 204, 210, 226
  effects, ice scour, 135
divergence, inter-population, 279
diversity, 341
  allelic, 288
  alpha, 130, 132
  alpha, latitudinal clines, 124–8, 141–2
  bathymetric profiles, 28, 101–4
  beta, 130
  coastal sediments, 25–9
  definition, 341
  evolution of, 95
  gamma, 130
  genetic, xvii, 65, 90, 91
  genetic, loss of, 289

Subject index

geographical range, 000
gradients, 207–8, 212, 260
'hotspots', 404
in Antarctic, 20
isopod, 99–108
latitudinal patterns, 104, 110
local, 111
macroalgae (sea-weeds), 20, 136, 325
of bivalve molluscs, 20, 183–6
of gastropod molluscs, 186
polychaete, 99
regional, 111
small-scale, 95
dominance, 30–1
bathymetric changes, 84
Drake Passage, 51, 135
dynamic equilibrium model, 22 102

East Australian Current, 227
echinoderms, 159
ecological determinism, 204
ecological processes, 150, 275
gradients, 208–9
economics, xix–xx
ecosystem function
and biodiversity, 319, 338
stability, 331–3
ecosystem management, 367
ecotones, 42, 338
ecotypes, 310
eddies, 52, 64
El Niño Southern Oscillation (ENSO), 85
endangered species, 398
endemics, island, 246
endemism, 207, 268
in Baltic, 326
in neritic species, 70
in reef fish, 221, 226–8
endocrine disruption, 396
energy flows, in diverse and simple ecosystems, 327–30, *328*
enzyme electrophoresis, 298
enzyme polymorphisms, 295
Eocene, 262
eradication of communities, 135
estuarine
benthic faunal zones, 345, *346*
biotic communities, 338
invertebrates, 344–7
morphometrics, 338, 353–4, *355*, *358*
nutrient inputs, 343
systems, *339*, *340*
ethical responsibility, xvi
euphausiids, 71, 73
evenness, 160
evolution, 193–4
in polar regions, 137–8
rates, 6

evolutionary
development, 112
histories, 275
processes, 150
exotic species, 71, 399, 405–6
expatriates, 53, 90
extinct species, 398, 401
extinction, 111, 276
and speciation, 46, 276
events, 49–50
crisis, 13
environmental controls, 194
rates, xiii, 6, 194, 223, 259, 261, 402
rates, variation with latitude, 267–8
species at risk, 420
of species, 327
extrinsic factors, 202

Fanning Island, 183, 185, 187
farmed
fish, 374–5
macroalgae (sea-weed), 374
molluscs, 374–5
shellfish, 374
faunal boundaries, *88*, 275
feedbacks, 341, 343
feeders
generalist, 182, 186, 189, 193
specialist, 182, 189, 193
feeding
algal, 191
suspension, 181, 182
fish assemblages, 212, 344
fisheries, xiv, 316
fishing, xxi
impacts, 399, 404
Muro–Ami method, 399
fitness, 279
food conversion ratios, 386
food in the deep sea, 110
food production, energy demands, 389
food supply, human, 372–3
food webs, 187–93, *188*, *190*, *192*, *324*
foraging efficiency, 232
French Polynesia, 183, 184, 185, 237
functional diversity, definition, 366
functional groups, 330
fynbos, 12, *125*, 126

gastropod
assemblages, 99
clines, 104–8, *106*
gene flow, 79, 277–80, 294, 295, 297, 311
generation times, 228, 260
genetic drift, 279, 282, 287
genetic heterogeneity, 79
genetic homogeneity, 305–6
genotypic variation, 311–13

geographical isolation, 104
geographical ranges, 103
glaciations, 154
Global Environment Facility, xviii
Gondwana, 135–6
Great Barrier Reef, 181, 209, 212, 229, 230
greenhouse gases, xv
growth rates, 182
Guam, 187, 189
Gulf Stream, 53, 156, 339

Habitat Directive of European Union, 417
habitat diversity, 138, 151
habitat loss, bird declines, 343
habitat selection, 231, 310
Hawaii, 181, 183, 184, 185, 187, 230
herbivores, 182
herring, 82
heterogeneity, 14, 79, 205
　large scale, 153
　small scale, 170
heterozygote superiority, 287
High Energy Benthic Boundary Experiment (HEBBLE), 155, 158, 160–2, 166, 168
hierarchical model of coastal man, 342
historical contingency, 204
homozygous loci, 287
Hong Kong, 184, 185, 186, 187, 191, 192, 193
Hurlbert's rarefaction Index, 98, 100, 106, 160
human population increases in coastal zones, 395
hybrid zone, 283, 284
hybridization, 294, 301, 311
hydrodynamic regime, 153–6, 169
Hydromedusae, 85
hypothesis
　area, 22
　competitive exclusion, 151
　disturbance, 22
　dynamic equilibrium, 151
　'Elm–Oyster' model, 286
　equal chance, 234
　equilibrium, 259
　intermediate disturbance, 152
　lottery, 234, 242–3
　multi-species equilibrium, 234
　null model of diversity distribution, 223
　reef fish community structure, 235
　spatial heterogeneity, 22
　stability–time, 21
　taxonomic diversity gradient evolution, 269
　vortex model, 223, 225

inbreeding, 286–9
　depression, 287

industrial revolution, xiii
information required for conservation, 416–24
Integrated Coastal Zone Management (ICZM), 407–10
interbreeding populations, 296
Inter-governmental Panel on Climate Change, xv
intrinsic factors, 202
introductions, 275, 327, 405, 424–5
　isolation, 280, 282, 294
　caused by sea level changes, 82
isopod, variation, 104, 106

k-dominance plots, 141
kelp forest, 32
keystone species, 337, 398

larvae, surpluses, 286
latitudinal clines, 45, 45
　alpha diversity, 141
　bivalve molluscs, 127, 140
　bryozoans, 139
　diversity, 124–8
　life cycles, 45, 82
　marine, 125–8
　mean population size, 46, 84
　shallow seas, 139–41
　terrestrial, 125–6
　vascular plants, 125
Law of the Sea, xxi
Leeuwin Current, 189
life history characteristics, 342
life history strategies, 204, 210–11, 367
lineages, evolutionary, 295
Lyellian curves, 267–8

Madagascar, 237
Mahalanobis distance, 304, 305
maintenance of richness, 14, 259
management strategies, 275
management, ecosystem, 338
mangrove restoration, 397, 404–5
mangroves (mangal), 32
mariculture, 373–89
　broodstock, 377–8
　destruction of terrestrial habitats, 376
　disease, 383–4
　environmental impacts, 376–84, 395
　environmental impact assessments (EIA), 386
　eutrophication effects, 385
　feed, 378–9
　impacts on biodiversity, 384–5
　management guidelines, 386
　predator control, 379–80
　release of exotic stock, 384
　technology, 374–75

Subject index

use of chemicals and drugs, 381–3, 387
wastes, 380–1, 387
wastewater treatment, 388
water demands
marine habitats, losses, 395, 397
Marine Nature Conservation Review (UK), 417
marine protected areas (MPA), 404, 417–23
  management, 423–4
  site selection criteria, 417–23
marine protection, 402–3
Mediterranean Sea, 61–2
meiobenthos, 150
metapopulation, 53, *137*, 247–8, 366
Milankovitch–Croll cycles, 36, 51
Miocene, 51, 136
mitochondrial DNA (mtDNA), 280–1, 284, 285, 295
mitochondrial RNA, 3, 259
molecular clocks, 260, 284
molecular markers, 280–2
monsoon regions, 181, 184, 220
mortality, 182, 362
  by entanglement, 397
mutation rate, 288
mytilids, 184–5

natural resources and low diversity, 330
natural selection, 277
nature conservation
  definition, 415
  main approaches, 416
nepheloid layers, 156, 157
neritic species
  ranges, 70, 79, 82
  species richness, 71
neuroactive peptides, 208
non-equilibrium ecosystems, 234
North Sea ecosystem, *328*, 329
Norwegian Sea, 99–*100*, *105*, 107–8, 115
nuclear DNA, 282, 284
nutrient
  control model, 179–82
  concentrations, 42, *42*
  fluxes, 102, 362
  input effects, *180*
  limitation, by iron, 44
  pulses, 182
  recycling, 180
  supply, 179, 183–6
  traps, 362

ocean circulation patterns, 223
Ocean Drilling Program (ODP), 50
ocean management, xxi–xxii
oceanic islands, 179
Okinawa, 183–4
ontogenetic migrations, 54, 60–1

organic matter, export, 47
ostracods, planktonic, 58, *62*
outbreeding depression, 278
over-fishing, 247, 364, 373, 396, 398
oviparity, 296
ovoviviparity, 297
oysters, 343

palaeontology, 36
Panama Isthmus, 51, 63
Pangea, break-up, *49*
panmictic populations, 280, 282, 294
panmictic species, 296
parapatric species, 301
particulate organic carbon (POC), 58
patchiness, mosaics, 152, 232, 338
patterns in time, 135–7
patterns, large-scale, 47
pelagic ecology, 37
pesticides, 396
phenotypic variation, 309–11
photographs, sea-bed, 162
Phuket, 186, 187
phyla, 70, 124, 149, 401, *see also* disparity
phytodetritus, 101, 152, 157
phytoplankton, 187
picoplankton, 43
planetary cycles, 36, 51
planktivores, 182
plastic waste, 397
plate tectonics, 49
Pleistocene, 339
Pliocene, 284, *285*
Polar Front, 40
pollution, xiv
polychaetes, 159, 168
polymerase chain reaction amplification, 300
polymorphic enzymes, 305, 311
polymorphic loci, 280
polymorphic species, 296
population biology, 38
population decline, 365
population genetics, 275–7
population size, 276
  effective, 286–9
  human, xiv
precautionary principle, 402
predation, effects, 101, 102, 234, 310
  role of, 39, 84, 241–2
predators, 187–93
Preston plots, *131*
primary production, 179, 329
productivity, 10
productivity regimes, 85
  high, 72
  nutrients, 87
prosobranch molluscs, 110–5

Subject index

proxies, 63
Pteropoda, 71, 73
public awareness, xviii

Quaternary glaciation, 115

r-selection, 182, 226, 228
radiation, 202
ranges, geographical, 178
Rapoport's Rule, 103, 146
rare species, 402
rarefaction, 14, 21, 22–3, 23, 26, 26, 27, 109, 167, 168, 169
  problems with, 29–31
rarity, 84, 242, 245–6, 418–20
recombinant genotypes, 279
recruitment, 161, 234–7, 242
  dynamics, 113
recycling, 184
Red Sea, 184, 185, 220, 230
Red Data Books, 419
redundancy, 331
reefs, 183
  artificial, 141, 144, 405
  coral, 32
  experimental, 236
  oyster, 337
  oyster, as critical habitats, 360–4
  oyster, biodiversity, 356–7
  oyster, ecosystem function, 348–51
  oyster, functional role, 363
  patch, 233
refugia, 103, 137, 222
regional comparison
  Arctic versus Antarctic, 130, 132–5
  Atlantic versus Indo-Pacific, 225–6
regional enrichment, 112
relicts, 285, 333, 418
  glacial, 52
reproduction, timing of, 87
reproductive isolation, 298, 299, 310
resilience, 364, 398
  to disturbance, 332
resource partitioning, 237–8, 259
resource limitation, 150, 233, 237–9
resources, living, xiv
ribosomal RNA, 284
Rockall Trough, 156, 159, 161, 162–3, 163, 164, 165, 166, 168
Rottnest Island, 189, 190
Russell cycle, 332

sampling effects, 19, 30, 95, 108–9, 130, 151
sand transport, 166
scale effects, 101, 217, 243–5
  importance of, 35–6, 337, 341
  large, 116

sea level oscillations, xvi, 51, 147, 220, 222, 227, 338, 400
sea otters, xv
sea-floor spreading, 38
sea-grass, 181
  beds, 32, 185, 360, 405
  fragments, 167
  seasonality, 64
sediment
  characteristics, 102
  grain size, 28
  heterogeneity, 28
  mounds, 152
  slides, 154
  stability, 157–9
  structure, 151
sedimentation, 343
self-fertilisation, 286
Setubal Canyon, 161, 164–7
sewage, 181, 191, 395
Shannon–Wiener index (H'), 161, 208
sibling species, 13, 63, 202, 314
size of ocean, 69
slope current, 34
Southern Ocean, 40, 51, 122–3, 128–38, 260
space, competition for, 204
spatial variability, 126
specialisation, 101, 151, 228, 237
speciation, 111, 221, 224, 276–7
  allopatric, 136, 220–1, 283, 294
  evolutionary processes, 116
  notothenoid fish, 137–8, 260
  rates, 50–2, 63, 70, 194, 211, 222, 223, 263–5, 259
  rates, in molluscs, 261–7, 266
  rate variations with latitude, 261–7
  serolid isopods, 137–8
species abundances, 160
species described
  associated with coral reefs, 203
  insects, 124, 149
  marine, 149
  plants, 124
species distributions, nutrient input, 183–6
  oceanic gyres, 128
species diversity indices, 161, 208, 417–8
  oyster reefs, 356–7
species losses, 337
species pool, 109
species richness
  abundance, 85
  area relationships, 126, 212, 218, 224
  bathymetric pattern, 71, 95
  birds, 3,7,9
  British insects, 4
  British marine Crustacea, 4
  Chaetognatha, 71–3

Subject index

Copepoda, 90–1
coral reefs, 203, 208
Crustacea, 70
deep-sea invertebrates, 4
discontinuity at 40°N, 43, *60*, 127
euphausiids, 55, 71, 91
Indo-Pacific maxima, 128
insects, 4–6, 7, 12
latitudinal patterns, 71, 73, 218
longitudinal patterns, 219–25, *221*
marine/terrestrial comparisons, 48, 124
mysids, 55, 90
neritic species, 71
Ocean mixing effects, 71
oceanic, 70
oceanic/neritic comparisons, 54–7
pelagic taxa, 48, 70
phytoplankton, 48, 70
plants, 4, 48, 70
polar, 132–5
Pteropoda, 71, 90
shallow-water communities, 127–8
Southern Ocean, 128–38
total marine, 123–4
tropical maxima, 127
species, polymorphic, 91
specificity, 12–13
stability, 19, 234
stochastic processes, 101
Stommel diagram, 36, *37*, 85, *85*
stone crabs, 283–6
Strangelove ocean, 50
substrate type, 209
succession, 101
Suez Canal, 405
surrogate, 341, *see also* proxies
sustainability, 403
sustainable exploitation, 373
Suwanee Straits, 284
symbionts, 185
symbioses, 180
systematics, 400

Tagus Abyssal Plain, 161, 164–7
Taiwan, 183, 184, 185
taxonomic diversity pumps, 136, *137*
taxonomic problems, 401
taxonomy, 38, 123, 150
tectonics, 178
temporal gradients, 338
thermohaline circulation, 40
tide-pool, 278–9
tides, internal, 154
toxic algae, 396
toxic compounds, 396
tracers, of ocean currents, 41
trade in wild species, 398–9
translocation, 277
trophic levels, 182
trophic structure, 204–5
tropical rain forest, 12
turbidity, 153
turnover
  faunal, 130, 194, 205, 273
  distance related, *206*

upwelling, 179, 184, 228
USA East Coast, 338–67
  estuarine fishes, 348–51

vicariance events, 36, *49*, 50–2, 64
vulnerability, 245

Wallace's Line, 52, 228
warm-core rings, 227
water masses, 41
waves
  internal, 154, 157
  topographical, 154
Weddell Sea, 108–9, 132, *134*
Western Boundary Undercurrent, 99
wood, 152

xenophyophores, 158